Building Information Modeling

Building Information Modeling: Framework for Structural Design introduces a structure and architecture synergy (SAS) framework that helps develop and enhance the understanding of the fundamental principles of architectural analysis using building information modeling (BIM) tools. Based upon three main components, namely, the structural melody, structural poetry, and structural analysis, along with the BIM tools as the frame enabler, this new framework allows users to explore structural design as an art while also factoring in the principles of engineering. The framework stresses the influence structure can play in form generation and in defining spatial order and composition. By highlighting the interplay between architecture and structure, this book emphasizes the conceptual behaviors of structural systems and their aesthetic implications and enables the readers to thoroughly understand the art and science of whole structural system concepts.

Key Features:

- Is updated throughout.
- Adds new exercises; cutting-edge information on the use of AI, Machine Learning (ML), and modelling; and issues of sustainability.
- Presents the use of BIM technology as part of a design process or framework that can lead to a more comprehensive, intelligent, and integrated building design.
- Places special emphasis on the application of BIM technology for exploring the intimate relationship between structural engineering and architectural design.
- Includes a discussion of current and emerging trends in structural engineering practice and the role of the structural engineer in building design using new BIM technologies.

Building Information Modeling

Framework for Structural Design

Second Edition

Nawari O. Nawari and Michael Kuenstle

CRC Press
Taylor & Francis Group
Boca Raton London New York

CRC Press is an imprint of the
Taylor & Francis Group, an **informa** business

Designed cover image: Shutterstock

Second edition published 2026
by CRC Press
2385 NW Executive Center Drive, Suite 320, Boca Raton FL 33431

and by CRC Press
4 Park Square, Milton Park, Abingdon, Oxon, OX14 4RN

CRC Press is an imprint of Taylor & Francis Group, LLC

First edition published by CRC Press 2015

Library of Congress Cataloging-in-Publication Data
Names: Nawari, Nawari O. and Michael Kuenstle, 1958- authors.
Title: Building information modeling : framework for structural design / Nawari O. Nawari and Michael Kuenstle.
Description: Second edition. | Boca Raton : CRC Press, 2026. |
Includes bibliographical references and index. | Summary: "Building Information Modeling: Framework for Structural Design introduces a structure and architecture synergy (SAS) framework that helps develop and enhance the understanding of the fundamental principles of architectural analysis using BIM tools. Based upon three main components: the structural melody, structural poetry, and structural analysis, along with the BIM tools as the frame enabler, this new framework allows users to explore structural design as an art while also factoring in the principles of engineering"—Provided by publisher.
Identifiers: LCCN 2025036338 | ISBN 9781032983165 (hardback) |
ISBN 9781032984001 (paperback) | ISBN 9781003598435 (ebook)
Subjects: LCSH: Building information modeling.
Classification: LCC TH438.13 .N39 2026
LC record available at https://lccn.loc.gov/2025036338

ISBN: 978-1-032-98316-5 (hbk)
ISBN: 978-1-032-98400-1 (pbk)
ISBN: 978-1-003-59843-5 (ebk)

DOI: 10.1201/9781003598435

Typeset in Times
by codeMantra

Contents

Preface

This book focuses primarily on building information modeling (BIM) technology and its application in the structural analysis and design of building structures. The material is presented using relevant case study BIM projects and provides example modeling techniques and exercise problems with solutions. An underlying goal for the material covered is to present the use of BIM technology as part of a design process or BIM framework that can lead to a more comprehensive, intelligent, and integrated building design—a design by which an optimized structural solution can be achieved in harmony with a building's intrinsic architectural concepts and spatial purpose. With this unique emphasis on the application of BIM technology for exploring the intimate relationship between structural engineering and architectural design, the material presented is well suited for students of engineering, architecture, and construction management, and it can also serve as a valuable resource for building design professionals, building contractors, subcontractors, and fabricators.

This book includes a discussion of current and emerging trends in structural engineering practice and the role of the structural engineer in building design using new BIM technologies. This new technology is significantly transforming twenty-first-century practice activities and is emerging as one of the most promising advances in the architecture, engineering, and construction (AEC) disciplines. Presently, the AEC industry continues to inform its association members and stakeholders about BIM adoption in a variety of ways, including this book. However, at the very core of the BIM evolution is education.

The BIM framework for structural design proposed and presented in this book provides a method to explore a structural design by blending various threads of knowledge that can inform a given building structure; some may seem contradictory and incompatible to arrive at structural beauty and correctness. This BIM framework is referred to as the *structure and architecture synergy framework* (SAS framework) or as the *buildoid framework* (students' preferred name). The SAS framework facilitates the exploration of a structural design as an art while emphasizing engineering principles and thereby provides an enhanced understanding of the influence structure can play in form generation and defining spatial order and composition. The proposed SAS framework will allow architects and engineers to applaud the fusion of art and science and cultivate professional qualities to meet the demands of today's as well as tomorrow's integrated practice requirements.

This key relationship between structure and architecture is therefore fundamental to the art of building. It sets up conflicts between the technical, scientific, and artistic agendas that architects and engineers must resolve. The method by which the resolution is carried out is one of the most critical criteria for the success of a building design. The SAS framework focuses on the resolution of such conflict by enhancing the understanding of the interplay between architecture and structure as well as expanding the design vocabulary.

Preface (2nd Edition)

This is the second edition of the *Building Information Modeling (BIM): Structural Modeling* book, published a decade after the first edition. Over the years, we have received tremendous support and encouragement from students, academics, colleagues, and professionals, urging us to release this updated edition. Sadly, in 2020, my co-author, dear friend, and colleague passed away, a profound loss for his family and for our school. In his honor, I made the decision to complete this second edition, dedicating it to his memory and in recognition of his invaluable contributions to the School of Architecture at the University of Florida.

In this edition, we have extensively updated several chapters and introduced new sections and exercises. In Chapter 1, we added a section on artificial intelligence (AI) and its connection to building information modeling (BIM). This section provides a concise introduction to AI, exploring current models and how they can enhance structural modeling. We also discuss generative AI and generative design, accompanied by practical application examples. The "BIM in Education" section has been revised to reflect recent developments in the field.

Chapter 2, which covers the structure and architecture synergy framework, now includes additional exercise problems to deepen understanding and improve its practical application. This chapter, "Architects + Engineers = Structures," delves into the intricate relationship between architecture and engineering, emphasizing how the two disciplines intersect to create functional and aesthetically pleasing structures. The main theme centers around the idea that architecture cannot exist without structural engineering, as structure is a fundamental aspect of every building and facility.

In Chapter 3, we updated the level of development (LOD) and IFC data schema and standard. The update covers the latest *LOD specification (2019)* that allows for more precise model definitions to support different phases, including design, construction, fabrication, and operation. It also encourages clarity in the exchange of model elements across the project's lifecycle. IFC (Industry Foundation Classes) is an open and standardized data model developed by *buildingSMART International* for exchanging and sharing BIM data across different software platforms, industries, and project stakeholders. The most recent version of the IFC standard is IFC4 (officially ratified as ISO 16739) and IFC4.3, which offer many enhancements over its predecessor, IFC2 × 3.

Moreover, in Chapter 3, a new section "BIM Process in Practice" has been added. The user interface section has been updated, and all the new icons and functions are described and explained. Also, we added new exercise problems to this chapter.

In Chapter 4, the modeling of structural elements has been updated to include steel connections, concrete, and steel reinforcements. Additionally, the structural analytical model has been revised with the introduction of the new "analytical automation" tool, which offers various methods for creating the analytical model. As with other chapters, more exercise problems have been added to enhance learning.

Chapter 5 focuses on updates to architectural elements, with revisions to site components and topography modeling. New topics have also been introduced, including sustainability analysis, energy analysis, solar and shadow studies, lighting analysis, and total carbon analysis.

The last chapter deals with the structural analysis tools. Many revisions have been made to this chapter. Specifically, the interoperability between Autodesk Revit and robot structural analysis tools has been expanded and revised. More examples, covering steel, wood, and concrete structures, are discussed and explained.

About the Authors

Dr. Nawari (Ph.D., P.E., F.ASCE) has more than 25 years of experience in design, teaching, and research in architectural structures, building information modeling (BIM), and artificial intelligence (AI). He currently instructs both graduate and undergraduate courses at the University of Florida, focusing on architectural structures, building technology, sustainability and resilience, and building information modeling (BIM).

Dr. Nawari has authored and co-authored over 200 publications and six books, and he holds more than eleven patents and patent-pending technologies. Additionally, he has mentored over 100 master's and Ph.D. students. As a regular participant and invited speaker at national and international conferences, Dr. Nawari has made significant contributions to the field.

Throughout his career, he has introduced several innovations, including the structure and architecture synergy (SAS) framework and the generalized adaptive framework (GAF). These works have established new paradigms in teaching and designing building structures using the SAS framework. Furthermore, his research in BIM and blockchain has greatly advanced the standardization process within the built environment.

Dr. Nawari has also made notable contributions in developing encoding building rules and regulations and automating building code conformance checking in BIM workflows with groundbreaking approaches like the generalized adaptive framework (GAF) and blockchain technologies (BCT). As a member of the Structural Engineering Institute's (SEI) BIM committee, he co-chaired the subcommittee on BIM in education.

A board-certified professional engineer in Florida and Ohio for over 20 years, Dr. Nawari was inducted as a fellow of the American Society of Civil Engineers (ASCE) in 2016, in recognition of his sustained contributions to the field.

Michael W. Kuenstle, AIA, received his Graduate Architecture degree from Columbia University in New York City, where he graduated with honors for excellence in design and was awarded the William Kinne Fellows Memorial Fellowship for postgraduate research. He holds a Bachelor of Architecture degree from the University of Houston, where he graduated with honors for his design thesis project. Prior to attending Columbia University, he worked as a research assistant at the Chicago Institute for Architecture and Urbanism. He served as adjunct associate professor at the New York Institute of Technology from 1990 to 1993. He has been an assistant and associate professor in the School of Architecture at the University of Florida since 1993.

Over the past 20 years, Michael Kuenstle's wide range of accomplishments as an educator include developing and teaching innovative architecture design studio courses at every level of the undergraduate and graduate curriculum, advancing and implementing new building structures courses as well as technology seminars that integrate digital modeling, analysis, and fabrication techniques as essential learning tools in an evolving technology curriculum. He has served as principal investigator for two significant funded interdisciplinary research projects for the Florida Department of Education, and he is a co-author of several important publications on school facilities design and construction.

Michael Kuenstle received his early professional training in the Chicago office of Skidmore, Owings and Merrill and is co-founder and principal partner in the research-based architecture firm of Clark+Kuenstle Associates, Inc., located in Gainesville, Florida. Parallel with his teaching accomplishments, his building design projects have received several AIA design awards and have been published and exhibited throughout the US and Canada. He is a licensed architect and currently serves as a member of the Board of Trustees to the Florida Foundation for Architecture.

1 Introduction

GENERAL

Historically, one of the oldest architectural structures dates to 2700 BC, when the step pyramid for Pharaoh Djoser was built by Imhotep, who is considered the first architect and engineer in history known by name. Pyramids were the most common major architectural structures built by ancient civilizations because the structural form of a pyramid is inherently stable and can almost be infinitely scaled linearly in size and proportion to increased loads. There is no record of any scientific or engineering knowledge being employed in the construction of pyramids during that era. The physical laws that underpin structural engineering began in the 3rd century BC, when Archimedes published his work, *On the Equilibrium of Planes*, in two volumes. He used the principles derived to calculate the areas and centers of gravity of various geometric Figures, including triangles, parabolas, and half-circles. Together with Euclidean geometry, Archimedes' work on this and on calculus and geometry established much of the mathematics and scientific foundation of modern structural engineering (Addis, 1992, 2007).

At the beginning of the 18th century, advances in mathematics were needed to allow structural designers to apply the understanding of structures gained through the work of Galileo, Hooke, and Newton during the 17th century. In that period, Leonhard Euler founded much of the mathematics and many of the principal methods that allow structural engineers to model and analyze architectural structures. Specifically, in about 1750, he developed the Euler–Bernoulli beam equation with Daniel Bernoulli (1700–1782); this equation is one of the fundamental theories used in structural analysis and design. A few years later, Euler (1757) was able to create the Euler buckling formula, which significantly advanced the ability of engineers and architects to design slender columns. His buckling equation is still one of the most fundamental equations used in various building codes to design columns and walls.

In the early 19th century, new construction materials, such as iron and Portland cement, played major roles in shaping the building design profession. Much of the previous century's practice tradition had to be discontinued or radically reconceptualized. This method did not fit well within the ancient norms of architecture and soon required a new type of training and education. By the middle of the 19th century, many engineering schools across Europe and the United States had been founded, and the modern engineering profession established. Hence, there was no split between architecture and engineering; rather, a new discipline emerged alongside an older one.

From 1854 to 1872, Eugene-Emmanuel Viollet-le-Duc published important contributions to the field of architecture: the *Dictionnaire raisonne de i'architecture francaise du Xle au XVle siecle* (1854–1868) [Dictionary of French Architecture from the 11th to the 16th century (1854–1868)], and the *Entretiens sur l'architecture* (1863).

DOI: 10.1201/9781003598435-1

Their impacts were enormous, both in Europe and in America. Viollet-le-Duc became the most prominent scholar to emphasize the importance of structures in architectural design. He asked the question (Viollet-le-Duc [1854], 1990, p. 28): "On what could one establish unity in architecture, if not on the structure, that is, the means of building?" He also said: "Construction is a science; it is also an art. The practice of architecture means adapting both art and science to the nature of the materials employed."

Based on Viollet-le-Duc's principles, Pier Luigi Nervi (1965), an architect and an engineer, published his book, *Aesthetics and Technology in Building*, in which he placed his design firmly on the tradition of Viollet-le-Duc's principles, with architecture and structure inseparable. In his book, he insisted that architecture cannot be based only on pure art and explained that structure, whether large or small, must be stable and lasting, must satisfy the needs for which it was built, and must be efficient (achieving maximum results with minimum means). He indicated that these are the criteria for good architecture. He also emphasized the idea of employing the materials "according to their nature." For instance, in discussing the advantages of reinforced concrete, he stated: "Reinforced concrete beams lose the rigidity of wooden beams or of metal shapes and ask to be molded according to the line of the bending moments and the shearing stress" (as cited in Sandaker, 2008, p. 28). He asserted his views on the necessity for design to take account of the particular properties of each material and to form or adapt it to a particular shape. These views can be magnificently seen in his design of the aircraft hangers for the Italian Air Force (1940); Stadio Flaminio, Rome (1957); Palazzetto dello Sport, Rome (1958); and the Cathedral of Saint Mary of the Assumption, San Francisco, California (1967).

Schueller (1995, 2007) approached the issue of the interplay between architecture and structure by emphasizing the engineering principles in architectural education alongside the application of software tools in training architects and engineers. Another viewpoint is the concept of "structural art," as described by Billington (2003). This perspective considers structural engineering as a new art form that is parallel to but independent of architecture in the same way that photography, that other new art of the 19th century, is parallel to but independent of painting. Billington explored structural art in the nineteenth and 20th century, specifically in Switzerland. He thoroughly reviewed the work of Swiss structural engineers such as Wilhelm Ritter (1847–1906) and Pierre Lardy (1903–1958) and four of their students: Robert Maillart (1872–1940) and Othmar Ammann (1879–1965), who studied with Ritter, and Heinz Isler and Christian Menn, who studied with Lardy. Addis (1990, 1998) and Addis (2001) shared similar viewpoints as Billington in considering structural art as a form of art that is parallel to but independent of architecture.

In the United States, the work of Khan (2004) in the 1970s and 1980s represents a remarkable contribution to structural art and innovation because of the introduction of trussed frames and tubes, tube within tube, and bundle tubes in high-rise structural systems. Structures such as the John Hancock Center, the Sears Tower, and One Magnificent Mile are important milestones in the history of buildings.

Sandaker (2008) considered structures as a part of the architectural context. Thus, the purpose of structure is not only the support function but also spatial harmony and enclosure organization. In his view, the main purpose of the structure is to establish architectural spaces physically. It follows that the form of structures must heavily consider the spatial functions and emphasize that an understanding and appreciation of structures needs to be taken into account.

The proposed building information modeling (BIM) framework, which is referred to as the structure and architecture synergy framework (SAS framework) in this book, shares some views similar to those proposed by Sandaker (2008), Schueller (2007), Nervi (1965), and Viollet-le-Duc (1854, 1990); however, the SAS framework introduces an innovative design to enable the resolution of the challenges related to the conceptual linking and integration between architectural and structural engineering principles. The framework hinges on BIM and the concepts of structural melody and poetry. This new framework focuses on how to engage the student's imagination and use it no less creatively than a musician or artist producing ideas; at the same time, it elaborates on structural analysis skills as well as on improving the ability to handle cross-disciplinary interests.

BIM IN EDUCATION

OVERVIEW

Building information modeling is a comprehensive information management and analysis technology that is becoming increasingly essential for academic education. Architecture, engineering, and construction (AEC) schools have implemented a variety of pedagogical methods for introducing BIM into their curriculums. These methods range from using BIM in everything from architectural studio, sustainable design, and construction management to civil engineering (Önür, 2009; Sharag-Eldin and Nawari, 2010; Barison and Santos, 2010; Sacks and Barak, 2010; Wong et al., 2011). For instance, Önür (2009) and Sharag-Eldin and Nawari (2010) described how BIM is integrated into the architectural curriculum. Sacks and Barak (2010) introduced BIM as an integral part of freshman-year civil engineering education.

Several academic institutions have integrated BIM in their curricula using different approaches; however, there is no commonly agreed upon methodology for teaching BIM in AEC programs (Barison and Santos, 2010). Most schools offer BIM in only one or two different courses. Many courses limit coverage to a short period (one to two weeks) (Becerik-Gerber et al., 2011). The BIM course is limited to a single discipline in 90% of the cases (Barison and Santos, 2010). The majority of schools introduce BIM on a basic level by teaching a specific software tool, limiting students' perspective on BIM to viewing it simply as another computer-aided design (CAD) productivity-enhancing tool for creating two-dimensional (2D) and three-dimensional (3D) drawings (Sacks and Pikas, 2013). However, BIM by nature goes far beyond digital drafting (Eastman et al., 2011). A comprehensive literature review on the subject can be found in the work of Barison and Santos (2010) and Sacks and Pikas (2013).

Because BIM is different from traditional CAD, it does require new ways of thinking and teaching. For instance, BIM facilitates cross-discipline and interdisciplinary collaboration and teamwork that must be incorporated in teaching BIM courses. Furthermore, BIM provides rich visualization of building elements and parametric modeling of behavior, which can enhance students' learning experience and understanding of virtual construction and how building elements fit together just as they must on a physical site.

BIM for Structural Engineering and Architecture

Building structures are, and have always been, essential components of building design. This is attributed to the roles and meanings of safety, economy, and performance of buildings to society at large. From early societies to the present, buildings have provided shelter, encouraged productivity, embodied cultural history, and represented an important part of human civilization. In fact, the roles of structures are constantly changing in terms of shaping certain quantities of materials and making them support the architecture against gravity and other environmental forces (Addis, 2007). Also, from the earliest times, a sense of beauty has been inherent in human nature; some buildings were conceived according to certain aesthetic views, which would often impose on structures far more stringent requirements than those of strength and performance. Thus, designing structures is becoming deceptively complex, as buildings today are also life support systems; communication and data terminals; centers of education, justice, health, and community; and so much more. They are expensive to build and maintain and must constantly be adjusted to function effectively over their life cycle (Prowler, 2012). Hence, for many, the subject is frequently marked by complexity.

Structural analysis courses at the undergraduate level focus mostly on computation and understanding the principles of statics and strength of materials, without stressing the importance of understanding the conceptual behaviors of structural systems and their aesthetic implications. Addis (1990) noted that at all times in architectural engineering history, there have been some types of knowledge that have been relatively easy to store and to communicate to other people, for instance, by means of diagrams or models, quantitative rules, or a mathematical form. At the same time, there are also other types of knowledge that, even today, still appear to be difficult to condense and pass on to others; they have to be learned afresh by each young engineer or architect, such as a feeling for structural behavior and aesthetic functions, for instance. Currently, in the education of young structural engineers, educators have tended to concentrate particularly on knowledge that is easy to store and communicate. Unfortunately, other types of knowledge have come to receive rather less than their fair share of attention (Addis, 1990; Rafiq, 2010).

On the other hand, architectural students in the design studios are concerned primarily with artistic expressions and philosophical descriptions, independent of the building as an organism and how it is constructed. Structure is not adequately discussed and presented in their work. They apparently are not motivated by the current way of conveying structural concepts and design processes (Schueller, 2007). The purely mathematical approach of the classical engineering schools is not effective in architectural and building construction colleges. Thus, students of these schools are driven to consider themselves artists or contractors, with less interest in scientific and engineering principles. However, all artists must acquire mastery of the technology of their chosen medium, particularly those who choose buildings as their means of expression.

The structure of a building is the framework that preserves its integrity in response to external and internal forces. It is a massive support system that must be incorporated into the architectural program. It must, therefore, be given a form that is compatible with other aspects of the building. Many fundamental issues associated with the function and appearance of a building, including its overall form, the pattern of its fenestration, the general articulation of solid and void within it, and even, possibly, the range and combination of the textures of its visible skins, are affected by the nature of its structure. The structure also influences programmatic aspects of a building's design because of its ability to organize and determine the feasibility of pattern and shape of private and public spaces. Furthermore, structures can be defined to control the inflow of natural light, improve ventilation, or fulfill many other functions that are needed by the architectural spaces.

The relationship between structure and architecture is, therefore, a fundamental aspect of the art of building. It sets up challenges between the technical, scientific, and artistic agendas that architects and engineers must resolve. The method by which the resolution is carried out is one of the most tested criteria of building design success. This issue has been recognized by many engineers and architects, such as Khan (2004), Addis (1990), Schueller (1995, 2007), Billington (2003), Schodek (2004), Sandaker (2008), and Nawari and Kuenstle (2011), among others.

With recent technological advancements, engineers and architects have smarter tools to create and analyze artistically efficient structural forms and to demonstrate how load combinations affect the stability and behavior of a structure. Specifically, BIM has the potential to provide solutions to issues related to the conceptual linking and integration between architectural and structural engineering principles and to advance different types of structural knowledge-sharing objectives without compromising their distinct requirements. BIM is a process that fundamentally changes the role of computation in structural design by creating a database of the building objects to be used for all aspects of the structure, from design to construction, operation, and maintenance. Based on this collaborative environment, a new framework is proposed to advance structural design education. This framework is referred to as the SAS framework, or, alternatively, as the "buildoid framework" (students' preferred name). The framework explores structural design as an art while emphasizing engineering principles; it thereby provides an enhanced understanding of the influence structure can play in creating form and defining spatial order and composition.

New Framework

The history of architecture intermixes with the history of mathematics, philosophy, and engineering at various levels. Designers have adopted concepts and language from these disciplines to assist in their own discourses. The term *synergy* refers to the collaboration of multiple objects in a system to produce an effect different from or greater than the sum of their discrete effects. In the context of the proposed framework, it also refers to the essence or shape of an entity's complete form.

In psychology, the term *Gestalt* is used in a similar sense, referring to theories of visual perception that indicate the human eye sees objects in their entirety (unified whole) before perceiving their individual parts. The phrase "The whole is greater than the sum of its parts" is often used when referring to synergy or Gestalt theories. Similarly, the SAS framework provides a useful language for understanding the structure as a whole in connection with its close relationship with architecture.

The SAS framework focuses on the interplay between architecture and structures and emphasizes a learning process that is highly creative. In this framework, the form of the structure is constrained not only by its function, the site, and the designer's vision, but also by how it works as a whole and by the need to provide a rational argument and calculations to justify expectations before the structure is built.

The proposed framework concept aims to advance other types of structural knowledge that center on how to engage the student's imagination and to use it no less creatively than a musician or artist producing ideas. On the one hand, structural correctness emphasizes the conceptual and quantitative engineering sciences of structural design. The framework combines various threads of knowledge (see Figure 1.1), which may seem, at first glance, conflicting and incompatible. These threads arise from many origins—an understanding of space and human activities, scale, proportions, engineering sciences, knowledge of the behavior of actual materials, and the construction process.

In structural design, the essential skill lies in choosing structural forms and arrangements that manage to satisfy, to varying degrees, many often-incompatible constraints. As with a musician when composing music, this skill relies on a mixture of precedent, experience, and inspiration. For this purpose, the vocabulary and methodology are introduced using the concepts of "structural melody," "structural poetry," and finally, "structural analysis." These are the main components of the proposed framework, along with BIM as the framework enabler. Figure 1.1 depicts an overview of this framework, without the traditional emphasis on first understanding

Beams, columns, footings, bearing walls, and so on, are considered two dimensionally, using the laws of statics and strength of materials; the framework emphasizes

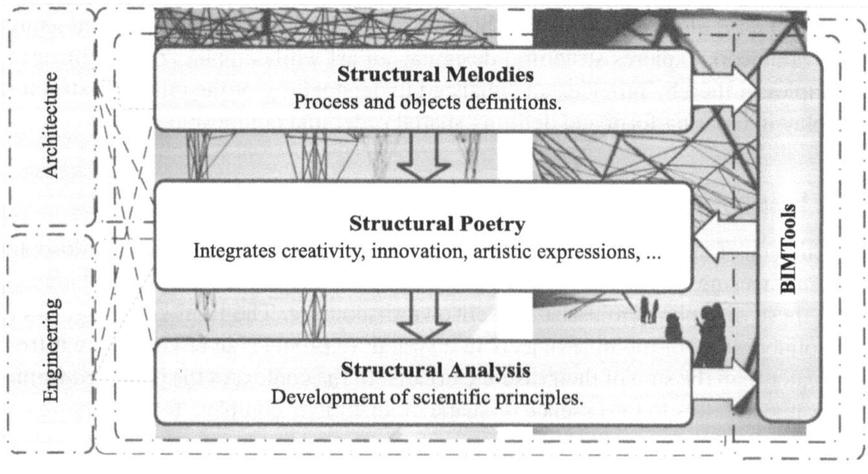

FIGURE 1.1 Structure and architecture synergy framework (SAS framework).

the building as a whole, and creates 3D structural systems using BIM tools and then develops them further into actual architectural solutions.

BIM Concept

Structural design in education is standing on the brink of a new technology that will transform the way structures are designed and constructed. The change is more significant and more profound than the transition from hand computation and drafting to CAD.

Building information modeling is a process that fundamentally changes the role of computation in the AEC industry (Autodesk, 2013). It involves new concepts and practices that are so greatly improved by innovative information technologies and business structures that they will radically reduce the multiple forms of waste and inefficiency in the building industry (National Institute of Building Sciences, 2007). In this concept, rather than using a computer to assist in producing a series of drawings that together describe a building, the computer is used to create a single, unified representation of the entire building, so comprehensive that it can generate all necessary construction documentations. The primitives from which the BIM software composes these models are not the same ones used in traditional CAD (points, lines, curves). Instead, the BIM application models with virtual building components that hold attributed information about actual elements and systems. Examples include trusses, columns, beams, walls, doors, windows, ceilings, and floors. The software platform that implements BIM recognizes the form and behavior of these objects, so it can ease much of the tedium of their coordination. Walls, for instance, join automatically, connecting structure layers to structure layers and finish layers to finish layers. Many of the benefits are obvious—for instance, changes made in one view propagate automatically to every other elevation, section, callout, and rendering of the project. Other advantages include the ability to use the same model to interact with other applications, such as structural and energy analysis software (Autodesk, 2013).

As a general concept modeling type, BIM deals with higher-level operations than traditional CAD does. It deals with placing and modifying entire objects, rather than placing drawings and modifying sets of lines and points. At the same time, BIM platforms allow you to do some standard drafting if needed. Consequently, the geometry is generated from the model and is, therefore, not open to direct handling (Autodesk, 2013).

Another important concept is that a BIM model encodes more than form; it encodes high-level design intent. Within the model, walls and floors are modeled not as a series of 3D solids, but as virtual walls and floors with material types and properties. That way, if a level changes height or walls change width, both of the objects automatically adjust to the new values. If the wall moves, any floor that has a relationship to that wall adjusts automatically.

Students must be introduced to these basics of BIM using one of the available BIM authoring tools. This introduction can take about 12 contact hours. The last phase of this introduction is an overview of the platform interface along with emphasizing the comprehension of new concepts, such as model element, categories, families, types, and instances (see Figure 1.2). Phases I and II in Figure 1.2 cover the introduction to

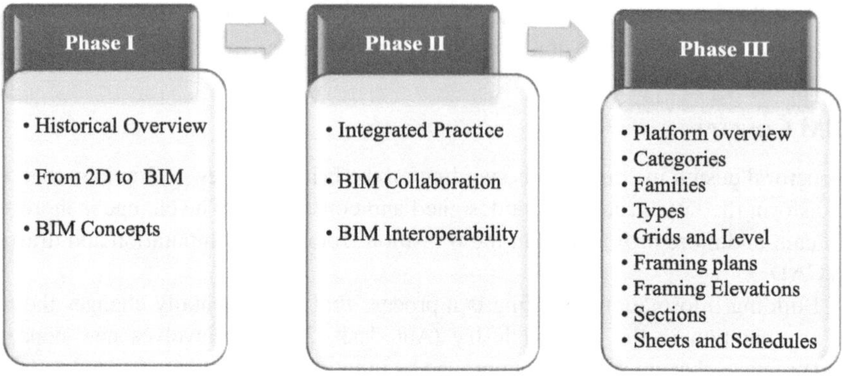

FIGURE 1.2 BIM introduction blocks. (From Sharag-Eldin and Nawari [2010]. BIM in AEC education. In *Joint Structures Congress with the North American Steel Construction Conference.*)

the basics of BIM. In Phase III, a specific platform is chosen, and students learn more in depth about object-oriented modeling techniques. This last phase of instruction normally takes a full semester.

Following the introduction, students can be engaged in learning about the various analysis tools that integrate with BIM platforms. Some of these tools are available as extensions to the basic versions of the software.

BIM AND AI

Nowadays, AI-powered tools are prevalent in everyday life. From your iPhone to banks and hospitals, we see AI applications used increasingly in our daily activities. Banks use AI algorithms to analyze your financial situation and approve loans. Hospitals use AI to analyze medical data, such as X-rays and MRIs. More examples include Facial recognition in Airports, mobile phones, and personal computers, human resources utilize AI tools to screen job applicants.

Introduction to AI

Background

Artificial intelligence (AI) is the field of computer science focused on creating systems and machines that can perform tasks typically requiring human intelligence, such as problem-solving, learning, recognizing patterns, summarizing, understanding natural language, and making decisions. A key technology behind many AI systems is the neural network—a computational model inspired by the human brain's structure, consisting of layers of interconnected nodes that process and learn from data. Neural networks enable AI systems to learn from experience and improve their performance over time without explicit programming.

Neural networks are inspired by the structure and function of the **biological neural networks** found in the human brain, though they are a much-simplified model. The human brain consists of billions of neurons, each connected to many others through synapses. Neurons in the brain receive signals, process them, and either fire (pass the signal along) or remain inactive, depending on the strength of the signal. This is similar to how artificial neurons in a neural network process inputs and decide whether to activate based on the sum of inputs and the activation function.

This biologically inspired computational model consists of processing elements (called neurons) and connections between them with coefficients (weights) bound to the connections, which constitute the neuronal structure, and training and recall algorithms attached to the structure. The connection weights are the "memory" of the system.

Even though neural networks have similarities to the human brain, they are not meant to model it. They are meant to be useful models for problem-solving and knowledge acquisition and processing in a humanlike way. The human brain is much more complex, and unfortunately, many of its cognitive functions are still not well known. Therefore, it's worth examining the key characteristics of the human brain from an information-processing perspective to understand how effectively they have been replicated in artificial neural networks.

At least three levels can be distinguished in the human brain from the point of view of the information processing performed in it: (a) the *structural level:* neurons, regions of neurons, and the connections between them; (b) the *physiological level:* the way the brain processes information as chemical and physical reactions and transmission of substances; and (c) the *cognitive level:* the way humans think.

Three points are briefly discussed here: (1) The brain's structure and organization, (2) The brain as a communication system, which is massively parallel and robust, and (3) The major information-processing functions and characteristics of the brain: learning, recall and generalization, associative storage, and chaotic behavior.

Let us look at *brain organization* first. The human brain contains about 10^{11} neurons participating in perhaps 10^{15} interconnections over transmission paths. Each of these paths could be a meter long or more. Neurons share many characteristics with the other cells in the body, but they have unique capabilities for receiving, processing, and transmitting electrochemical signals over the neural pathways that make up the brain's communication system. Figure 1.3 shows the structure of a typical *biological neuron.* The neurons are composed of a *cellular body,* also called the *soma,* and its one or several branches. According to the traditional definition, the branches conducting information into a cell (*stimulus*) are called *dendrites,* and the branch that conducts information out of the cell (*N"action*) is called an *axon.* An activation of a neuron called an *action potential,* is transmitted to other neurons through its axon at the first instance. A signal (spike) emitted from a neuron is characterized by *frequency, dW'll-tion,* and *amplitude.*

The brain consists of different types of neurons. They differ in shape (pyramidal, granular, basket, etc.) and in their specialized function. There is a correlation between the function of a cell and its shape; for example, pyramidal cells are excitatory, and basket cells are inhibitory. In terms of variety, over 50 kinds of functionally different neurons are found in the cerebellum (one of the major parts of the brain). This fact may influence the creation of heterogeneous neuronal computational models, which consist of different types of specialized artificial neurons.

The human brain is an intricate communication system that transmits signals through electrical impulses, which are consistent and repetitive in nature. A single

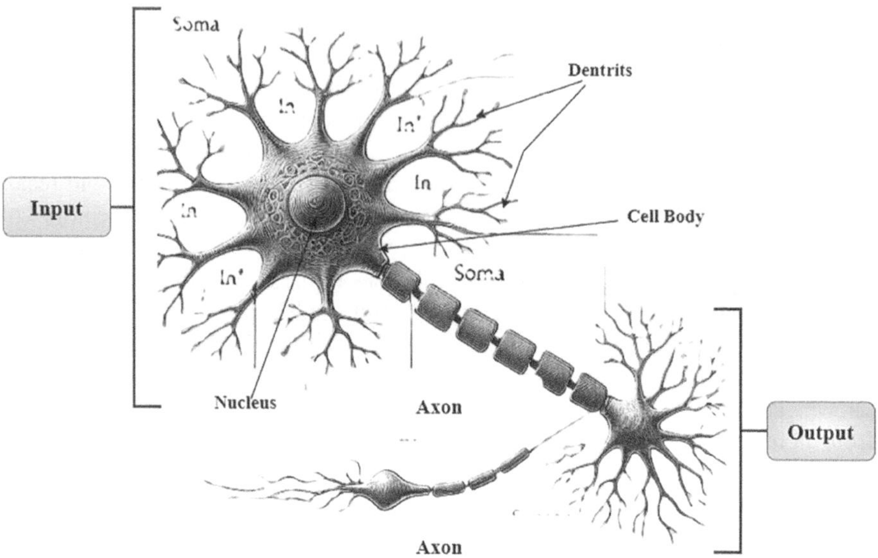

FIGURE 1.3 A typical structure of biological neuron.

nerve impulse contains very little information, so processing complex data is only possible through the interaction of vast networks of neurons and nerve fibers. These networks are massive in both number and size, with the total length of neural connections in the human body reaching approximately 10^{14} m. This extensive network enables a high degree of parallel processing, a hallmark of brain function. While individual neurons respond to external stimuli relatively slowly (around 200 ms per reaction), the brain as a whole can solve complex problems in fractions of a second to several seconds, thanks to its enormous capacity for parallelism.

The brain's ability to solve complex problems and respond effectively to unfamiliar situations stems from its heuristic capacity to make decisions based on previously acquired knowledge and its adaptability to new circumstances. It can learn and generalize by storing information in synapses, where concentrated chemical substances preserve the knowledge we gain through experience. The brain receives input through receptors, which are connected by dendrites to neurons in the corresponding areas of the cortex—the region responsible for processing signals from our sensory organs. Thus, from an information-processing perspective, two of the brain's key functions are learning and recall (or generalization).

The human brain stores patterns of information in an associative mode, that is, as an associative memory. *Associative memory* is characterized in general by its ability to store patterns, to recall these patterns when only parts of them are given as inputs, for example, part of u face wo have seen

The question of how information is stored in the human brain touches on the debate between local and distributed representation. Biological evidence indicates that certain brain regions, and even individual neurons, respond to specific stimuli or are responsible for particular functions. A neuron, or a group of neurons, representing

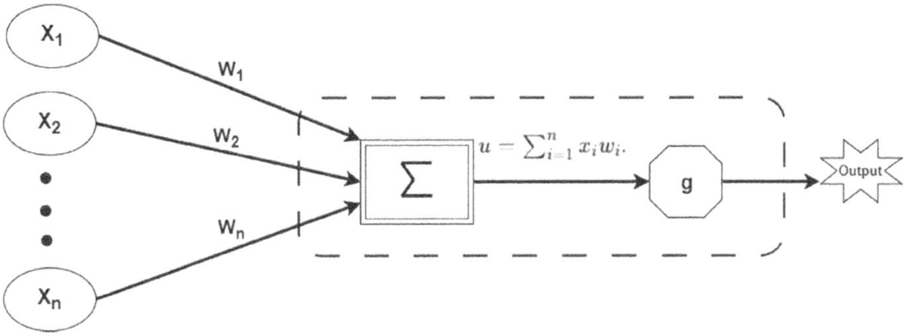

FIGURE 1.4 A model of an artificial neuron (neural network processing unit).

a distinct concept is often referred to as a "grandmother cell." However, there is also evidence indicating that brain activity is a collective effort, with many neurons working together to encode patterns, features, functions, tasks, and goals, reflecting the brain's highly distributed processing system.

The collaborative efforts of neuroscientists, biologists, physiologists, computer and data scientists, mathematicians, physicists, chemists, engineers, artists, and others have resulted in new insights into brain processes and the development of updated neural computational models. These models build upon and refine previous ones. In the following section, some basic artificial neural models will be explored.

The first mathematical model of a neuron was proposed by McCulJoch and Pitts in 1943. It *was* a binary device using binary inputs~binary output, and a fixed activation threshold. In general, a model of an artificial neuron is based on the following parameters, which describe a neuron processing unit (see Figure 1.4):

- **Inputs**: $x_1, x_2, \ldots, x_n$. There are weights attached to each input parameters: $\mathbf{w}_1, \mathbf{w}_2, \ldots \mathbf{w}_n$;
- **Input Function** *f*: calculates the aggregated net input signal to the neuron **u = f(x, w)** where **x** and **w** are the input and weight vectors correspondingly, $= u = \sum_{i=1}^{n} x_i \; w_i$
- ***An activation (signal) function s*** calculates the actitation **level** of the neuron ***a*** **= s(u)** (see Figure 1.5).
- ***An output function*** calculates the output signal value emitted through the output (the axon) of the neuron: $o = g(a)$ the output signal is usually assumed to be equal to the activation level of the neuron, that is, $o = a$ (see Figures 1.3 and 1.4)

Explanation

$\sum_{i=1}^{n}$ Block represents the sumation of weighted inputs given by $u = \sum_{i=1}^{n} x_i w_i$

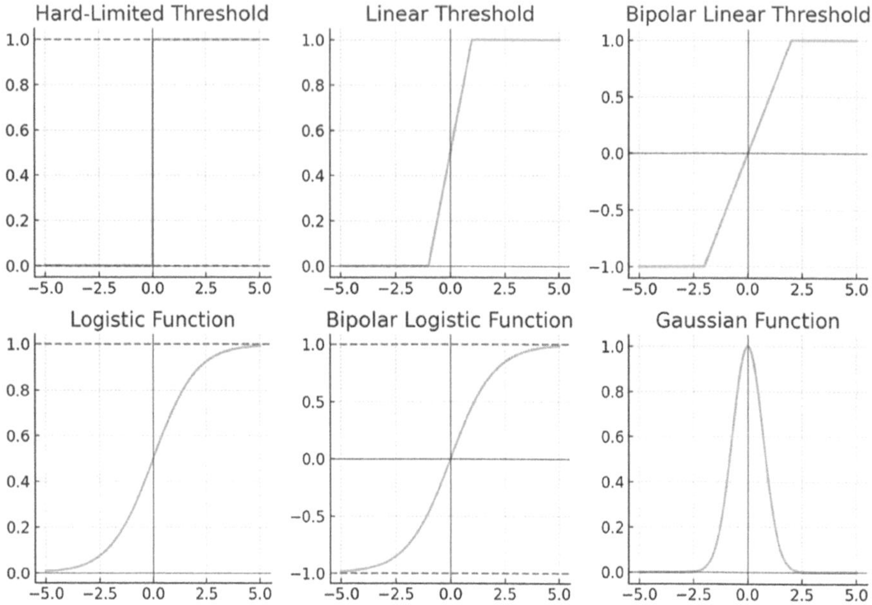

FIGURE 1.5 Examples of activation functions.

g Block (Activation Function): A non-linear transformation applies to the sum. Common activation functions include:

- ReLU (Rectified Linear Unit): $g(u) = \max(0, u)$
- Sigmoid: $g(u) = \dfrac{1}{1+e^{-u}}$
- $\tan h : g(u) = \tan h(u)$

Key parallels between biological and artificial neurons include:

- **Neurons and Nodes**: Both biological neurons and artificial nodes receive multiple inputs, process them, and transmit output to other neurons or nodes.
- **Synapses and Weights**: In the brain, synapses control the strength of signals between neurons, while in neural networks, weights determine the importance of inputs from one node to another.
- **Learning Mechanism**: Just as the human brain adjusts its synaptic connections to strengthen certain pathways (learning from experience), neural networks adjust weights and biases through training to improve performance.

However, the complexity of the human brain far exceeds that of artificial neural networks. The brain's neurons are highly specialized and connected in intricate ways that current AI systems cannot fully replicate. Neural networks in AI are inspired by this biological process but are much simpler, focusing on specific problem-solving tasks rather than the vast range of functions managed by the brain.

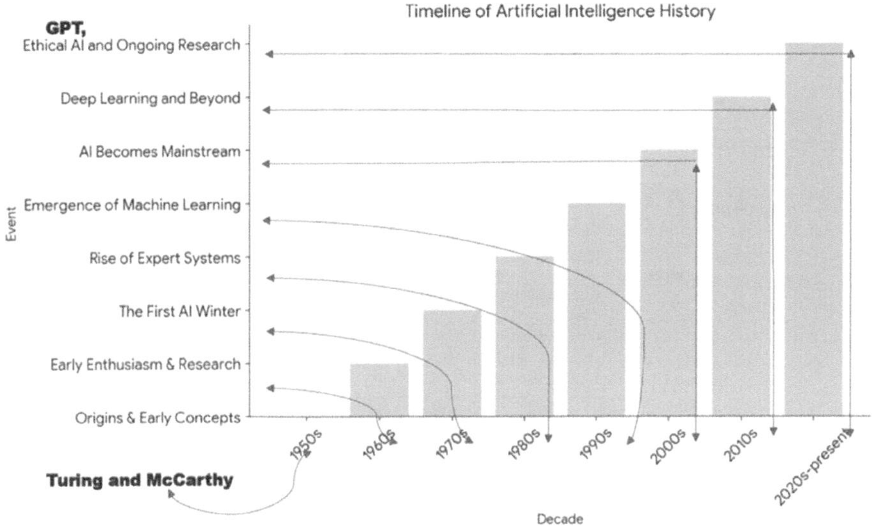

FIGURE 1.6 AI milestones.

Artificial Intelligence (AI) is a multidisciplinary field of science and engineering aimed at building intelligent agents, which are systems capable of performing tasks that typically require human intelligence. These tasks include, but are not limited to, **reasoning**, **problem-solving**, **understanding natural language**, **speech recognition**, Design, and **visual perception.**

A I's conceptual foundations were laid in the late 1940–1950s. Alan Turing's 1950 paper, "Computing Machinery and Intelligence," posed the famous question, "Can machines think?" and introduced the Turing Test. In 1956, John McCarthy coined the term "***Artificial Intelligence***" at the Dartmouth Conference, marking the formal start of AI as an academic discipline. Figure 1.6 depicts a brief history of AI development.

AI is a transformative technology that is reshaping various aspects of our modern world. For example:

- **Economic Transformation**: AI is a major driver of the Fourth Industrial Revolution, significantly impacting economies globally. It's creating new industries and transforming existing ones, leading to increased productivity and economic growth. Understanding AI is crucial for future leaders and innovators in virtually every field.
- **Technological Innovation**: AI is at the forefront of technological innovation, pushing the boundaries of what's possible in areas like Design, healthcare, transportation, and communication.
- **Data Analysis and Insight**: With the explosion of big data, AI's role in extracting meaningful insights from vast datasets is indispensable. AI techniques are essential in fields like genomics, climate science, and market analysis, making skills in AI highly valuable for researchers and analysts (Figure 1.7a).

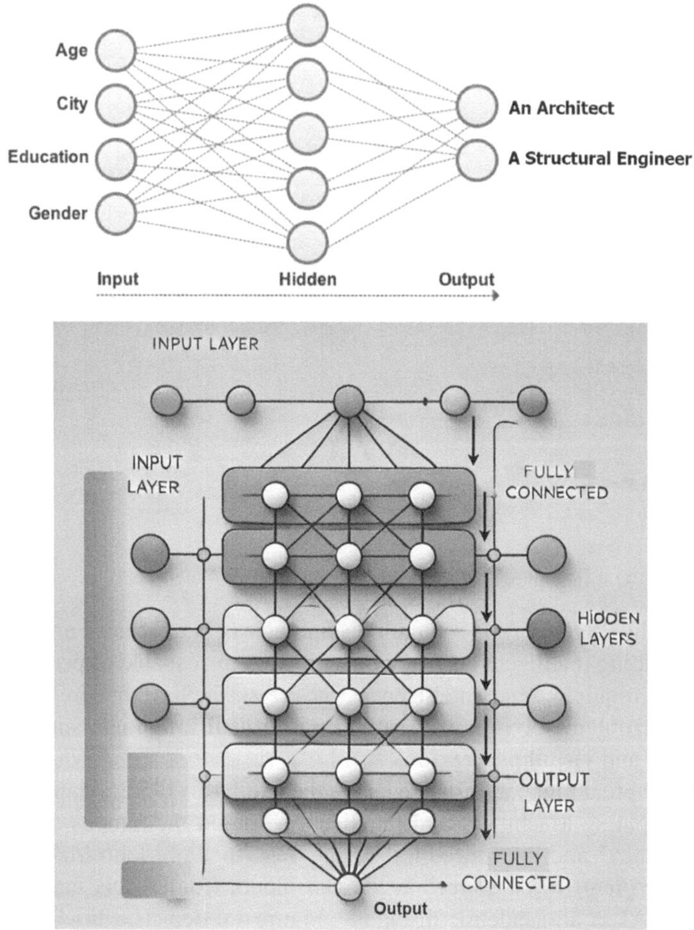

FIGURE 1.7 (a) Multilayer neural network to classify students data professional preference. (b) An example of deep neural networks (DNNs).

- **Creative Collaboration**: AI acts as a creative collaborator in the fields of Design and art. It offers artists and designers new tools and methods for creating innovative and unique works. This collaboration can lead to unforeseen and inspiring results, expanding the boundaries of creative expression.
- **Enhanced Creativity**: AI algorithms can analyze vast collections of art and design, learning styles and patterns. Artists and designers can leverage these insights to push their creative boundaries, incorporating elements they might not have considered before.
- **Generative Art**: AI has given rise to a new genre called generative art, where artists use algorithms to create art. This can include anything from digital paintings to sculptures, often with stunning and complex results that would be difficult or impossible to achieve by human hand alone.

- **Architectural Innovations**: In architecture, AI is used for optimizing designs, simulating environmental impact, and even generating creative design alternatives based on set parameters, leading to innovative architectural solutions.
- **Automated and Enhanced Design Processes**: In the design industry, AI can automate mundane tasks, such as resizing graphics or generating base models, allowing designers to focus more on the creative aspects of their work.
- **Interactive Art**: AI enables the creation of interactive art installations that can respond to audience behavior, environment changes, or social media trends, creating immersive and dynamic experiences.
- **Problem Solving and Decision Making**: AI's ability to process complex information and provide solutions to intricate problems is unparalleled. This has applications in areas ranging from optimizing supply chains to developing new pharmaceuticals, and even in addressing societal challenges like poverty and education.
- **Personalized Experiences and Services**: In the consumer sector, AI's ability to personalize experiences – from shopping recommendations to content creation – is transforming customer expectations and business models.

AI works through a combination of various algorithms, models, and techniques that enable machines to perform tasks that would typically require human intelligence. To understand how AI works, it's essential to delve into the core components and processes involved. Namely these are data collection and processing.

- **Data Collection:** The process of gathering information from various sources is to be used for training AI models. Sources can be Databases, sensors, web scraping, user-generated content, etc. Examples include Collecting building records for an architectural design AI system.
- **Data Preparation:** The data cleaning includes removing errors and inconsistencies from the data. Normalization: Adjusting values measured on different scales to a common scale. Splitting: Dividing data into training, validation, and test sets. Example: Standardizing the site requirements, program, and scale in a design dataset.

Model Selection

Choosing an appropriate AI algorithm or model architecture based on the problem type and data characteristics. The following are the major AI techniques and algorithms.

Key Algorithms and Techniques

1. **Supervised Learning**: The model learns from labeled data, which means each training example is paired with an output label. The common Algorithms include:

- Support Vector Machines (SVMs): Classify data by finding the optimal hyperplane.
- Random Forest: An ensemble of decision trees to improve classification accuracy.

2. **Reinforcement Learning**: The model learns by interacting with an environment to maximize a reward signal. The main components of this model are
   - Agent: The learner or decision maker.
   - Environment: The external system the agent interacts with.
   - Reward Signal: Feedback from the environment to evaluate actions.
3. **Neural Networks and Deep Learning**

   The model learns by interacting with an environment to maximize a reward signal. The fundamental elements include:
   - Neural Networks: Composed of layers of neurons
   - Activation Functions: Functions like ReLU, Sigmoid, and Tanh that introduce non-linearity into the model.
   - Training: Uses algorithms like Stochastic Gradient Descent (SGD) for optimization.
   - Deep Neural Networks (DNNs): Neural networks with many layers (deep architectures) (Figure 1.7b).
   - Specialized Architectures, such as
     - CNNs: For image processing, capturing spatial hierarchies in images.
   - Recurrent Neural Networks (RNNs): For sequence data, capturing temporal dependencies.
4. **Unsupervised Learning**: Deriving insights from unlabeled data. Includes clustering algorithms (like K-means, hierarchical clustering), dimensionality reduction techniques (like PCA, t-SNE), and association rule mining.
5. **Semi-Supervised and Transfer Learning**: Using a combination of labeled and unlabeled data or leveraging pre-trained models for new tasks.

Practical Applications

- **Natural Language Processing (NLP)**: Applications like text classification, machine translation, sentiment analysis, and question answering.
- **Computer Vision**: Image recognition, object detection, and other applications involving image and video data.
- **Predictive Modeling**: In finance, healthcare, marketing, etc., for predicting future trends, customer behavior, disease outbreaks, etc. (Figure 1.8).

Challenges and Ethical Considerations

The rapid advancement of artificial intelligence (AI) has revolutionized industries and reshaped how we approach complex problems, but it comes with significant challenges and ethical considerations. As AI becomes increasingly integrated into everyday life, it raises concerns about bias in algorithms, data privacy, transparency, and accountability. Ethical dilemmas, such as the displacement of jobs, the potential

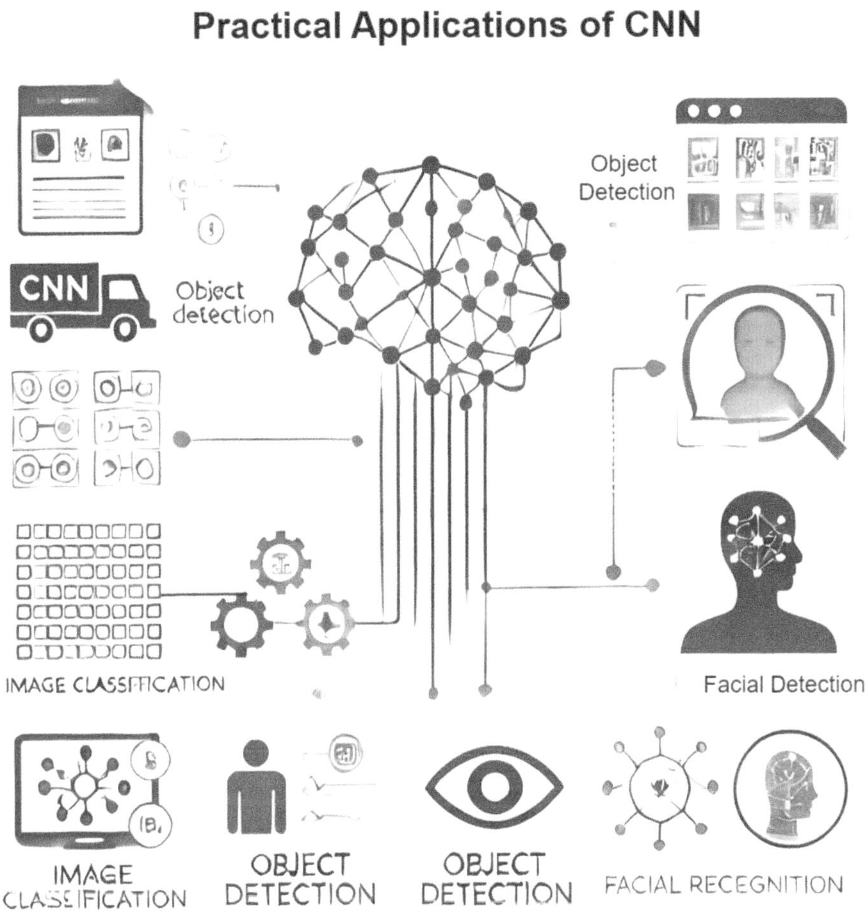

FIGURE 1.8 Samples of AI and ML applications.

for surveillance overreach, and decision-making in critical areas like healthcare and criminal justice, highlight the urgent need for a balanced approach. Addressing these challenges involves not only developing more responsible AI technologies but also creating policies and frameworks that ensure fairness, inclusivity, and human-centered outcomes. The conversation around AI must move beyond technological achievements to address its broader societal impact and the moral responsibilities of its creators and users.

Here are the main factors to consider when addressing the ethical and fairness of AI applications:

Bias and Fairness: Addressing biases in data and algorithms and ensuring fairness in model predictions.

- **Explainability and Transparency**: Developing interpretable models, especially important in sensitive areas like healthcare and criminal justice.

- **Privacy**: Understanding the importance of data privacy, techniques like differential privacy, and General Data Protection Regulation (GDPR) compliance.

Generative Design

Generative Design is a collaborative design process between humans and AI. During this process, the designer defines the design parameters, and the computer produces design studies (alternatives), evaluates them against quantifiable goals set by the designer, improves the studies by using results from previous ones and feedback from the designer, and ranks the results based on how well they achieve the designer's original goals. It is a specific application of the computational design approach, with the following distinctions:

- The designer defines goals to achieve a design (rather than the exact steps).
- The AI algorithms help the designer to explore the design space and generate multiple design options (not just one).
- The computer enables the designer to find a set of optimal solutions that satisfy multiple competing goals (Figure 1.9).
- The designer compares multiple design scenarios to find a set of design options that fits the design goals.

Generative Design can help find better designs for a project more quickly by leveraging what computers are good at: computation and repetition. Computers can generate and evaluate a huge number of design variants in only a fraction of the time it would take an individual designer, allowing you to learn what does and doesn't work at an accelerated pace. Using generative Design, an algorithm can generate thousands of variants in a few minutes.

Generative Design Process

In Generative Design, this workflow involves the following stages: *generate*, *analyze*, *rank*, *evolve*, *explore*, and *integrate* (Figure 1.10). Each of these stages can be further broken down into **Define**, **Run** and **Results** steps. The Define step is the responsibility of the designer, while the Run and Results steps are performed by computational algorithms.

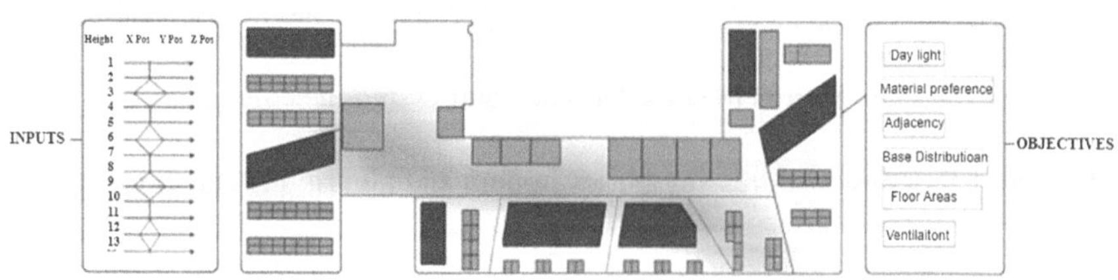

FIGURE 1.9 Generative design.

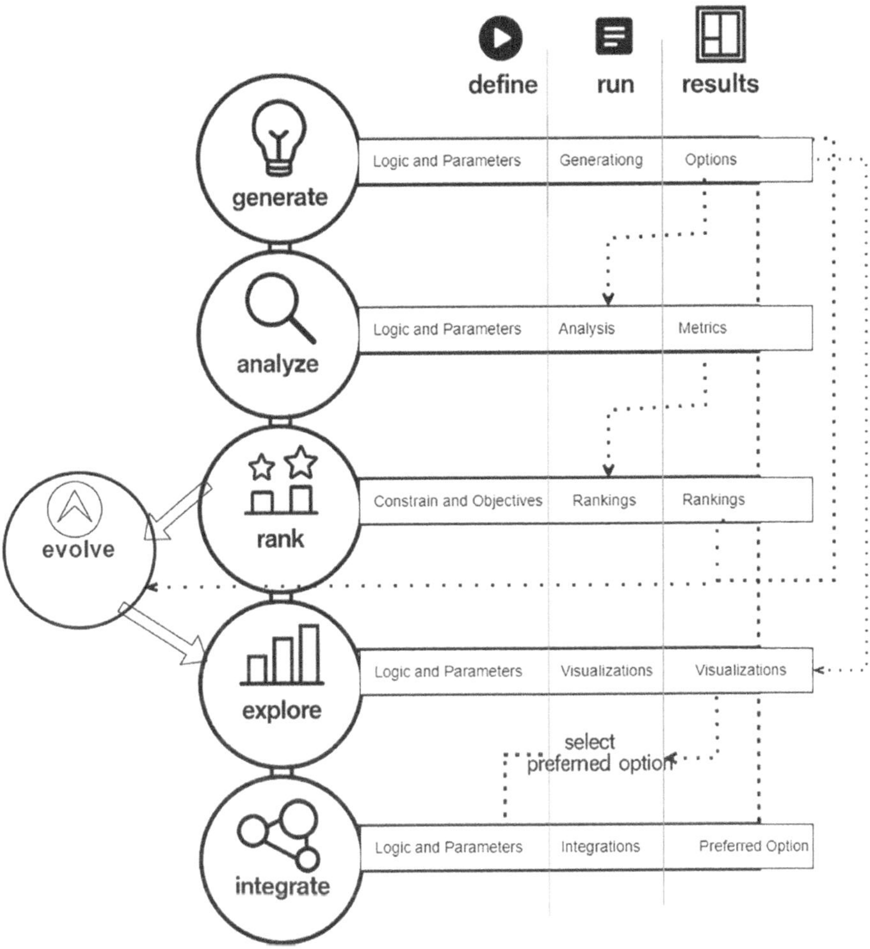

FIGURE 1.10 Overview of the key steps.

The **Define** step is essential to the entire generative design process, as the accuracy and relevance of the outputs depend heavily on the quality of the designer's input at this stage. This step involves setting up the generation algorithm—the logic that dictates how designs are created, including any constraints and rules. Additionally, it requires specifying the generation parameters, which are the input variables necessary for guiding the design process.

The **Run** stage occurs once the algorithm and its parameters have been fully defined. At this point, the computer initiates the process, generating various design options. This operation can be performed locally on the designer's computer, or, for more complex and resource-intensive calculations, it may be executed using cloud computing.

Results represent the outputs that are generated during the **Run** step are the final yields from each stage. These are then used as inputs or parameters in subsequent phases. For example, the designs created in the generate phase will be used as one of input parameters in the analysis phase.

Structural Design Fundamentals

Common Attributes of Architecture

Throughout the United States, accredited schools of architecture and design are influencing and educating the future generation of architects, who may go on to create the next masterpiece. Their knowledge of many branches along with their judgment is the practice and theory of architecture (Waldrep, 2006). It is not this issue that is being called into question; rather, the question considers the current role of the architect.

To understand the role of architects, it is imperative to acknowledge their focal points in design. For example, Salingaros and Mehaffy (2006, p. 30) suggest that architects may consider "order on the smallest scale that is established by paired contrasting elements, existing in a balanced visual tension; large scale order occurs when every element relates to every other element at a distance in a way that reduces entropy; the small scale is connected to the large scale through a linked hierarchy of intermediate scales with various scaling ratios." One of the overall objectives is to give rise to different experiences that users of a building undergo. The practical functions, such as the entry and exit, and circulations are also influenced by the structural form and order.

The basic practices an architect of today would follow are appraisal, design brief, concept, and design development (Chappell and Willis, 2000). These actions encompass understanding the needs of the client, an outline of the preparatory work agreed on by the architect and client, a sketch design to illustrate the external public and private spaces, internal public and private spaces and appearances, as well as a final version showing a clear representation of the entire building, including components, materials, and layout.

Common Attributes of Engineering

Engineering students are trained in understanding advanced calculus and numerical methods for analyzing and designing buildings and other structures—they know how to set up the analytical model and solve equations to obtain solutions to verify safety and stability. However, they can lack understanding of overall structural behavior of the building and its connection to other architectural and construction aspects and thus may use an abstract mathematical and analytical model that imperfectly simulates reality. Furthermore, engineers rarely have the opportunity to entertain engagement in aesthetic matters of buildings (Addis, 1990; Billington, 2003). Their focus is relative to their discipline, whether it is structural, civil, or mechanical.

Also, engineers have their own set of preferred geometric forms that have their origins in the mathematical models found in structural science. A wide-flange I section or an inverted T shape is an efficient cross section for a beam; depending on the material and how it is manufactured, efficient cross sections for a column might be a solid circle, a tube, or an H section. To use the minimum amount of material, beams and columns should taper as a parabola or paraboloid from their centers to the end support points. Trusses need to be made up of triangles, sometimes of identical shape and size, sometimes changing. Suspension structures (cables and arches) feature catenaries or parabolic curves. Shells are usually made in the form of paraboloids, hyperboloids, or hyperbolic paraboloids, but may also be elliptical, spherical, cylindrical, or even have the form of logarithmic spirals and epicycloids.

In a nutshell, through their education, structural engineers are taught a fundamental understanding of applied mathematics and the knowledge of behavior and the science of materials under various loading conditions, as well as theories of structural analysis, which normally guide their motives in making design decisions. However, structural design is concerned not only with science, mathematics, and techniques but also with space enclosure, scale and proportions, nature and the environment, communication links for people and objects, circulations, and—after all—aesthetic values and innovation.

Differences and Oversight between Architects and Engineers

Simply stated, architects and engineers have differing goals when approached with a design. On one hand, architects are concerned with what they have been taught, space organization and order, flow, circulation, comfort of occupants, cultural and social issues, and aesthetic impact. On the other hand, engineers deal with structural planning, safety, and economy. As can be seen, there is no clear overlap or platform that would facilitate communication and coordination between the disciplines. Accordingly, if a suitable method is utilized by both professions to understand the interplay between these fields, a more fluent, cohesive, and efficient design process could be achieved.

Along with the structural Gestalt framework, BIM will assist in minimizing the oversight between architecture and engineering because of its ability to specify the interaction of architectural forms and features, structural stresses, section properties, material strength, deformation, and performance based on type of connections and boundary conditions using a single BIM model.

Summary

To advance other types of structural knowledge, structural melodies and poetry focus on how to engage the student's imagination and to use it no less creatively than a musician or artist producing ideas. On the other hand, structural analysis concentrates on the conceptual and quantitative aspects of structural performance.

The proposed BIM framework seeks to emphasize the value of these knowledge domains, specifically the interplay between architecture and structures, as well as the qualitative understanding of structural behavior. Furthermore, the framework elaborates on improving the student's ability in handling cross-disciplinary interests through the use of BIM knowledge and other related digital tools. By approaching structural design education and training in this manner, several objectives can be achieved:

- First, students are exposed to the fundamentals of BIM at an earlier stage of their current core curriculum.
- Second, students will explore structure as an art and thereby gain an understanding of the influence structure can play in creating form and defining spatial order and composition.

- Third, it introduces students to the vocabulary and hierarchy of a structural system, enabling structural decision making to be integrated early in the design thought process.
- It establishes the notion that structural analysis computation is primarily a tool to verify structural decisions rather than a design strategy.
- Finally, it initiates an attitude of understanding the interplay between architecture and structural systems that should continue into the student's remaining education and forward into her or his professional career.

In summary, during the **Generate** step, the created designs are evaluated based on how well they meet the goals established by the designer in the **Analyze** phase. Following this, the **Rank** phase involves organizing or prioritizing the design options based on the analysis results. This ranking helps guide the direction for further development in the **Evolve** phase. In the subsequent **Explore** step, the designer reviews and compares the generated designs, examining both their geometry and evaluation results. Finally, in the **Integrate** phase, the designer selects the preferred design option and incorporates it into the broader project or design work.

2 Structure and Architecture Synergy Framework (SAS Framework)

INTRODUCTION

The history of architecture intermixes with the history of mathematics, philosophy, and engineering at various levels. Designers have adopted concepts and language from these disciplines to assist in their own discourses. The term *synergy* refers to the collaboration of multiple objects in a system to produce an effect different from or greater than the sum of their discrete effects. In the context of the proposed framework, it also refers to the essence or shape of an entity's complete form. In psychology, the term *Gestalt* is used in a similar sense, referring to theories of visual perception that the human eye sees objects in their entirety (unified whole) before perceiving individual parts. The phrase "The whole is greater than the sum of its parts" is often used when referring to synergy or Gestalt theories. Similarly, the structure and architecture synergy framework (SAS framework) provides a useful language for understanding the structure as a whole in connection to its close relationship with architecture.

Structure is the main ingredient of architecture. In the contemporary sphere, it has acquired an independent personality, so that its own intimate aesthetic quality is highly valued. At the same time, structure must obey scientific and engineering laws to be correct. The separation between these aspects in practice has continued since the beginning of the Industrial Revolution, when structural engineering became a specialized field, separate from architecture.

The SAS or, as preferred by students, "buildoid framework" focuses on the interplay between architecture and structures and emphasizes a learning process that is highly creative. In this framework, the form of the structure is constrained not only by its function, the site, and the designer's vision, but also how it works as a whole and by the need to provide a rational argument and calculations to justify expectations before the structure is built.

The proposed framework aims to advance other types of structural knowledge that center on how to engage the student's imagination and to use it no less creatively than a musician or artist producing ideas. On the other hand, structural correctness emphasizes the conceptual and quantitative engineering sciences of the structural design. The framework combines various threads of knowledge (see Figure 2.1), which may seem at first glance contradictory and incompatible. These threads arise from many origins: an understanding of space and human activities, engineering sciences, knowledge of the behavior of actual materials, and the construction process.

DOI: 10.1201/9781003598435-2

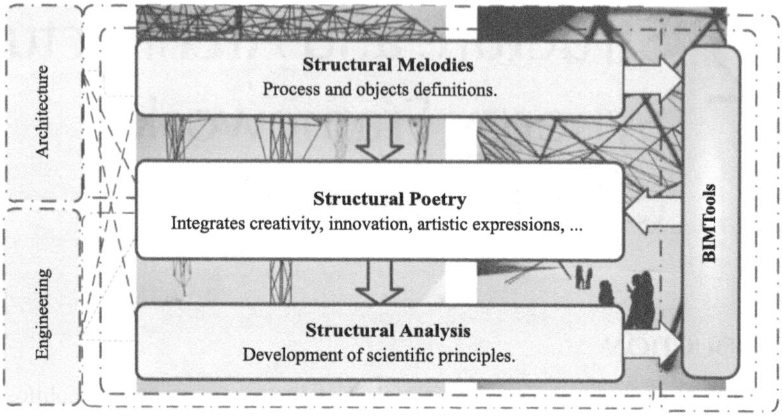

FIGURE 2.1 Structure and architecture synergy framework (SAS framework).

In structural design, the essential skill lies in choosing structural forms and arrangements that manage to satisfy, to varying degrees, many often-incompatible constraints. As with a musician when composing music, this skill relies on a mixture of precedent, experience, and inspiration. For this purpose, the vocabulary and methodology are introduced using the concepts of structural melody, structural poetry, and finally structural analysis. These are the main components of the proposed framework, along with building information modeling (BIM) tools as the framework enabler. Figure 2.1 depicts an overview of this framework. Without the traditional emphasis on first understanding beams, columns, footings, bearing walls, and so on two-dimensionally, using the laws of statics and strength of materials, the framework emphasizes the building as a whole and creates a three-dimensional (3D) structural system using BIM tools and then develops it further into an actual architectural solution.

VOCABULARY AND OBJECTIVES

The proposed (SAS) framework refers to the collaboration of multiple objects in a system to produce an effect different from or greater than the sum of their discrete effects. It also refers to the essence or shape of an entity's complete form, as in the sense of Gestalt or the whole as being greater than the sum of its parts. Similarly, the SAS framework provides a useful language for understanding the structure as a whole in connection to its close relationship with architecture. This includes, for example, human activities, spatial systems, forms, circulations, enclosure systems, structural systems and elements, load diagrams, lateral stability, force equilibrium, support reactions, shear force, and bending moment diagrams.

To advance other types of structural knowledge, structural melodies and poetry focus on engaging and having the student use imagination creatively. Conceptual and quantitative aspects of the structural performance are addressed by structural analysis.

The proposed framework seeks to emphasize the value of these knowledge domains, specifically the interplay between architecture and structures as well as the qualitative understanding of structural behavior. Furthermore, this framework elaborates on improving a student's ability to handle cross-disciplinary interests through the use of BIM knowledge and other digital tools. By using such vocabulary and methodology, several objectives are realized:

- Students have the chance to learn the fundamentals of BIM at an earlier stage of their current core curriculum.
- Students will explore structure as an art and thereby gain an understanding of the influence structure can play in creating form and defining spatial order and composition.
- Students are introduced to the vocabulary and hierarchy of a structural system, enabling structural decision-making to be integrated early in the design thought process.
- The notion is established that structural analysis computation is primarily a tool to verify structural decisions rather than a design strategy.
- Finally, an attitude of understanding the interplay between architecture and structural systems is initiated, which should continue into the student's remaining education and forward into a professional career.

In the following sections, the SAS framework's three main components (structural melody, structural poetry, and structural analysis; Figure 2.1) are described in detail along with their interrelationships.

Structural Melodies

Structural melody is the first basic component of the SAS framework. Its initial goal is to understand how linear, nonlinear, planar, and volumetric structural elements can be orchestrated to create spatial order in architecture using BIM tools. It is also the intention to develop this idea as a holistic vehicle to introduce structural vocabulary, the hierarchy of structural members, and the interplay between architectural concepts and structural systems, such as exoskeleton, endoskeleton, stratification, transition, hierarchy, and the heart of spaces. Structural melodys also deal with the structural vocabulary, such as element names and their purpose, order, and hierarchy (Figure 2.2a and b).

Structural melody introduces the language of structural design and therein introduces students to relationships between systems and details. It aims to provide students with the basic vocabulary and grammar for expressing design ideas. Structural melodies start incrementally from a simple 3D system and then evolve into a complete architectural solution with lateral and vertical stratification.

Furthermore, using BIM tools, the following vocabularies are introduced:

- **Beams**: These are mostly horizontal, spanning linear prismatic elements to support vertical loads from floors or roofs. They are the secondary and sometimes tertiary level of framing.
- **Girders**: Girders are a specific type of beam that spans between columns. They can resist vertical and horizontal applied loads. They normally represent a primary level of structural framing.

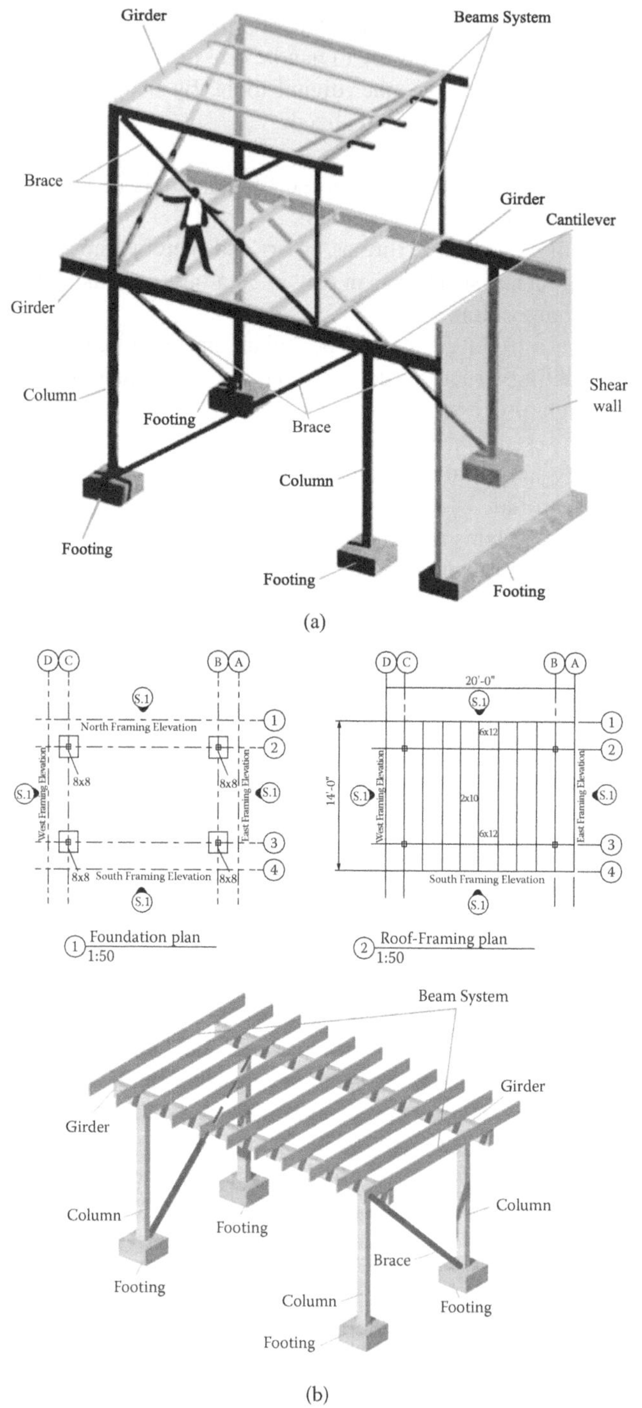

FIGURE 2.2 Structural melody: basic definitions. Structural melody: basic definitions.

(Continued)

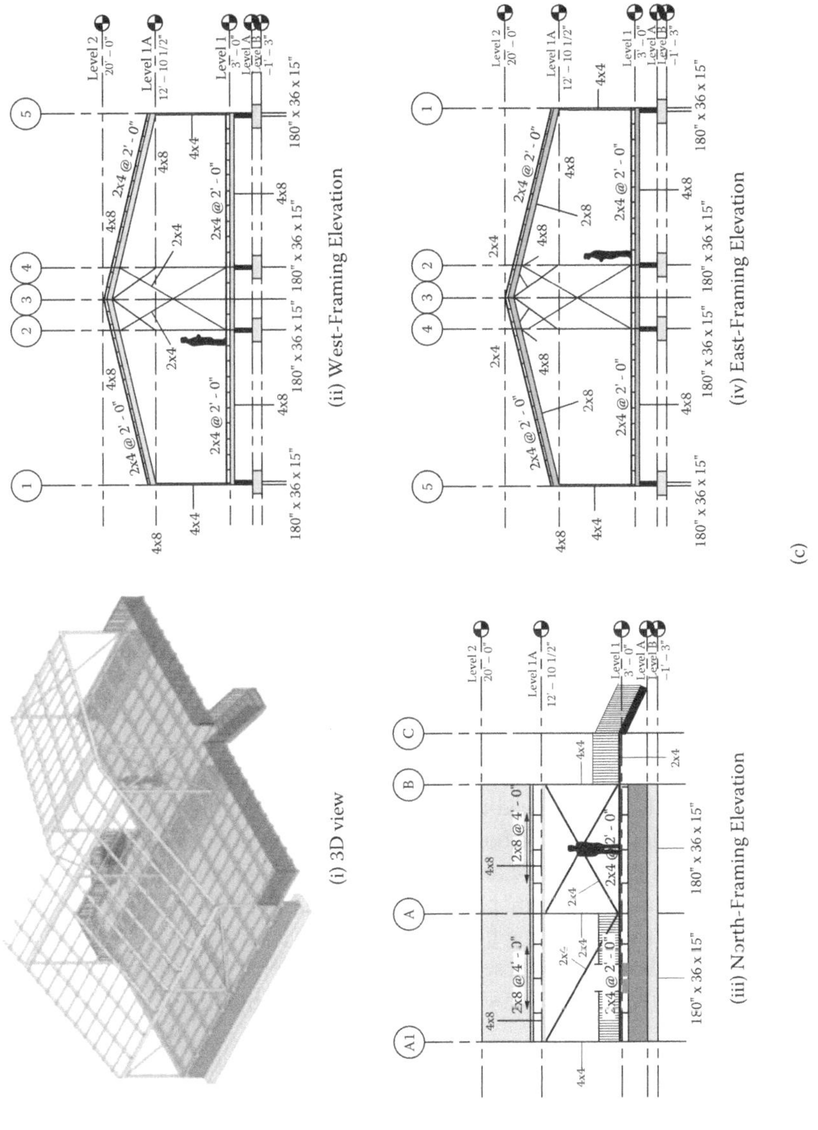

FIGURE 2.2 (*Continued*) Structural melody: basic definitions. Structural melody: basic definitions.

- **Truss**: Generally, a truss is a framework arranged in triangular patterns using slender, straight elements connected at each end joint with pins. The versatility of the truss in terms of its form and how it is manufactured makes it one of the few structural elements useful for a wide range of spatial characteristics (i.e., small, intermediate, or large spaces).
- **Columns**: These are lineal compression members having a length substantially greater than their least lateral dimension. They generally support other structural members such as girders, beams, trusses or slabs, and transmit loads to the foundation.

- **Walls**: These are planar elements used for vertical or lateral support of the structure.
- **Braces**: Braces are structural members responsible for carrying lateral loads. They are normally connected to columns or columns/beam members.
- **Arches**: An arch is a structural system that distributes loads to supports through a curvilinear form within a single plane. The forces developed within an arch are primarily compressive in nature.
- **Cables**: Structural cables are steel members that are inherently flexible and form-active, thus they can only transmit tensile forces. Cables adjust their suspended form to the respective loading conditions so that they can respond in tension.
- **Foundation**: Types of foundations are single footing, combined footing, mat, and piles.

Structural melody introduces the language of structural design and therein introduces students to relationships between space and support systems (i.e., how to deal with different types of spaces, such as small or large architectural spaces). It aims to provide students with the basic vocabulary and grammar for expressing structural design ideas for different architectural programs. Structural melodies start incrementally from one element and then evolve into a complete 3D architectural solution. Furthermore, using BIM tools, the following vocabularies are introduced:

- **Grid Lines and Reference Planes**: These are essential lines in structural melody, used to define the structural layout and boundaries. Grid lines are represented by dotted lines with a bubble at one or both ends. The bubbles are utilized to number grid lines using digits in one direction (y direction) and letters in the other orthogonal direction (x direction) (see Figure 2.2b).
- **Foundation Plans**: These are plan views at the foundation level.
- **Framing Plans**: These are plan views of the roof or floor showing the structural support system at the floor or roof level.
- **Framing Elevations**: These are special elevation views that depict structural elements at a specific section line across the building.

These definitions are illustrated in Figure 2.2. This facilitates the understanding of the relationships between the 3D models and their two-dimensional representations. The stability and hierarchy of the structural elements are also introduced using 3D models similar to those given in Figure 2.2.

In addition, structural melody includes understanding various support patterns derived from basic linear, planar, and nonlinear elements and how these patterns respond to the functional organization of buildings. For instance, columns and walls can be utilized to separate and reinforce spaces to allow for different activities. The primary structural instrument to determine the appropriate degree of fit between the functional spaces and the structural support patterns (Figure 2.3) is provided by the patterns in Figure 2.4a and b in conjunction with the rules of thumb (defined next). Figure 2.4a and b illustrate some examples of structural support patterns and their spatial characteristics. For instance, introducing overhangs in Figure 2.3 helps reduce the stresses on the beam support system as well as serves as cantilevers

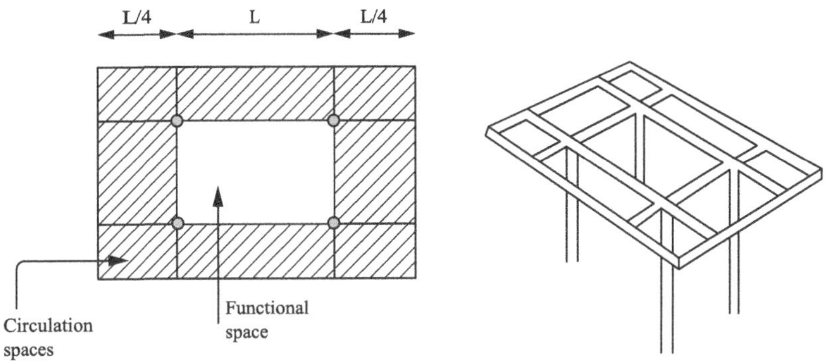

FIGURE 2.3 Relationship between the structural support and the building spatial program.

covering the circulation zones. The architectural implication of using such overhangs includes the visual overthrow of the vertical support elements along the façade (i.e., the façade design does not have to include these vertical columns).

In general, for smaller spaces, it is usually recommended to choose a flat system with fewer vertical support elements. As the spatial areas increase, specifically the functional spaces, the flat system is not going to be the most effective structural solution; another structural framework needs to be designed. For example, creating structural support for 20 ft (6 m) × 20 ft (6 m) would be efficient using flat beam-and-joist systems rather than support utilizing cable or arch structures.

The density of the vertical support pattern is a critical consideration in structural design. The choice between a concentrated and a widespread support system needs to be addressed in relation to the resulting consequences on the organization of the overall architectural program. Most of the time, the structural support density will significantly shape the form of the building as a whole and often have an impact on the spatial characteristics of the building.

The vertical support pattern in a structure does require stabilization to resist lateral loads caused by forces such as wind and earthquakes. This is generally achieved by the introduction of triangulation patterns, rigid connections between the vertical support and horizontal elements, or structural shear walls. The triangulation is attained using brace elements in various configurations (Figure 2.5).

Another important part of structural melody is the rules of thumb for the relationship between the sizes of the structural elements and the space they define. The following are some examples of rules that are introduced to determine the initial sizes of linear and planar elements (Schueller, 2007): The ratios of the overall depth of a beam *d*, a girder, or a planar element thickness *t* to the span of the space *L* are as follows:

$\frac{L}{L} = 60$ for a cantilever beam and trusses d

$= 20$ for a non-cantilever roof or floor beam and trusses d

$= 30\,\frac{L}{t}$ for a planar element (t = thickness of a roof or a floor slab)

$\frac{L}{t} = 60$ for arches and vaults

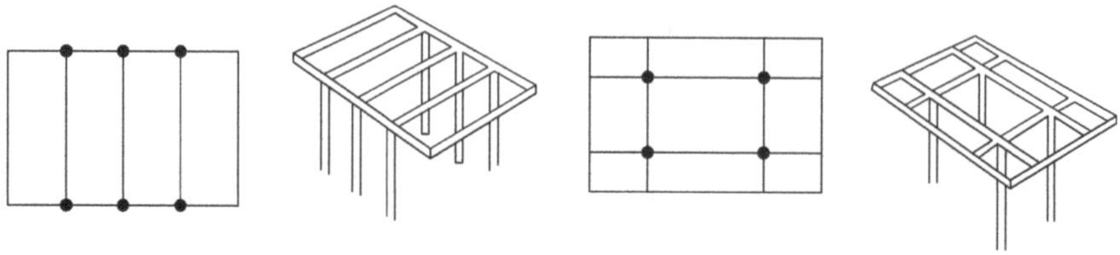

(i) Concentrated

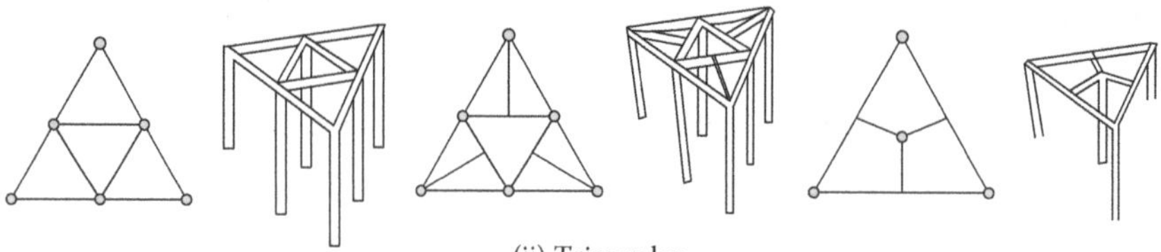

(ii) Triangular

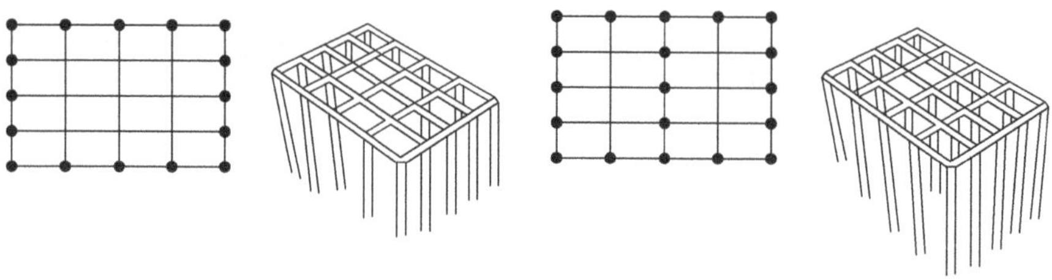

(iii) Perimeter (iv) Perimeter with transitional

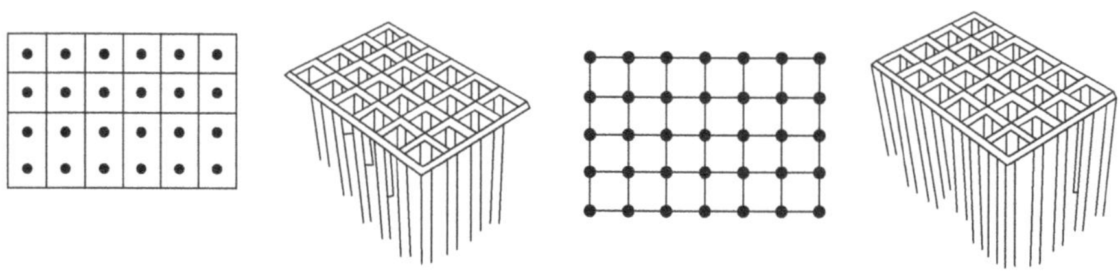

(iv) Uniformly distributed

(a)

FIGURE 2.4 Structural melody: (a) linear vertical support patterns. (b) Surface support patterns.

(Continued)

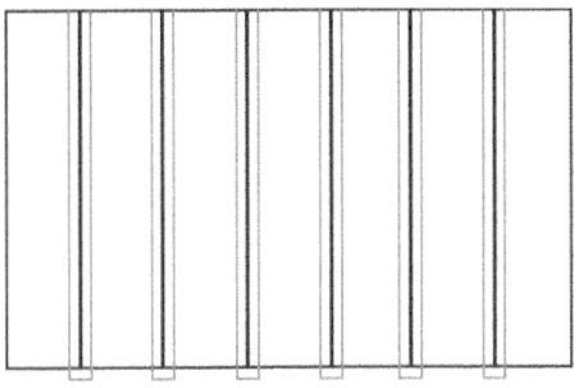

(i) Parallel: one-way

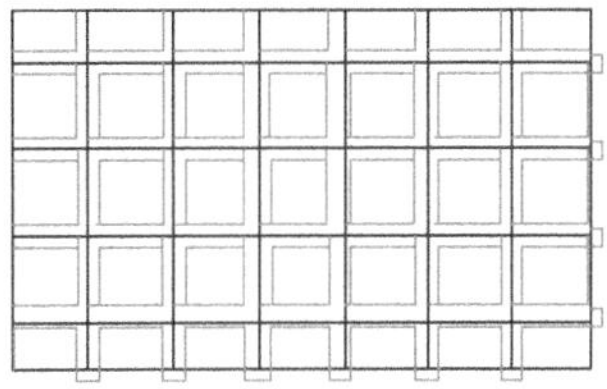

(ii) Parallel: two-way

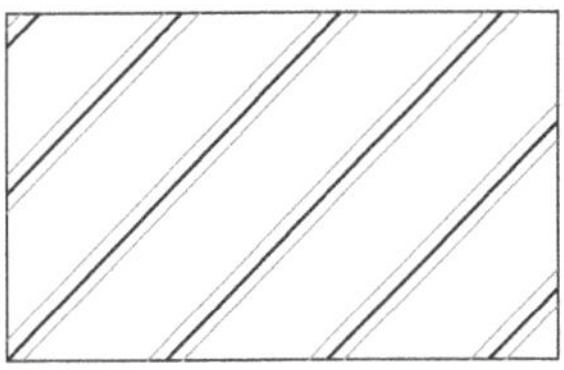

(iii) Diagonal: one-way

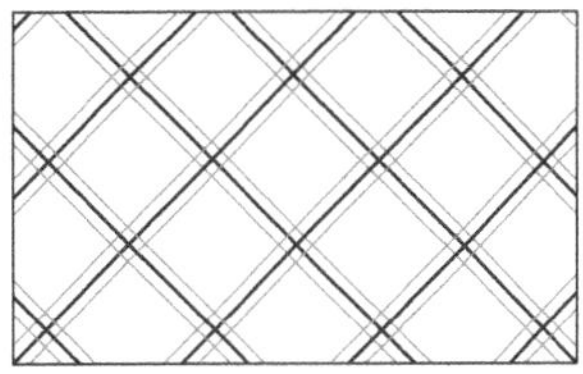

(iv) Diagonal: two-way

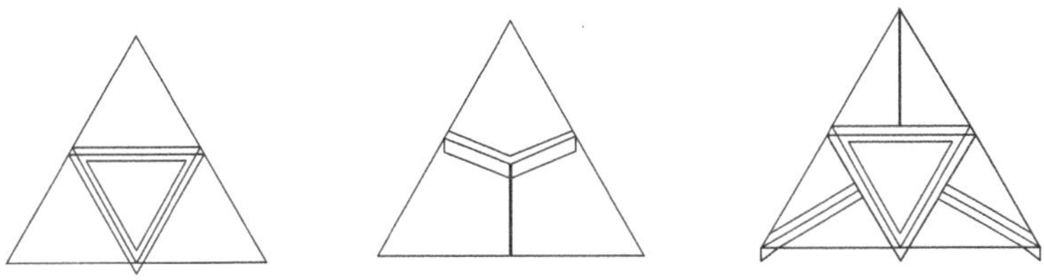

(v) Triangular

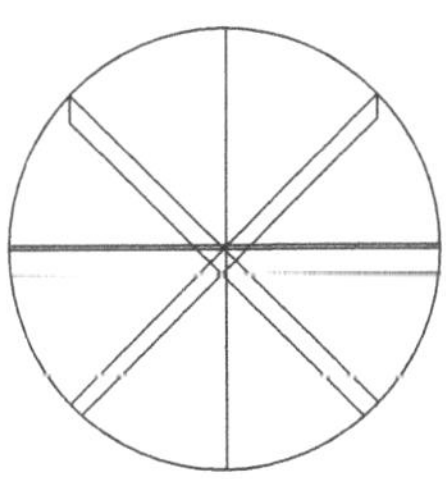

(vi) Radial

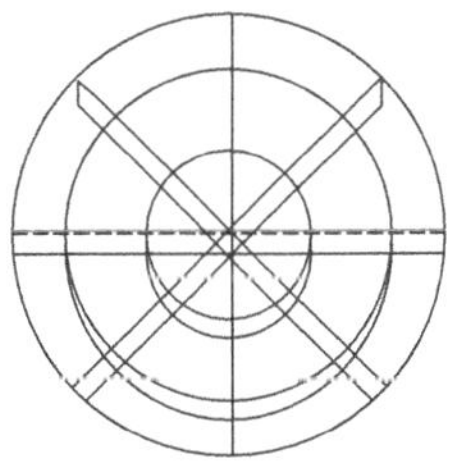

(vii) Circumferential

(viii) Reciprocal

(b)

FIGURE 2.4 (*Continued*) Structural melody: (a) linear vertical support patterns. (b) Surface support patterns.

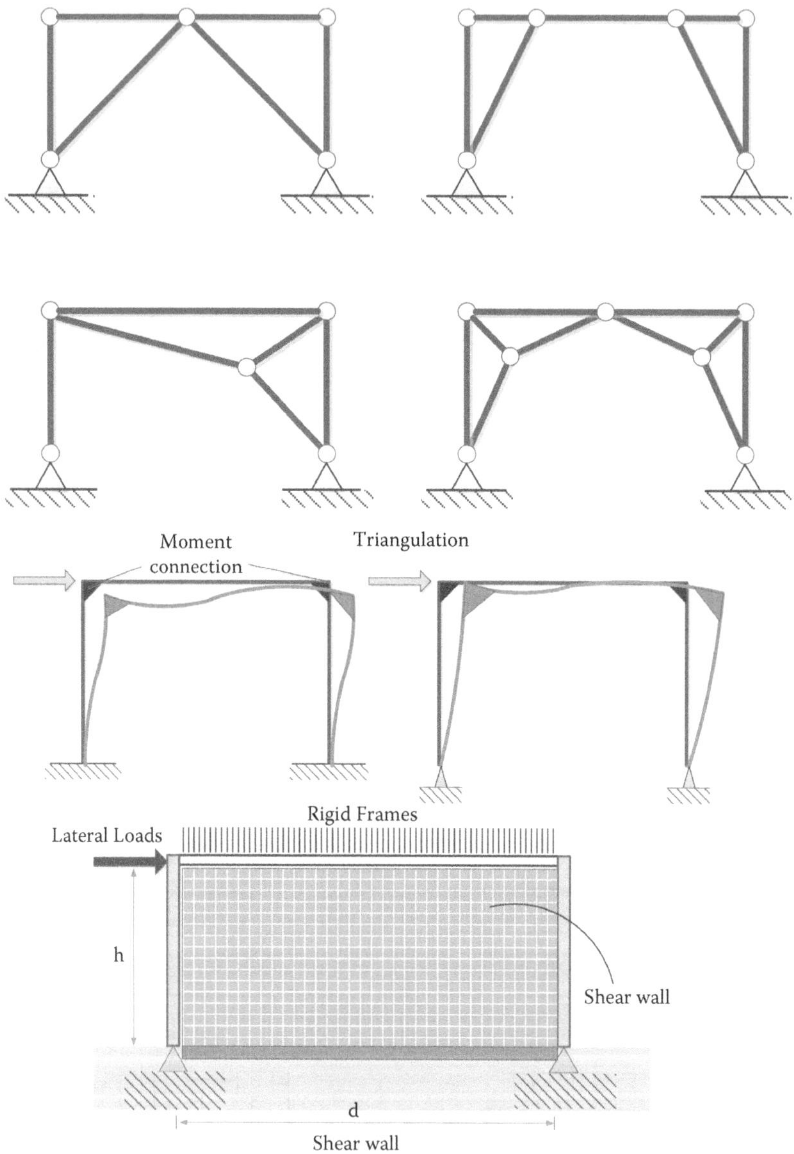

FIGURE 2.5 Stabilization of the structural support patterns.

Structural Poetry

Structural poetry is the second component of the SAS framework. It is based on structural melody and aims to develop structural creativity and spatial thinking as well as enhance the conceptual design abilities using BIM tools.

The word *poetry* has its origin in Greek and means a making: forming, creating, or the art in which language is used for its aesthetic and evocative qualities, in addition to, or instead of, its notional and semantic content (New Oxford American Dictionary, 2010). In other words, poetry is a fundamental creative act using language. Similarly, structural poetry is a creative exercise to provide structural systems using structural vocabulary and melodies to organize and stabilize architectural spaces.

In a more general sense, structural poetry strives to develop structural creativity and spatial thinking, as well as enhance the conceptual design abilities. This allows you to develop an imaginative, complex structural system without a thorough understanding of its individual components at the initial design stages. Without the traditional emphasis on first understanding beams, columns, bearing walls, and so on two-dimensionally, using the laws of statics and strength of materials, structural poetry utilizes the power of BIM to create 3D structural forms to satisfy spatial, aesthetic, and other programmatic requirements.

In this approach, a parallel can be drawn to language poetry to enhance the student's comprehension. For instance, poetry may use a condensed or compressed form to convey emotion or ideas to the reader's or listener's mind or ear; structures can be formed using a few members in different forms to provide a certain aesthetic and framework for spaces. Poetry may also use devices such as assonance and repetition to achieve musical or incantatory effects; similarly, structures can be orchestrated by repeating the same pattern of supports to achieve simplicity, optimization, and elegance. Poems frequently rely on imagery, word association, and the musical qualities of the language used for their effect. Similarly, structures can use the form, orientation, type, and quality of materials to have an impact on the final design. The interactive layering of all these effects to generate meaningful spaces is what marks structural poetry.

Figure 2.4 shows the basic structural models, or simply buildoids. These buildoids can grow horizontally and vertically to fulfill the desired programmatic objectives using BIM tools. This process is similar to the natural growth of living objects. All living organisms are composed of one or more cells. In these living organisms, the cell is the basic unit of structure, function, and organization. Furthermore, biological forms are hierarchical structures, made of materials with elusive properties that are capable of change in response to variations in local conditions. In a similar fashion, structural poetry (Figure 2.6) aims to learn from such a natural growth process by designing systems that are self-assembled, using small, primarily units (buildoids) to make a whole architectural solution. This resembles growth in most multicellular organisms, in which growth is not only about the volume increase of a single cell, but also about the multiplication and rearrangement of cells (Figure 2.7) (Sinnott, 1960; Bard, 1990).

This process is demonstrated in Figure 2.8 by depicting buildoids' plan configurations to generate various spatial growth expressions. Figure 2.9 illustrates the manifestation of this growth concept using the buildoid shown in Figure 2.6c.

The development of biological forms begins with the growth of individuals from a single cell to a fully developed adult. The evolution and development start from simple components that are assembled together to form larger structures. Most of these living forms are hierarchical structures made from simple members to create large and strong structures. The geometrical and hierarchical organization of these members is critical to their structural capacity, and that capacity emerges from the way in

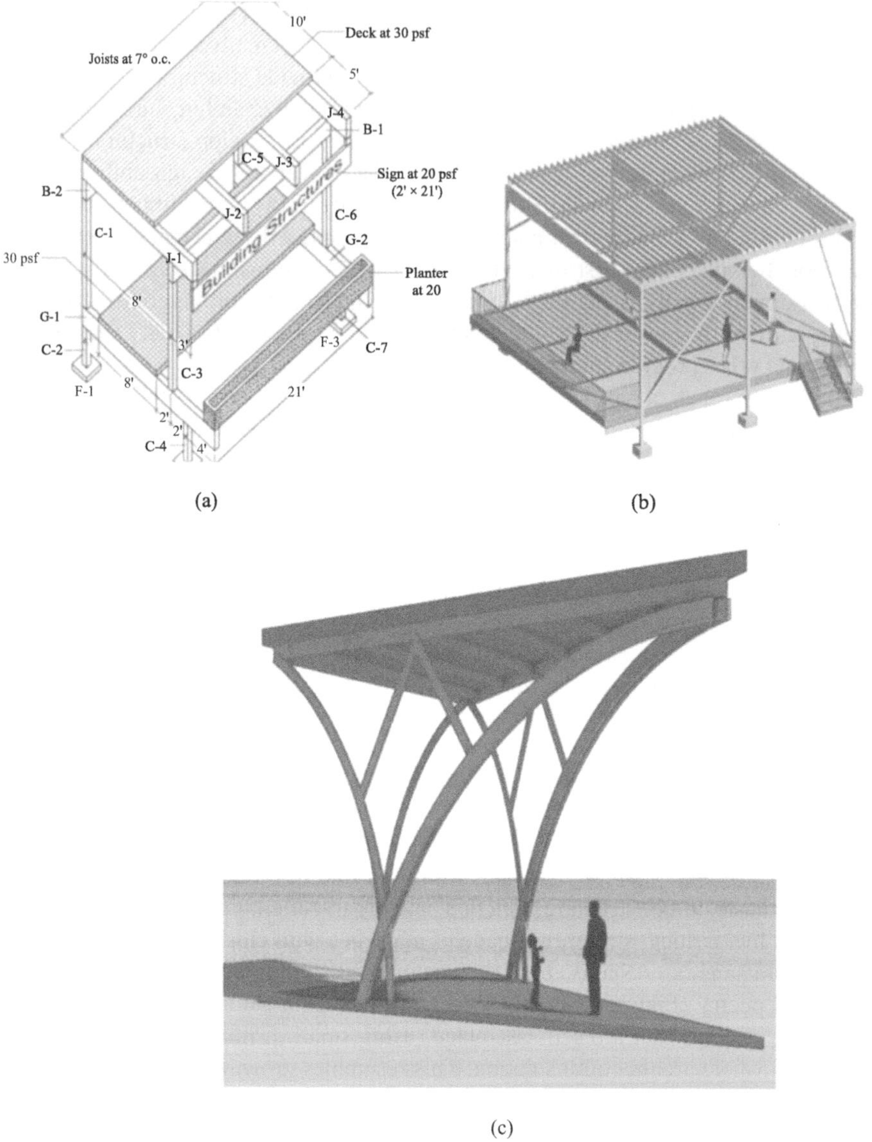

FIGURE 2.6 Basic units (buildoids). Examples of structural poetry. (a) Buildoid showing linear and planar elements; (b) wood framing buildoid; and (c) buildoid created using linear, nonlinear, and planar elements.

which they are assembled together (Hensel et al., 2010). This high level of integration of form, structure, and function is inherent in all biological forms (Figure 2.7).

Structural poetry has a similar process that creates, elaborates, and maintains structural forms emerging from a starting buildoid. Figure 2.8a depicts the plan organizations and growth process as one of the expressions in structural poetry. Examples of the full models for such a process are displayed in Figures 2.9 and 2.10.

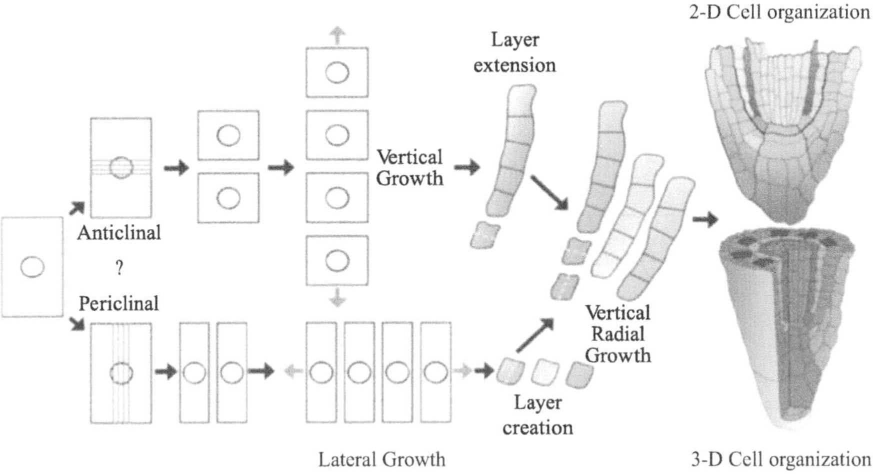

FIGURE 2.7 Profound influence of oriented cell divisions on plant morphogenesis as a result of absence of cell migration that allows straightforward analysis of the relations between cells, tissues, organs, and final plant architecture. (From Dhonukshe, 2011.)

Figures 2.8–2.10 depict the development of the basic BIM model (buildoid) to define larger spaces by expanding progressively in the horizontal and vertical directions. This illustrates how structural poetry provides various architectural programmatic solutions using BIM tools. The ability of the BIM tools to capture and analyze various architectural and structural attributed data associated with the building components, and their object-oriented modeling nature, allows structural poetry to achieve these solutions. Furthermore, in addition to using the same model data to generate numerous expansion solutions, the BIM model can be sent to other structural and architectural analysis software platforms without having to remodel the project. These content-driven and object-oriented modeling capabilities, along with the interoperability feature, make BIM tools superior to traditional computer-aided design (CAD) tools in applying structural poetry. The last phase in this growth progression shows complete structural and architectural components of the building (Figures 2.11 and 2.12).

Structural poetry is thus an art that is an integral part of building design that flourishes with engineering knowledge. The variations of structural forms shown in Figures 2.7–2.10 depict various architectural and structural fundamentals, such as the idea of elevated floors; cantilevers; simple trusses; frames; arches; shear walls; bracing; two levels of framing and three levels of framing; linear and nonlinear frames; spaces established (interior, exterior, private, and public spaces); lateral and vertical circulation; structural hierarchy and organization; spatial order; and aesthetics.

To explore further creative activities, interaction with existing iconic structures can be conducted. Analogy is made here to musical composers who study variations on musical themes by others, or to poets or philosophers who memorize and learn about other great works to test their own contributions. Such creative interaction with the work of other existing creative models can be a source of ideas and can develop an understanding of how architecture interacts with structural components.

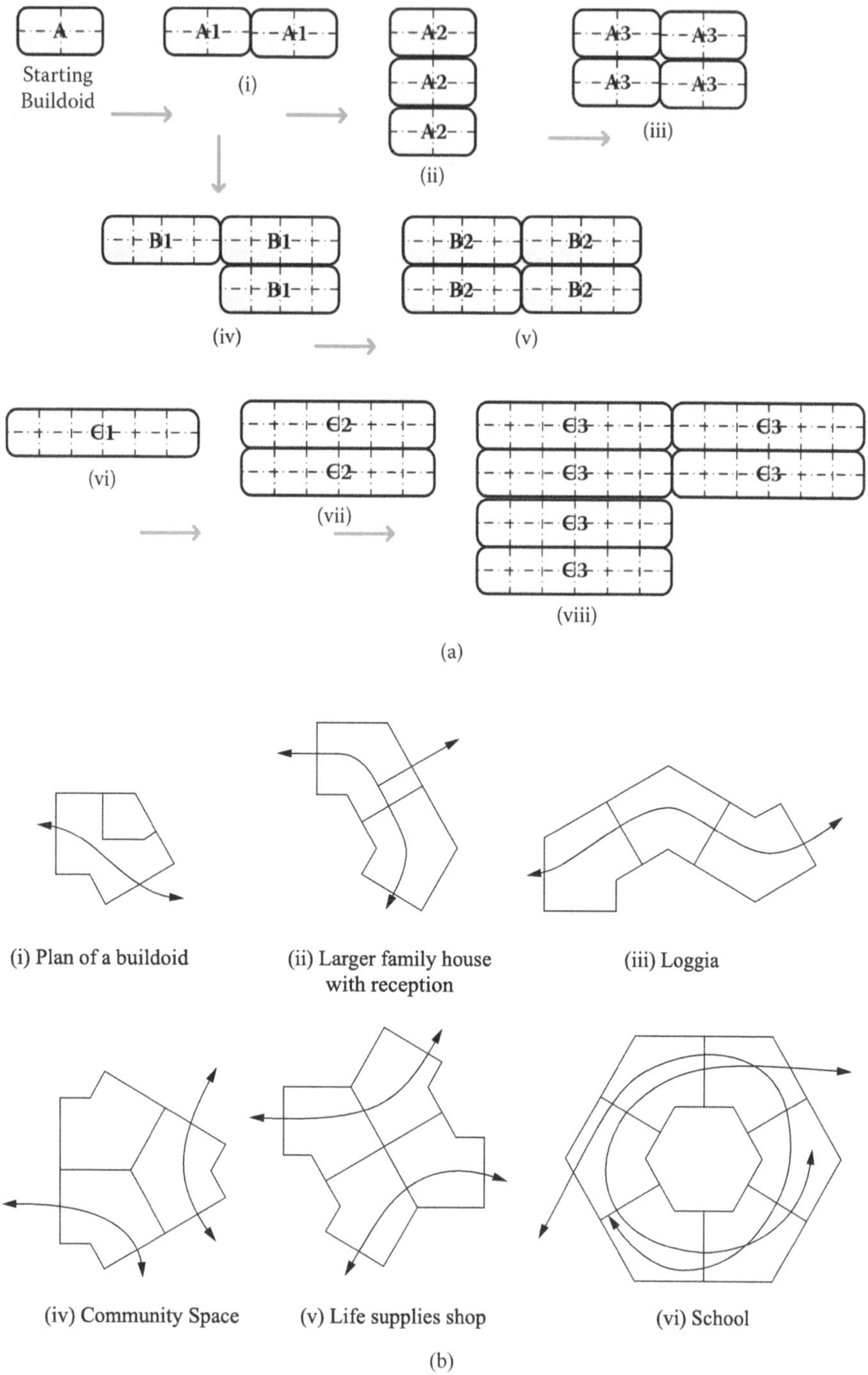

FIGURE 2.8 (a) Growth patterns of a buildoid having a rectangular plan. (b) Growth and organization of a buildoid having an irregular plan.

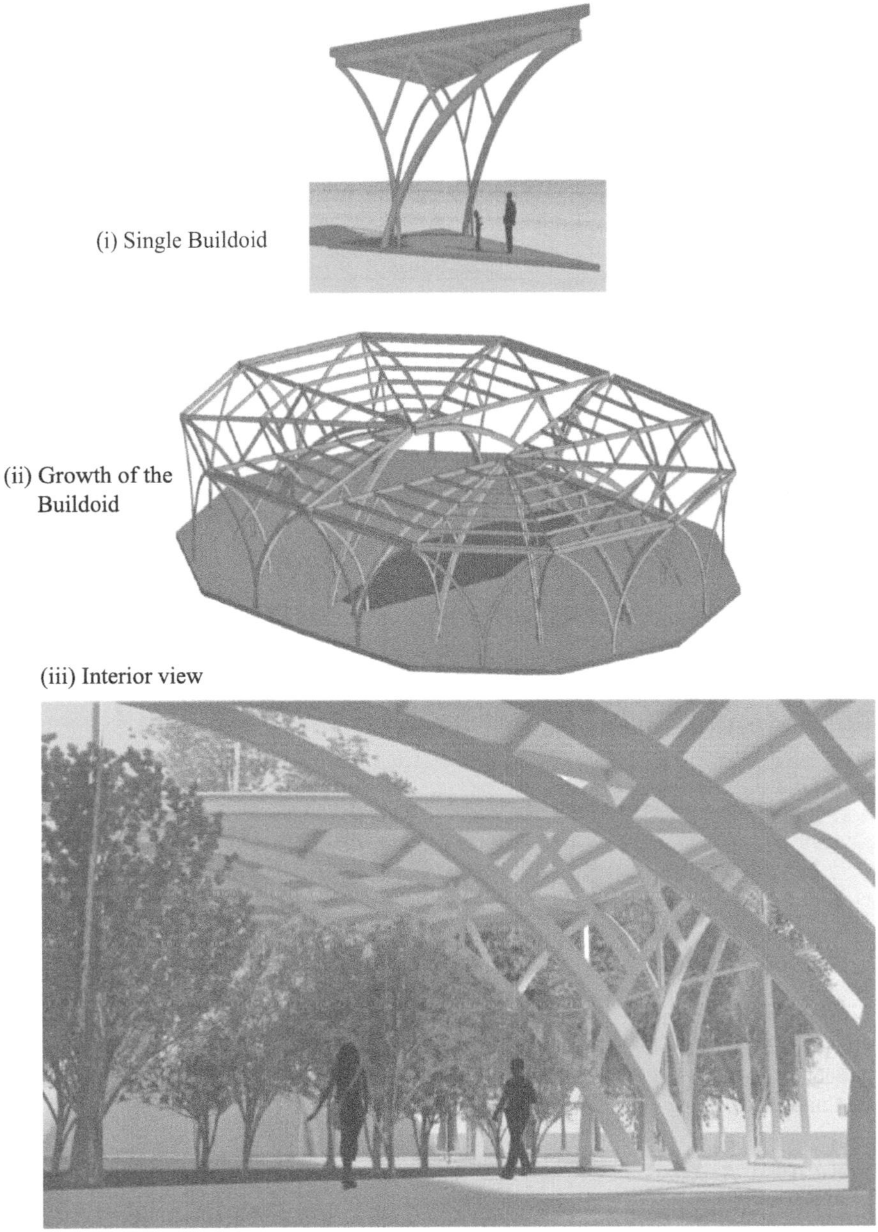

FIGURE 2.9 Examples of structural poetry.

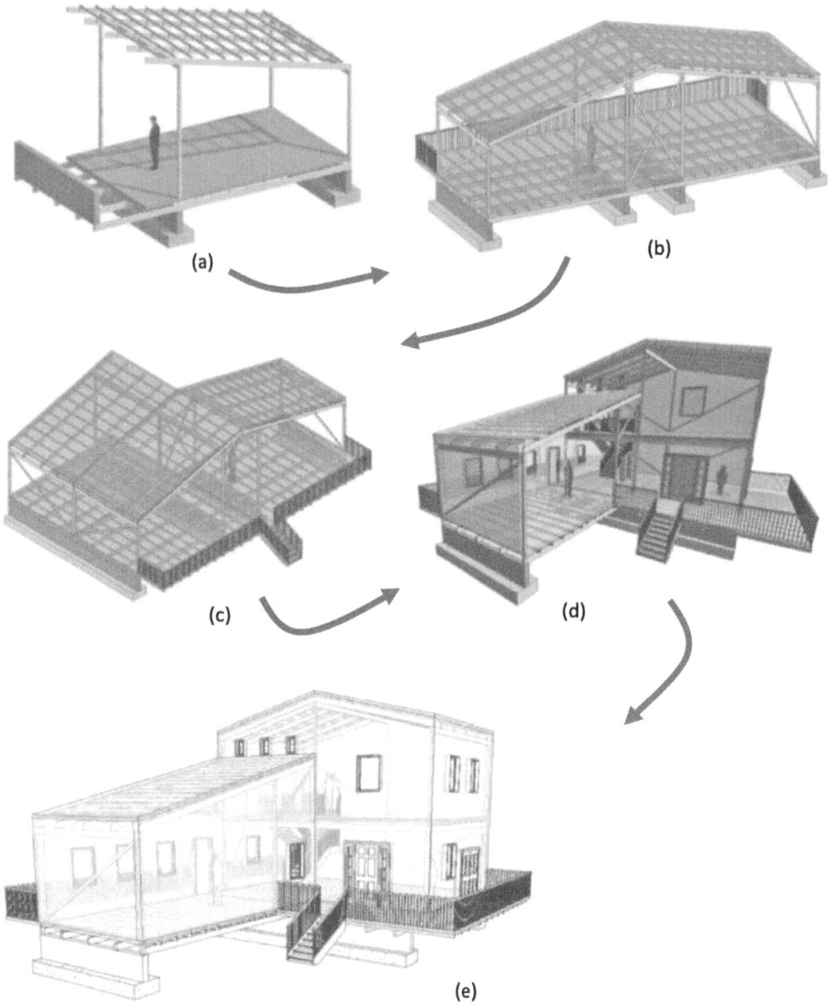

FIGURE 2.10 Growth in structural poetry. (a) Starting buildoid; (b) Combining two buildoids; (c) Lateral growth: combining three buildoids; (d) Vertical growth of (c); and (e) Complete building with all architectural elements.

Structural Analysis

After completing structural melody and poetry phases, BIM models are subjected to structural analysis. Various analysis tools within BIM platforms will be introduced.

BIM tools used in this phase are principally the beam, truss, frame simulation, the load takedown, and the integration with robot structural analysis. The load takedown played an important role in introducing load path, load tracing, reactions, and constraints in building structures (see Figure 2.11). Students will be able to understand

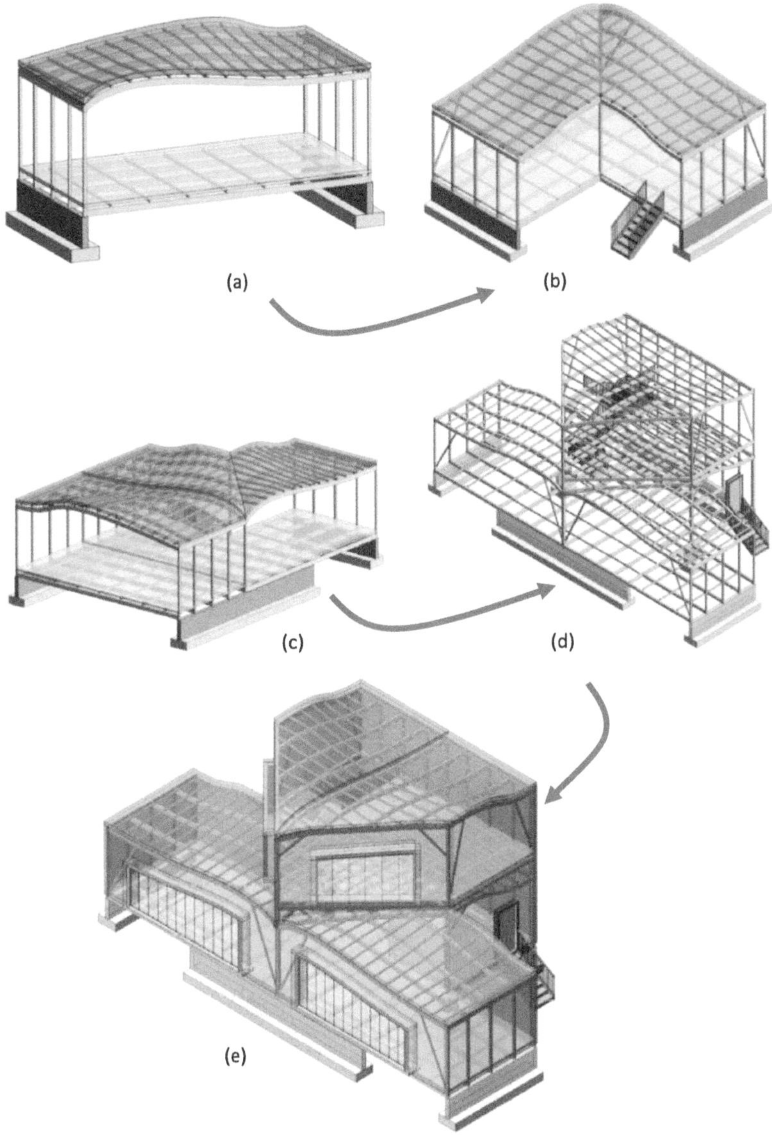

FIGURE 2.11 Progression in structural poetry. (a) Starting buildoid with nonlinear elements; (b) Arranging two buildoids; (c) Lateral growth: three buildoids; (d) Vertical progression of (c); and (e) Final stage: all main structural and architectural elements.

concepts such as tributary areas for beams, girders, and columns in a visually interactive manner, which can greatly stimulate interest and motivation to explore other analysis capabilities of the tools (Nawari et al., 2011). Furthermore, advanced topics such as nonlinear analysis and dynamic behavior can be investigated using structural analysis tools that are fully integrated with BIM platforms.

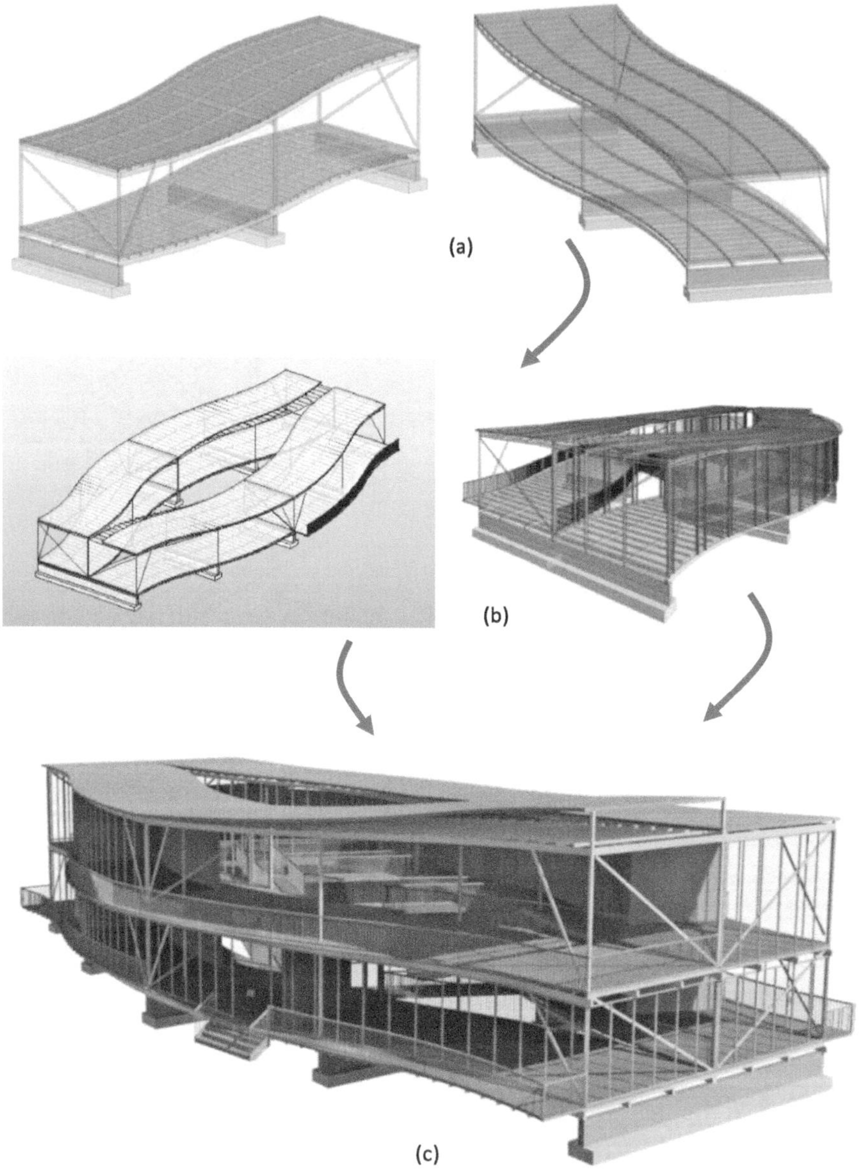

FIGURE 2.12 Progression in structural poetry: (a) starting buildoid; (b) lateral growth of buildoids; and (c) vertical progression.

Chapter 6 addresses BIM tools for structural analysis and how they can be applied to various structural buildoids to promote the understanding of fundamentals of structural analysis, such as the force equilibrium, support reactions, shear force, and bending moment diagrams; frame and truss analysis; and steel, wood, and concrete design. Figure 2.13 depicts an overview of the various BIM structural analysis tools utilized in this book.

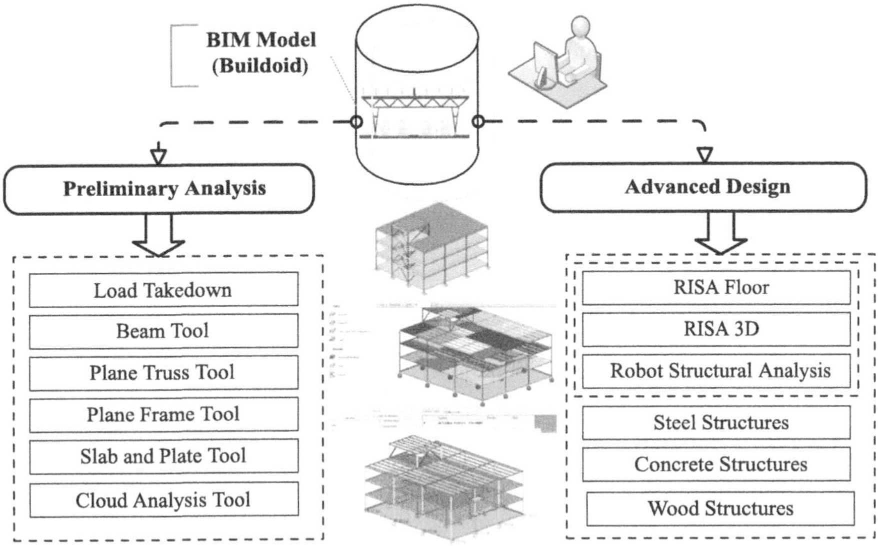

FIGURE 2.13 BIM tools for structural analysis.

EXERCISES

(Chapters 1 and 2)

1.1. What purpose does the structure serve in architectural design?

1.2. What is the oldest architectural structure recorded in history? Who was the designer of that structure and is considered to be the first architect/engineer known by name?

1.3. Describe briefly the role of the following scientists in advancing architectural structures:
   a. Galileo Galilei
   b. Robert Hooke
   c. Isaac Newton

1.4. Briefly summarize the advancement that took place in structural analysis during the 18th and 19th centuries.

1.5. Pier Luigi Nervi and Viollet-le-Duc shared the same principles in terms of the relationship between architecture and structure. Describe these principles.

1.6. What are the main ideas behind the SAS framework?

1.7. Describe briefly the main components of the SAS.

1.8. What is the role of structural analysis within the SAS?

1.9. What is the role of BIM in the SAS?

1.10. How does the structural system influence the spatial organization of a building?

1.11. In what ways do structural elements contribute to the aesthetic expression of architecture?

1.12. How do different structural systems (e.g., frame, shell, membrane, beams and columns) shape the architectural form and experience?
1.13. Discuss how technological innovations in materials have expanded the purposes and possibilities of structure in architecture.
1.14. In what ways does a structure act as both a physical and symbolic element in cultural or iconic architecture?
1.15. Describe the role of structure in enabling adaptability and flexibility within architectural design.

3 Building Information Modeling

INTRODUCTION

The building information model (BIM) provides the three-dimensional (3D) objects library of the physical building. In essence, BIM is a way to construct a building virtually before building it in the real world. It is a digital simulation of the physical and functional characteristics of a building. As such, it functions as a shared knowledge resource for information about a building, forming a dependable basis for decisions during its life cycle from inception onward.

Creating a BIM is different from making a drawing in two-dimensional (2D) or 3D computer-aided design (CAD). To create a BIM, a modeler uses intelligent objects to build the model. Unlike 2D or 3D CAD drawings, when you make a revision or change in any element in the model, you have to change it only once, and all the views and details in the model are automatically updated. Not only does this feature make revising a structural design almost effortless, but it also virtually eliminates the possibilities of errors associated with uncoordinated drawings. It is important to note that BIM is entirely unlike the CAD tools that emerged over the past 50 years and are still in wide use in today's practice. BIM methodologies, motivations, and principles represent a shift away from the kind of assisted drafting systems that CAD offers.

Building information modeling is a process that fundamentally changes the role of computation in building design. It is the human activity of using BIM software and other related technologies to create and share BIM models. It means that rather than using a computer to help produce a series of drawings and schedules that together describe a building, the computer is used to produce a single, unified representation of the building so complete that it can generate all necessary documentation. The primitives from which these models are composed are not the same ones used in CAD (points, lines, curves). Instead, you model with building components such as walls, columns, beams, doors, windows, ceilings, and roofs. The software used to do this recognizes the form and behavior of these components, so it can ease much of the tedium of their manipulation. Walls, for instance, join and miter automatically, connecting structure layers to structure layers and finish layers to finish layers. Many of the advantages are obvious; for instance, changes made in elevation propagate automatically to every plan, elevation, section, callout, and rendering of the project. Other advantages are indirect and take some investigation to discover. Some of these benefits are illustrated next.

Building information modeling expands significantly the design decision matrix, facilitating collaboration among the various disciplines and systems and identifying clashes before construction begins. Incorporating constructability into the structural

DOI: 10.1201/9781003598435-3

and architectural design decision matrix has the potential to help the design team make smarter decisions and to have a positive impact on overall project efficiency and quality. BIM can deliver projects that integrate design and construction insights in a highly collaborative and owner-friendly format.

In a nutshell, BIM represents the hub of data and applications that streamline the delivery process of design, detailing, manufacturing, construction, and operation. Its value as an integrator between the technology tools used to perform various functions of the architecture, engineering, and construction (AEC) industry and the ability of computational and simulation software to manipulate the model directly, with or without human intervention , is significant. In a typical BIM-enabled design environment, the data model serves as the principal means for communication between activities and professionals (Figure 3.1).

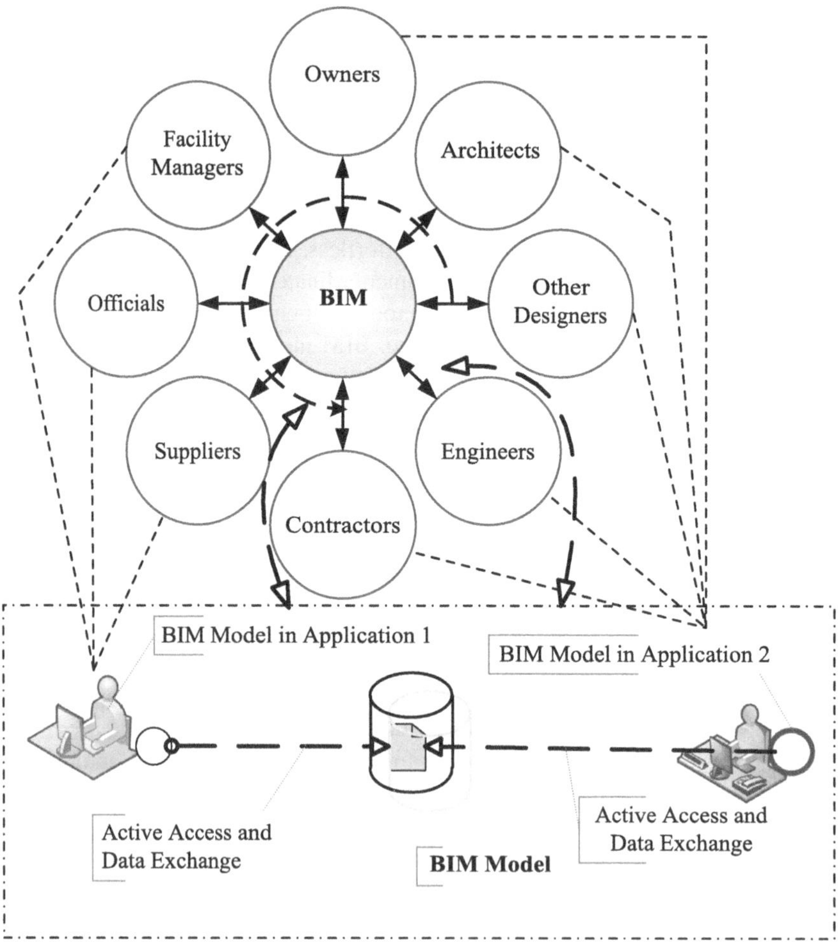

FIGURE 3.1 BIM concept and process.

According to the National Building Information Modeling Standard (National Institute of Building Sciences, 2007), a BIM is a digital representation of the physical and functional characteristics of a facility and its related project/life-cycle information, and it is intended to be a repository of information for the facility owner/operator to use and maintain throughout the lifetime of the structure. A fundamental premise of BIM is collaboration by different stakeholders at different phases of the life cycle of a project to insert, extract, update, or modify data in the BIM process to support and reflect the roles of that stakeholder. Thus, a BIM is a shared digital representation of a facility founded on open standards for practical interoperability (see Figure 3.1).

The following sections examine the key principles and concepts behind BIM.

LEVEL OF DEVELOPMENT (LOD)

The Level of Development (LOD) Specification serves as a vital reference tool designed to enhance communication among users of Building Information Models (BIM). It builds upon the LOD schema established by AIA Contract Documents in the E201 2022 BIM Exhibit for sharing models with project participants. The LOD Specification offers detailed definitions and visual representations of BIM elements across various building systems, illustrating their characteristics at different stages of development and application throughout the design and construction process.

Users of the LOD Specification should not dictate the required levels of development for various phases of the construction process; this decision is left to the discretion of each project team. However, it is anticipated that the clearer definitions provided by the specification will help minimize miscommunication among project team members. This clarity aids in establishing expectations for different stages of Design and construction, allowing team members to better understand their deliverables and enhancing the predictability of the effort needed to produce them.

The Level of Development (LOD) framework in Building Information Modeling (BIM) has seen significant updates to enhance clarity and consistency in model specifications. In December 2023, the BIMForum released the 2023 LOD Specification, marking the tenth anniversary of this essential document. This edition focuses on resolving conflicts and improving the readability of text and graphics, ensuring a more user-friendly experience for practitioners. Notably, the update maintains the core requirements of previous versions, with only minor relaxations where necessary.

Additionally, the 2023 draft introduced a major update to the landscape section (G2080), incorporating elements such as irrigation distinctions between LOD 200 and 300 and detailed modeling requirements at LOD 350, including canopy clearance at maturity and staking details. This expansion aims to provide comprehensive guidance for landscape architecture within the BIM framework.

These advancements reflect the industry's commitment to refining BIM practices, promoting effective communication among stakeholders, and ensuring that models accurately represent project requirements at various stages of development.

The LOD Specification is organized by CSI (Construction Specification Institute) Uniformat 2010 and Omniclass2, with the subclasses expanded to Level 4 (and in a few cases to Level 5) to provide detail and clarity to the element definitions.

See BIMForum LOD Spec Part II, tab 3, for a Uniclass-to-Omniclass/Uniformat cross-reference table. UniFormat™ and Omniclass are from UniFormat™, published by CSI and Construction Specifications Canada (CSC).

Here are definitions for reference tables according to the Construction Specifications Institute (CSI):

UniFormat: UniFormat is a classification system developed by CSI that organizes construction information based on systems or assemblies rather than individual products or materials. It is structured around major building systems (e.g., foundation, superstructure, mechanical, electrical) to assist with early project stages, such as schematic Design, cost estimating, and performance specifications. By grouping information based on functional systems, UniFormat facilitates better coordination in the early phases of Design and planning.

OmniClass: OmniClass (OmniClass Construction Classification System, or OCCS) is a comprehensive classification system for the construction industry, developed to organize and manage information throughout a building's lifecycle. OmniClass is designed to accommodate a broad range of information, covering areas such as construction entities, elements, spaces, and roles. It integrates with other classification systems, such as MasterFormat and UniFormat, providing a flexible, standardized framework for organizing project information across various phases, from planning to demolition.

UniClass: It is originally developed in the United Kingdom, is a comprehensive classification system for organizing information across all sectors of the construction industry. It covers various levels, including elements, work sections, spaces, and products, to facilitate interoperability and consistency in project data management. While UniClass is widely used internationally, it is compatible with U.S.-based systems like OmniClass and has been incorporated into digital design and construction workflows to support information management and BIM processes.

The following table summarizes and presents examples of these classification systems (Tables 3.1 and 3.2).

The LOD Specification serves as a reference standard and is designed to evolve alongside the advancement of BIM practices. Updated annually, previous versions are archived and accessible on the BIMForum website (www.bimforum.org/lod).

The major updates to the 2023 LOD specifications focus on the following aspects:

- **Elimination of Nebulous Terms:**
- **Simplification:** An example of this is the Interior Finishes section. In looking at the section as a whole, it was realized that the modeling needs were dictated by the thickness of the finish rather than the material. Thus, requirements based on thickness are defined for "C20 Interior Finishes", and these requirements apply to all children of that heading.

TABLE 3.1
Uniformat, OminiClass, and UniClass Definitions Using Single Concrete Footing

UniFormat	OminiClass	UniClass
UniFormat organizes building information based on systems and assemblies. A single concrete footing would be classified under: **A - Substructure** **A10 - Foundations** **A1010 - Standard Foundations** **A1010.20 - Column Foundations** **A1010.20 - Column Foundations**: This classification refers to individual footings that support structural columns, which is typical for single concrete footings.	OmniClass provides a comprehensive classification system with multiple tables. For a concrete footing, you would use: **Table 21 - Elements (Uniformat):** **21 - 01 20 20 - Standard Foundations** **21 - 01 20 20 12 - Column Foundations** **21 - 01 20 20 12 - Column Foundations**: This code specifies foundations designed for columns, aligning with the single concrete footing used to support a column.	UniClass is a unified classification system used primarily in the UK. For a concrete footing, refer to the Elements/Functions table: **EF - Elements/Functions** **EF_30 - Substructure** **EF_30_10 - Foundations** **EF_30_10_25 - Pad Foundations.** **EF_30_10_25 - Pad Foundations**: This category is for isolated (pad) foundations, which are individual footings like single concrete footings used to support columns or pillars.

TABLE 3.2
Summary of Table 3.1

Classification System	Code	Description
UniFormat	A1010.20	Column Foundations
OmniClass	21-01 20 20 12	Column Foundations
UniClass	EF_30_10_25	Pad Foundations

- **Defined Terms:** This section has been added to succinctly define terms that have specialized meaning in this Spec. The defined terms are in boldface where they appear in the LOD narrative descriptions.
- **LOD Requirements:** In the narrative descriptions of LOD requirements, only items required to be included are listed. No reference is made to items that are not required and no items have been listed as prohibited.
- **Narrative Descriptions:** These have been organized into lists of specific required items to facilitate use of the Spec as a checklist.
- **Non-Geometric Information:** On the principle that any number of non-geometric attributes can be attached to a model element of any LOD, non-geometric information is not addressed in Part I.
- **Duplicated Content:** The Spec was scrubbed for duplicated content – where found it was deleted or replaced with cross references.

BASIC LOD DEFINITIONS

LOD 100

BIMForum Interpretation: LOD 100 elements do not need to be geometric representations. Examples include information linked to other model elements, such as symbols indicating the presence of a component without specifying its shape, size, or exact location, or placeholder volumes for reserved spaces. Essentially, if information about an element can be inferred from the model but lacks the detail of LOD 200, it is classified as LOD 100. Any data derived from LOD 100 elements should be considered approximate.

LOD 200

BIMForum Interpretation: LOD 200 elements serve as generic placeholders that are identifiable as the components they represent (e.g., a pump, light fixture, or beam). Any information derived from LOD 200 elements should still be regarded as approximate.

LOD 300

BIMForum Interpretation: LOD 300 elements are developed in enough detail to clearly communicate the design intent of the represented item. While the LOD definitions and this Specification do not dictate who is responsible for modeling each element, it is uncommon for designers to produce model elements beyond LOD 300. For further details, refer to the interpretation of LOD 350 below.

LOD 350

BIMForum Interpretation: LOD 350 establishes requirements for model elements detailed enough to enable construction-level coordination. Achieving this level typically requires discipline-specific expertise; hence the note that designers rarely produce elements beyond LOD 300. However, it's important to note that neither the LOD definitions nor this Specification dictate who should model each element—design teams with access to discipline knowledge may choose to develop elements to LOD 350 or even higher.

Uniformat	Omniclass	Uniclass

B1010.10.30 21-02 10 10 10 30
Floor Structural Frame, Steel Framing Columns
Associate Masterformat sections: 05 10 00

LOD	Detail	Graphic
100	See B1010	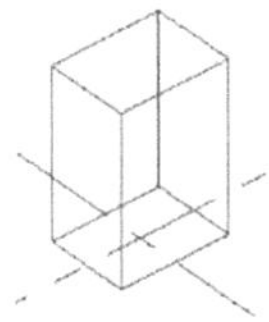
200	See B1010	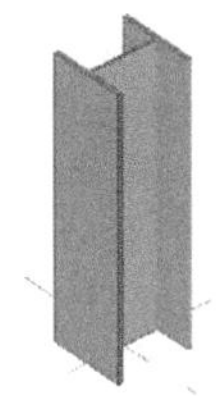
300	Presences: • Specific section type and size	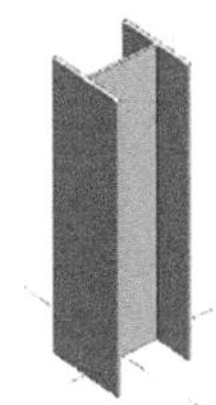
350	Presences: • Member connections such as base plates and gusset plates, anchor rods. • Connection details with correct and reliable dimensions. • Steel structural reinforcement and stiffeners (e.g. for penetration). • Cap plates.	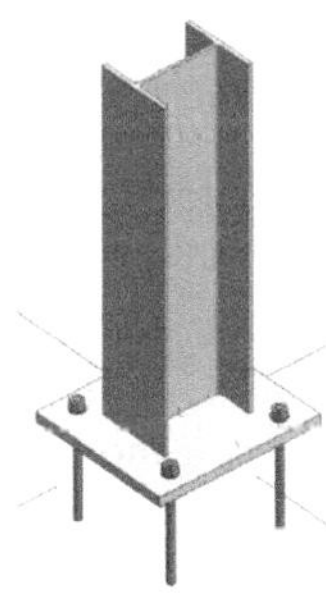
400	Presences: • Welds • Coping of members • Washers, nuts, etc.	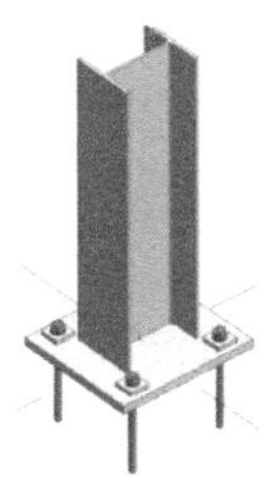

CUSTOMIZATION AND REUSE

In the process of design, we often need to draw an element multiple times. For instance, you are always required to draw things multiple times to see them in different views and representations. Drawing a door in a plan does not automatically place a door in your elevations or sections. Therefore, the traditional CAD program requires that you draw the same door object several times. In a BIM environment, you model objects only once and reuse them multiple times.

What are the Issues with Doing Anything More than Once?

There is more to the idea of not drawing anything more than once than just saving time in the initial work of design representation. Assume that someone has drawn a door or a beam in a plan and added that same object in two sections and one elevation. Now, if someone decides to move that door or beam, then suddenly every other representation of that element needs to be found, and its location also changed. In a complicated set of drawing details, the likelihood that someone can find all the instances of that element and make the necessary changes the first time is low.

TRACKING AND REPRESENTATION

Reference tracking and storing are two of about three fundamental operations that a computer does efficiently. In fact, computers are exceedingly good at them. The fundamental principle of BIM design is that you are designing not by drafting the forms of objects in a specific representation but by placing a reference to an object into a holistic, unified model. When you place a beam into your plan (Figure 3.2), it automatically appears in any section, elevation, or other plan in which it ought to be visible. When you move it, all these views update because there is only one instance of that beam in the model. However, there may be many representations of it.

IT IS NOT ONLY ABOUT DRAFTING

As one creates a model in a BIM software platform, one is not making a drawing. Normally, you have to specify only as much geometry as is necessary to locate and describe the building elements that need to be placed into the model. For instance, once a wall exists, to place a door into it, you need to specify only what type of door it is (which automatically determines all its internal geometry) and how far along the wall it is to be placed. That same information places a door in as many drawing views as there are. No further specification work is required.

Therefore, the act of placing a door or a truss into a model is not at all like drawing a door in a plan or elevation or even modeling it in 3D. You do not make solids or draw lines. You simply choose the type of door or truss from a library and select the location in a wall. You do not draft it.

IT IS NOT JUST A TRADITIONAL 3D MODEL

In a BIM environment, you will be creating a BIM. That means making detailed data sets that define a building. This should not be confused with making a classical 3D drawing of a building. A traditional CAD 3D model is just another representation of a building model with the same incompleteness as a plan or section.

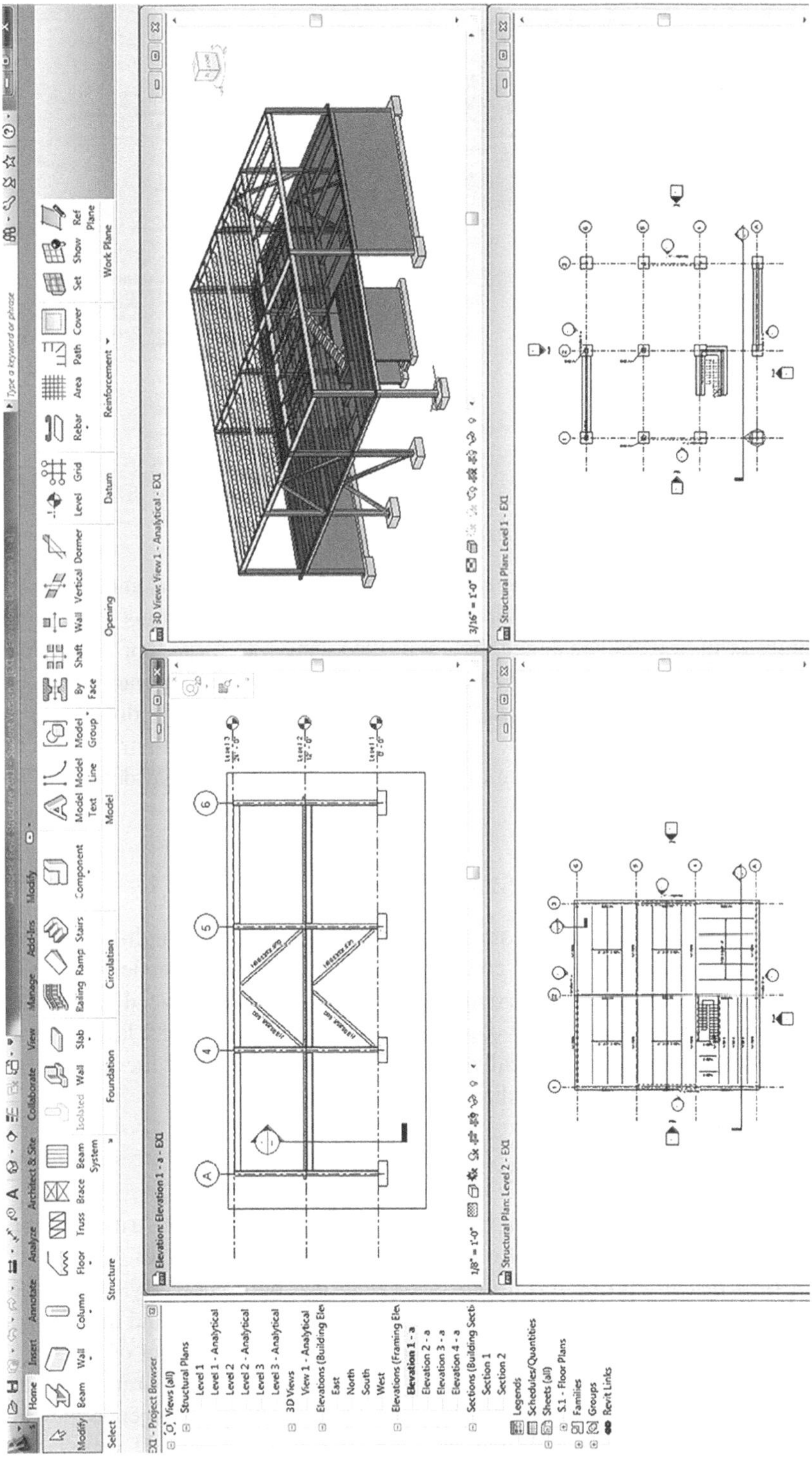

FIGURE 3.2 A beam modeled in a plan will appear in various views.

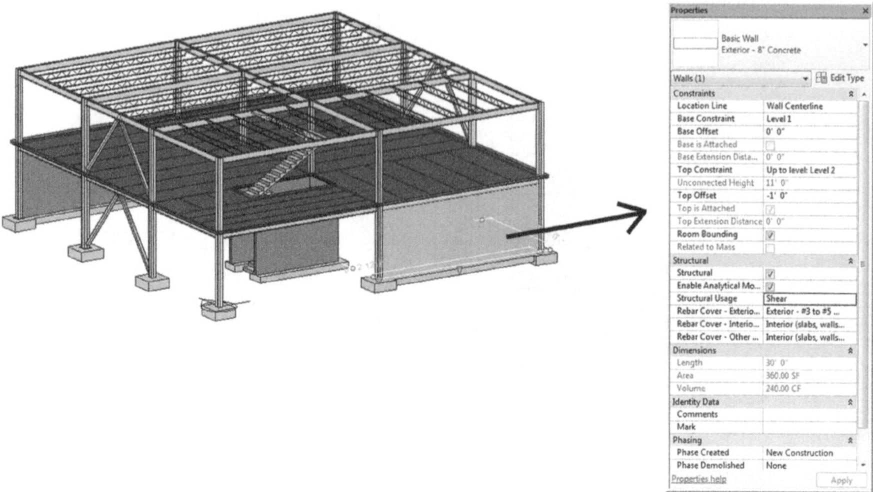

FIGURE 3.3 Reinforced concrete shear wall properties.

A full 3D model can be cut to reveal the basic outlines for sections and plans, but there are drawing conventions associated with these representations that cannot be captured this way. For instance, how will the swing line for a door be encoded into a 3D model? For a system to intelligently place a door swing into a plan but not into a 3D model, you would need a high-level description of the door building model elements separate from the geometric 3D description of their form. This digital data intelligence of storing and retrieving data sets is what describes the BIM environment.

MODEL CONTENT AND DESIGN INTENT

The BIM model encodes more than geometric forms; it encodes high-level design intent. Within the model, beams, columns, walls, and roofs are modeled not as a series of 3D solids, but as beams, columns, walls, and roofs that have all the properties and characteristics of physical objects (Figure 3.3). That way, if a level changes height, all of the objects automatically adjust to the new criterion. If the wall moves, any roof that has a relationship to that wall adjusts automatically.

OBJECTS AND PARAMETERS

Every object in a BIM environment has parameters: beams, columns, trusses, doors, windows, walls, ceilings, roofs, floors, and even drawings themselves. The objects used in BIM design encode much more data than just pure geometry, such as the ability to have much more information about the objects themselves and their associated objects. Parameters are quantities that are fixed under a given set of conditions but may be different under other conditions. Thus, some of these parameters have constant values, and some are modifiable. To allow flexibility in objects, they are created with a set of parameters that can take on arbitrary values. If you want to

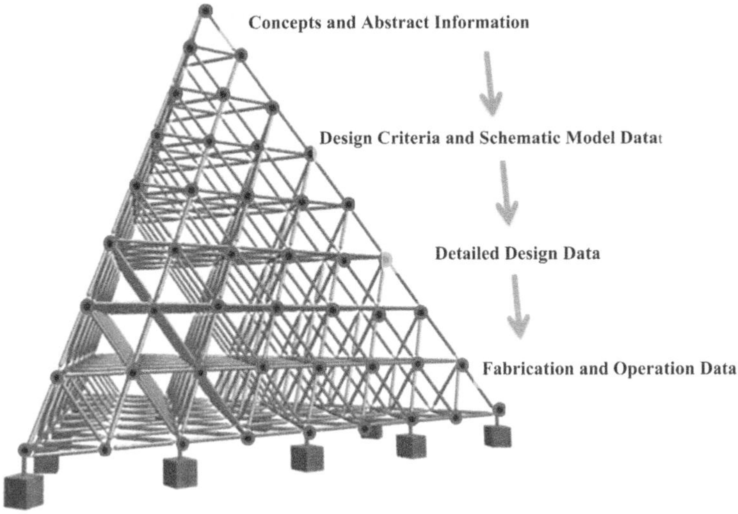

FIGURE 3.4 BIM model data development.

create a truss that is 8 ft high, it is only necessary to modify the height parameter of an existing truss. Then, all members of the truss will adjust according to the new height of the truss. In advanced modeling, you also can create custom object types with parameters of your choice.

Building information modeling design basically proceeds by "placing" objects into a model and then adjusting their parameters. These objects are fully developed building elements like walls, doors, and windows. Although their parameters may vary, the placed objects retain their fundamental identities. A wall that is made wider remains a wall no matter what. As the design process proceeds, more parameters and values will be added to the model.

In these processes, information in a BIM model is produced while information in the actual space is being established. Conceptual design information becomes concrete over a period of time as the design proceeds forward toward construction and operations. Information in a BIM process grows in a pyramid-like form (Figure 3.4), expanding continuously from the abstract to detailed information in a coherent manner to ensure the efficiency and quality of the construction and operation process of a facility.

DATA SHARING AND COLLABORATION

Unlike traditional CAD systems, in BIM, building information data are attached to each building object, creating comprehensive architectural and structural content libraries as well as mechanical, electrical, plumbing, landscape, and other libraries. BIM systems normally allow for various operations on the data created and stored. These include, for example, rich rendering and simulation capabilities, quantity takeoffs, scheduling, collaboration, and more.

Most BIM applications provide ways for seamless sharing of the virtual building data between the project team members. These can be further set up as collaboration environments. For example, internal collaboration between members of one company can be established as follows:

- Create a central database file storing the complete virtual building information.
- Team members work on local copies.
- Team members must have dedicated workspaces.
- Team members send and receive changes regularly from the centralized database file.

External collaboration between various companies participating in a project can be established by sharing the BIM data via different data formats that most BIM systems support. These include, for example,

- IFC (Industry Foundation Classes; http://www.buildingsmart-tech.org/ [publication of the IFC specification])
- DXF/DWG (AutoCAD Drawing Exchange Format/Drawing)
- PDF (Portable Document Format)
- XML (Extensible Markup Language)

IFC DATA SCHEMA AND STANDARDS

The Industry Foundation Classes (IFC) schema is an open, standardized data model used in the construction industry to facilitate interoperability and data exchange among various software applications. It is a key data format of Building Information Modeling (BIM) and is developed and maintained by **buildingSMART International**.

The main features of the IFC data schema are:

a. **Interoperability**: IFC serves as a universal data format to exchange information across different software platforms, eliminating silos in Design, construction, and facility management.
b. **Open Standard**: It is an open and vendor-neutral standard, ensuring transparency and broad adoption without proprietary constraints.
c. **Rich Data Model**: It encompasses geometry, spatial relationships, materials, schedules, and other attributes to fully represent buildings and infrastructure throughout their lifecycle.
d. **File Formats**: The IFC schema supports various file formats, such as IFC-SPF (.ifc), IFC-XML (.ifcxml), and IFC-ZIP (.ifczip), making it adaptable for different needs.

The IFS is now an ISO Standard. It has been recognized as an international standard under ISO 16739, demonstrating its global acceptance and reliability for BIM data exchange. Furthermore, the IFC schema supports various phases of a project, including conceptual Design, schematic Design, detailed Design, construction, operation, and maintenance.

By providing a structured and standardized schema to manage complex construction data, the IFC data format promotes efficiency, reduces errors, and enables better collaboration across stakeholders in the AEC industry.

For information on any construction project, IFC offers detailed and comprehensive specifications. This means IFC represents object definitions that are internationally standardized. The fundamental concept of IFC exchanges is that they are object-based. These objects have a 3D description geometry, with properties such as height, position, finish, and cost of their element name. Some objects are "abstract," such as calculation, cost, and quantities, while others are "real" like columns and beams. With IFC, the knowledge about an object can be shared with the various software programs from different members of the AEC industry and the many related subsidiary participants.

An example of IFC hierarchal schema is depicted in (Figure 3.5) (Nawari, 2018).

There are several versions of the IFC data schema. These versions are developed by building SMART and are regularly updated. It is preferable whenever possible to use the latest version. The newest version IFC4.3 provides better representation of complex geometries in addition to other features. Currently there are three versions supported by Autodesk Revit which are:

- **IFC4.3:** Builds upon IFC4 with enhanced geometric descriptions and interoperability. It Introduced to expand the IFC schema to better support infrastructure projects, such as roads, bridges, and railways. IFC4.3 is expected to play a critical role in bridging gaps between BIM and infrastructure information modeling. It is anticipated to be certified as ISO 16739-2. By bridging BIM and infrastructure modeling, IFC4.3 represents a significant leap forward in addressing the complexities of large-scale building construction and civil engineering projects.
- **IFC4:** It Introduced significant improvements in data definitions, geometry modeling, and classification systems. Also, it enhanced support for model-based workflows and extended the schema for infrastructure domains. IFC4 is recognized as ISO 16739:2013, establishing it as a global standard.

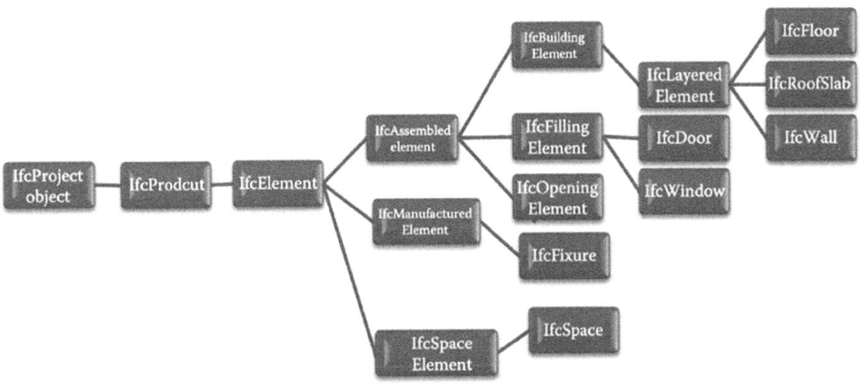

FIGURE 3.5 Example showing simplified part of the IFC classes hierarchy. (Nawari, 2018.)

- **IFC2 × 3:** It introduced better support for material definitions, spatial hierarchies, and property sets. It is currently the most supported and stable format, certified in Revit and recommended for production.
- **IFC2 × 2:** Early version of IFC and focused on basic interoperability with limited scope and data structures. (Recommended if the recipient of the file does not have software with IFC2 × 3 or IFC4 support).

BIM PLATFORMS

Currently, a number of software platforms support BIM concepts and principles. These include Autodesk Revit, ArchiCAD, Bentley Architecture, and Tekla Structures. In this book, Revit is used in most of the examples and exercises. A brief description of these BIM systems is given next.

Autodesk Revit

The Revit platform for BIM is a design and documentation system that supports the design, drawings, and schedules required for a building project (Autodesk, 2013).

BIM delivers information about project design, scope, quantities, and phases when you need it. It is a software platform with multidimensional capabilities (nD), with tools to plan and track various stages in the building's life cycle, from concept to construction and later demolition. Thus, Revit provides a practical collaborative environment between different disciplines in the building construction industry.

The platform also offers cloud integration, which helps users create and find the custom objects and components they need to make their BIM models complete. Revit also supports integrated model exchange management using neutral data forms such as IFC, gbXML (Green Building XML), and DXF/DWG round-trip conversion solutions between different applications. It offers extension engines that support structural and energy analysis.

ArchiCAD

ArchiCAD is a BIM software system offered by GraphiSoft® Incorporated. ArchiCAD creates a 3D BIM and all the necessary documentation and visualization. Based on profound knowledge of the architectural process, ArchiCAD's BIM simulates the way a real building is constructed. ArchiCAD's BIM tools cover everything from town planning to intricate details, from functional studies to complex designs (GraphiSoft Inc., 2014).

ArchiCAD enables combining architectural freedom with the BIM with a comprehensive set of tools that supports the design process. It has direct modeling capabilities in the native BIM environment, with its end-to-end BIM workflow using priority-based connections and intelligent building materials, as well as improved analysis tools. The platform also offers cloud integration that helps users create and find the custom objects and components they need to make their BIM models complete.

ArchiCAD supports integrated model exchange management using neutral data forms such as IFC and DXF/DWG round-trip conversion solutions between different applications. The platform offers has an energy evaluation engine that supports multiple thermal blocks.

Bentley Architecture

Bentley Architecture is a BIM application offered by Bentley Systems Incorporated. The software provides architects and designers with the tools to explore design options, to make better-informed design decisions, and to predict costs and performance. Design and construction documents are automatically coordinated, eliminating errors and omissions. It supports all phases of the architectural workflow, from conceptual design to construction documentation, and integrates design, visualization, drawing production, and reporting of quantities and costs (Bentley Inc., 2014). The software has a full range of advanced solids modeling tools, which allow the creation of virtually any feasible form.

The Bentley BIM platform integrates well with other Bentley building engineering, analysis and facilities management applications, such as the Structural Modeler; Bentley Building Mechanical Systems; Bentley Building Electrical Systems; Bentley Facilities; and more to facilitate a shared multidisciplinary model for team collaboration and coordination.

Tekla Structures

Tekla Structures is a software platform for BIM. The software enables users to create and manage 3D structural models in concrete, wood, or steel from concept to fabrication. The platform supports the automatic process of producing shop drawings, along with the creation of computer numerical control (CNC) files for controlling fabrication machines. Tekla Structures is available in different configurations and localized environments to suit different segment- and culture-specific needs.

Tekla offers a free tool known as Tekla BIMsight, which is a software application for BIM model-based construction project collaboration. It can import models from other BIM software applications using the IFC data format and other formats such as DWG and DGN.* With this free tool, you can perform spatial coordination, such as clash detection or conflict checking, to avoid design and constructability issues and communicate with others in a construction project by sharing models and review notes. This enables project participants to identify and solve issues in the design phase before construction.

BIM PROCESSES IN PRACTICE

The BIM process involves the generation, management, and sharing of digital representations of buildings and infrastructure throughout their lifecycle. It is a collaborative design process that integrates information about geometry, spatial relationships, materials, schedules, cost, and more. Here's how the BIM process is typically applied in practice:

Planning and Conceptual Design

The goal in this preliminary design phase is to Establish project requirements, feasibility, and initial design concepts.

- **Activities:**
  - Collecting client requirements and site data.
  - Creating preliminary BIM models to explore design concepts.
  - Using BIM tools for simulations like energy performance, structural analysis or environmental impact.
- **Deliverables**: Conceptual BIM models, feasibility reports, and initial cost estimates.

Schematic Design

In this phase, detailed BIM models that encompass all disciplines (architecture, structure, MEP, etc.) are developed.

- **Activities:**
  - Collaborative Design between architects, engineers, and consultants.
  - Integration of detailed geometry, materials, and specifications into the BIM model.
  - Clash detection to resolve conflicts between systems (e.g., structural beams and HVAC ducts).
  - Iterative design revisions based on stakeholder feedback.
- **Deliverables**: Coordinated BIM models, technical drawings, and refined cost estimates.

Construction Documentation

In this phase, the detailed Design is translated into actionable construction documents.

- **Activities:**
  - Producing shop drawings, schedules, and work packages directly from the BIM model.
  - Planning construction sequencing (4D BIM) and logistics.
  - Generating precise quantities for procurement (5D BIM).
- **Deliverables**: Construction plans, schedules, cost breakdowns, and procurement lists.

HANDOVER AND OPERATION

The completed BIM model is transitioned to the owner or facility manager.

- **Activities:**
  - Providing a complete digital twin of the asset, including all data about equipment, systems, and maintenance schedules.
  - Training facility management teams on using BIM for operations and maintenance.
- **Deliverables**: Digital as-built models, operation manuals, and maintenance schedules.

MAINTENANCE AND LIFECYCLE MANAGEMENT

In this phase of BIM process, BIM data is used for efficient operation, maintenance, and future renovations.

- **Activities include:**
  - Using BIM models to monitor system performance and schedule maintenance.
  - Making informed decisions for renovations or upgrades using digital records.
  - Planning end-of-life decommissioning using embedded lifecycle data.
- **Deliverables**: Updated lifecycle BIM model, maintenance logs, and decision-making reports.

THEORY OF MODELING

GENERAL

Building information modeling design is concerned with constructing a building model complete enough that representations can be automatically generated from it. The objects used in BIM design encode much more data than pure geometry.

Each object in Autodesk Revit belongs to a hierarchy that helps organize the objects in the building model. The terms used to describe this hierarchical classification, from broad to specific, are categories, families, types, and instances. This is the fundamental organization of the building model database. Most of the aspects of the building model, including the views, have this organizational structure. This concept is important because each of the objects has parametric control at these different levels of organization.

CATEGORIES

All objects in the building model belong to categories. A category is a group of elements that you use to model or document a building design. For example, categories of model elements include columns, beams, trusses, and foundations. Categories of annotation elements include tags and text notes. This broad category is further broken down into families. Figure 3.6 depicts the main categories used in Autodesk Revit.

FAMILIES

Families are subclasses of elements in a category. A family groups elements with a common set of parameters (properties), identical use, and similar graphical representation. For instance, the steel wide-flange column is a family within the columns category.

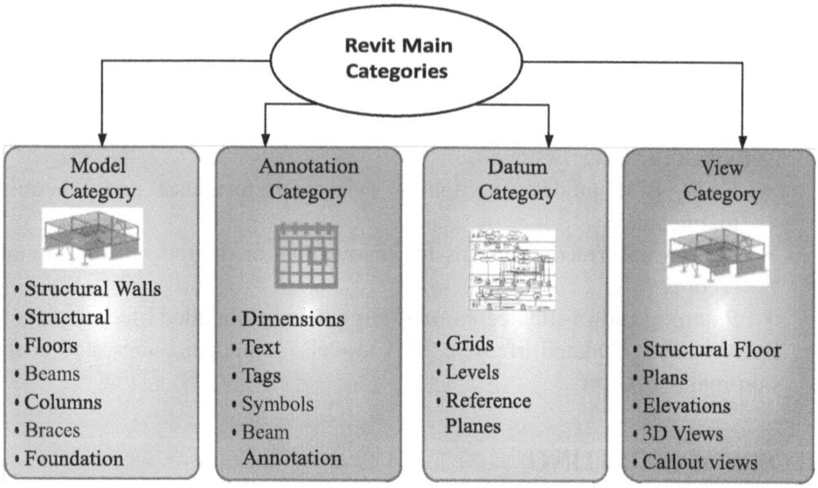

FIGURE 3.6 Main categories in the Revit platform.

Types

Families are divided into types. A family may have one or more types. The type defines what properties an object possesses, how it interacts with other objects, and how it can be represented into different forms. For example, the steel wide-flange column family has many sizes, such as W12×26 and W10×33. These are different types of one family.

Instances

An instance is simply a single object of a type in the building model. For example, the W8×31 column on the north side of the building is one instance of the column family and type, and the W8×31 column on the south side is a different instance of the same column family and type.

Instances of family types are thus the ingredients of a model. Whereas a family is an abstract description of what parameters a class of objects must have and how they relate, an *instance* is a concrete application of that type. They are the manifestation of the parameters and geometry that a type establishes. Types are unique; however, there can be various similar instances of any type. These instances can be positioned at different locations in the model. Some may have parameters set to different values, but fundamentally, they share a type and family.

Example 3.1

Explain the relationship between categories, families, types, and instances using structural column category.

Figure 3.7 illustrates the relationship between the column category, its families, types, and instances. The column category has different families based on their respective material and geometries. In this Figure, the column category has families of rectangular concrete, steel wide-flange, and timber columns. The Figure depicts also an example of types present in each family. For instance, the rectangular concrete family is represented by 8×24 ft columns; the wide-flange family by a type W10×49; and the wood family is represented by 12×12 ft timber columns. Instances of each of these types are depicted in the actual location of the building structure.

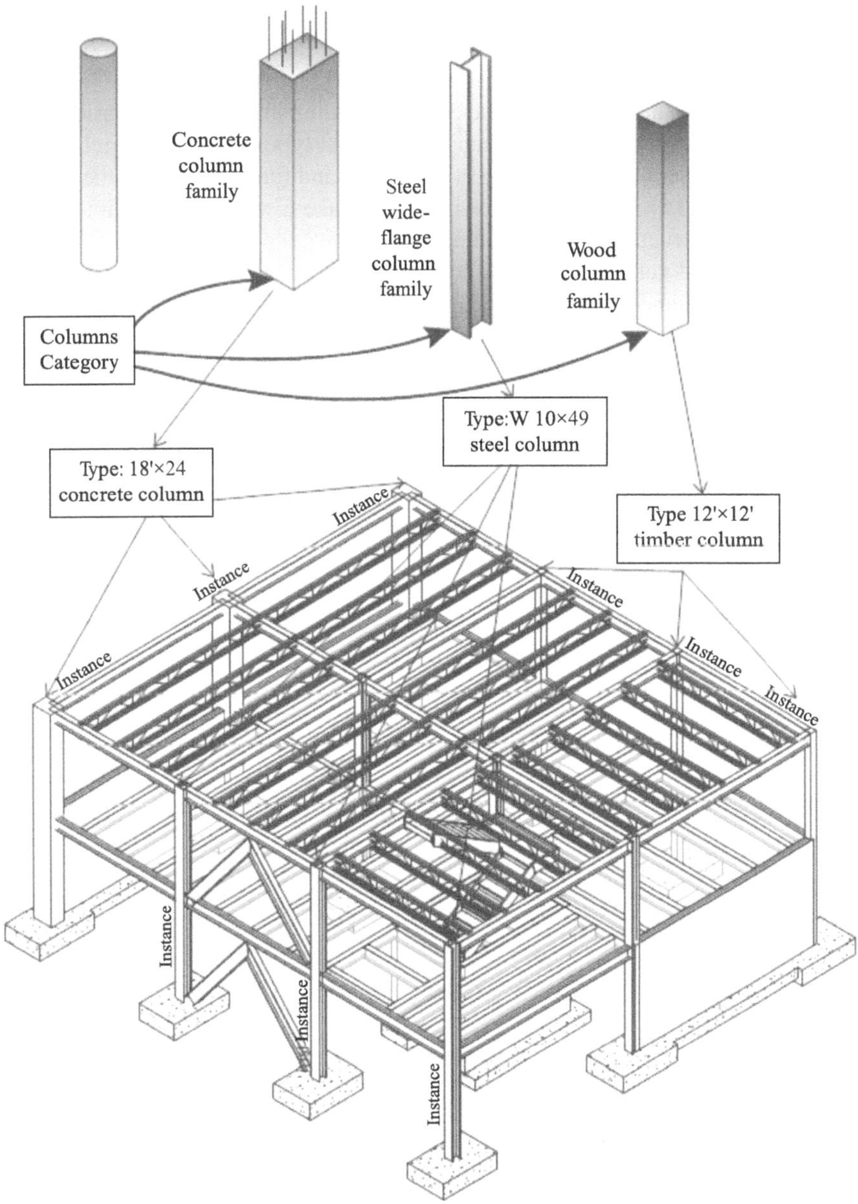

FIGURE 3.7 Relationship between categories, families, types, and instances.

Model Creation

You start modeling by placing objects. Actually, when you place an object into a model, you are making an instance of a family, as shown in Figure 3.7. Most families have multiple types. A type normally defines default values for a family's parameters. A type of a column has a particular geometry. On many occasions, types can be exchanged. For instance, one column type can easily be swapped for another column type once the object has been created. For example, a wide-flange steel column type can easily be exchanged for a column of a different family, such as a wood or concrete column. Revit handles all necessary graphical representations.

You can easily start with a specific concrete column and change its various properties later. You cannot, however, change one object type to a totally unrelated type because there is no internal mapping of properties and functionality for the platform to track (Figure 3.8). You cannot, for instance, make a wide-flange beam and then change it into a structural wall or column.

Designers deal normally with the concept of types and instances, but not always explicitly. For instance, when they create a schedule of columns, they must list every instance grouped by its type (e.g., 18 × 24 ft reinforced concrete column). Traditional CAD platforms have no notion of type versus instance; thus, they have no mechanism to assist with this issue. Conversely, BIM platforms have a full object-oriented database in which each object uses and retains the element category, family, type, and instance information that can be easily utilized to generate a graphical column schedule based on the instance or type. A similar schedule can be generated for any other type or instance of building elements (e.g., doors, windows, etc.).

The BIM platforms have the intelligence to realize the various relationships between building elements. For instance, when you move one end point of a wall, you would like both sides of the wall to move. You might also like objects embedded in the wall, such as windows and doors, to move with the wall as you edit it.

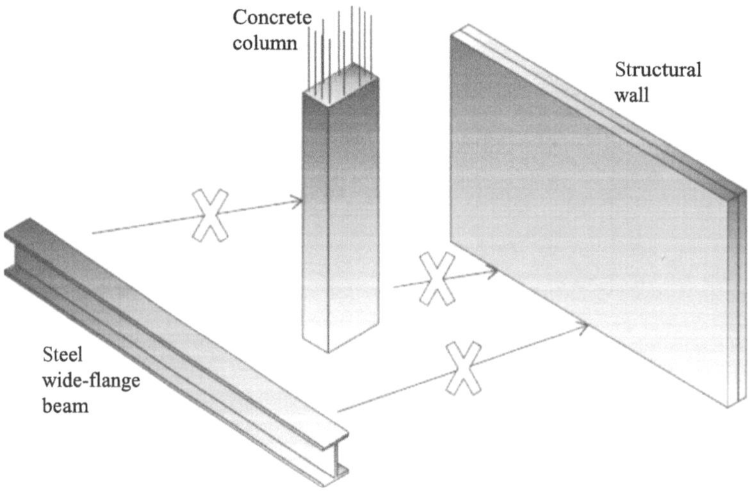

FIGURE 3.8 It is not possible to exchange an object from one type to an unrelated type.

That is exactly what BIM platform software does. However, in traditional CAD, you may have drawn a wall as a long, skinny rectangle. You might also have drawn a long, straight walkway in a similar manner. How would the software recognize that one was a wall and one was a walkway? Moving walls does not necessarily adjust other hosted or connected objects automatically. In summary, traditional CAD drawings lack the intelligence and content that BIM platforms possess in encoding a designer's intention.

In terms of the workflow in the BIM design environment, you begin with specifying what type of object is about to be placed into the model. Then, the designer identifies the necessary information about that building object so that it can be established properly. What that information is depends in large part on what kind of object is being established. For a wall, one must at a minimum specify its height and start and end points. Designers also have to select ahead of time what sort of wall it is (structural, architectural, partition, exterior, and so forth) and its type (masonry brick, concrete, wood stud, metal stud, etc.); however, you can change these at any later time.

Thus, there is no vagueness. Any time you are creating the building model, you are actually specifying some parameter of an object that you are establishing in the model. A floor would be a floor and a wall would be a wall from its inception. Notice that this methodology releases designers from unnecessary work. You do not have to draw four lines to create a wall in a plan view; picking two points for a straight wall or three points for a curved wall now is usually sufficient. You always have the opportunity later to change parameter values of any object you placed in the model, such as an arc's center or the diameter of a wall.

In summary, building models are made of design objects—columns, beams, walls, doors, windows, stairs, and so forth. Not only physical building elements that are "objects" in the Revit platform but also anything with properties that can vary is a Revit object, including visual objects such as views. A view has properties that specify what is visible. You can choose whether to show furniture, for instance, in a plan view. A section view specifies the location of the cut and direction of viewing along with the details exposed by the cut.

The objected-oriented approach is a powerful concept that has made its way from information technology to building design in the form of a BIM design method. BIM design enables designers to postulate design intention at a higher level. They are free to compose models from established digital building components and then vary parameters.

EXPLORING THE USER INTERFACE

To introduce the user interface of the Revit platform, it is crucial to understand the terms discussed next.

PROJECT

A project is the single database of information for the design (i.e., the BIM). The project file contains all information for the building design, from geometry to construction data. This information includes components used to design the model, views of the project, and drawings of the design. By using a single project file, it is easy to alter the design and have changes reflected in all associated areas (plan views, elevation views, section views, schedules, and so forth). Having only one file to track also makes it easier to manage the project.

Level

Levels are infinite horizontal planes that act as a reference for level-hosted elements, such as roofs, floors, and ceilings. Most often, levels are used to define a vertical height or story within a building. A level is created for each known story or other needed reference of the building, for example, first floor, top of wall, or bottom of foundation. To place levels, you must be in a section or elevation view.

In Figure 3.9a, the user interface is labeled. In the sections that follow, we navigate and become familiar with the user interface items.

Ribbon

The ribbon displays when you create or open a project file. It provides all the tools necessary to create a project or family.

As you resize the Revit window, you may notice that tools in the ribbon automatically adjust their size to fit the available space. This feature allows all buttons to be visible for most screen sizes.

Expanded Panels

An arrow next to a panel title or item indicates that you can expand the panel or item to display related tools and controls (see Figure 3.9c).

By default, an expanded panel closes automatically when you click outside the panel. To keep a panel expanded while its ribbon tab is displayed, click the pushpin icon in the bottom-left corner of the expanded panel.

Dialog Launcher

Some panels allow you to open a dialog to define related settings. A dialog launcher arrow on the bottom of a panel opens a dialog (see Figure 3.10).

Contextual Ribbon Tabs

When you start using certain tools or select model elements, a contextual ribbon tab displays tools that relate to the context of that tool or element. In many cases, the contextual tab merges with the **Modify** tab. A contextual ribbon tab closes when one exits the tool or clears the selection. For instance, when you select a floor, such as concrete on a metal deck, the contextual ribbon tabs shown in Figure 3.11 appear.

You can specify whether a contextual tab automatically comes into focus or the current tab stays in focus. You can also postulate which ribbon tab displays when you exits a tool or clears a selection.

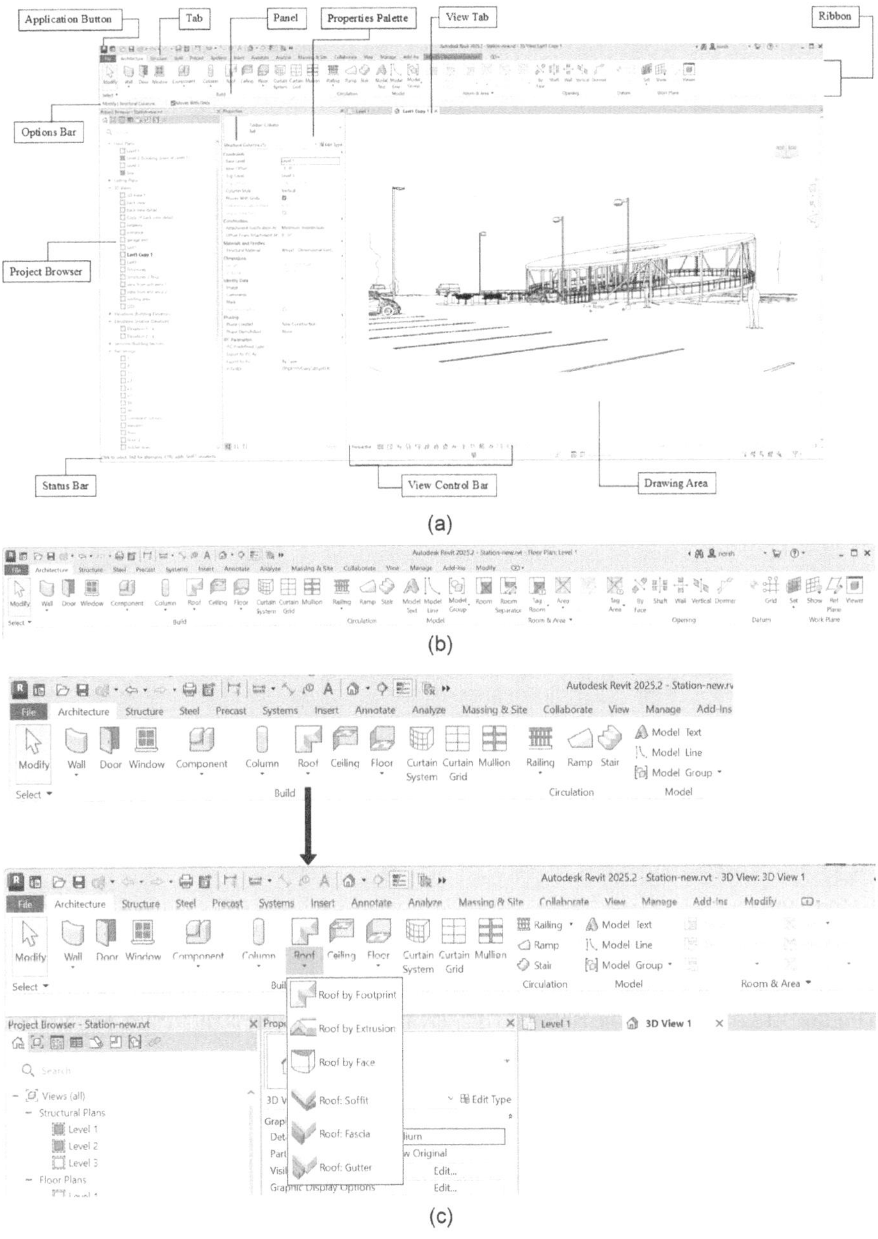

FIGURE 3.9 (a) Revit platform user interface. (b) Revit ribbon and (c) Expanded panels in Revit.

Quick Access Toolbar

The **Quick Access** toolbar contains a set of default tools (Figure 3.12). This toolbar can be customized to display the tools used most often. The **Quick Access** toolbar can display above or below the ribbon. To change the setting, on the **Quick Access** toolbar,

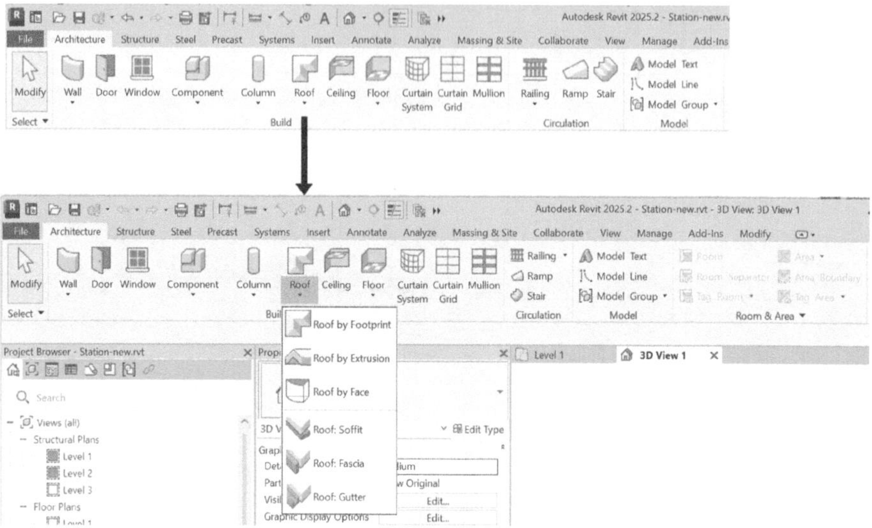

FIGURE 3.10 Dialog launcher.

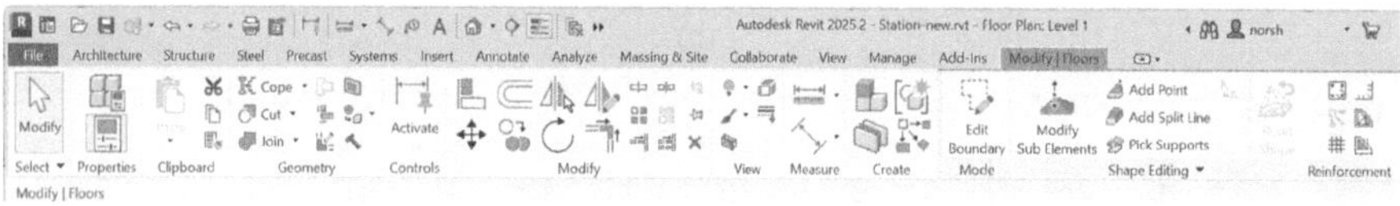

FIGURE 3.11 Contextual ribbon tabs.

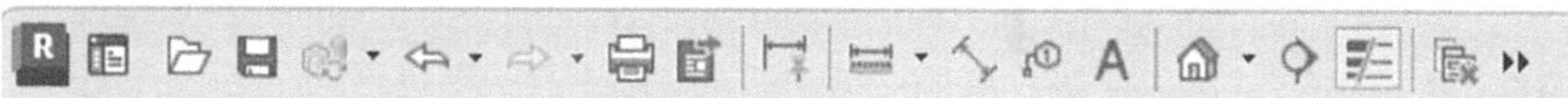

FIGURE 3.12 **Quick Access** toolbar.

click **Customize Quick Access Toolbar** drop-down ➤ **Show below the Ribbon** (Figure 3.13a). Navigate the ribbon to display the tool that you want to add. Right-click the tool and click **Add to Quick Access Toolbar** (Figure 3.13b).

To Customize the Quick Access Toolbar

To make a quick change to the **Quick Access** toolbar, right-click a tool on the **Quick Access** toolbar and select one of the following options:

- Remove from Quick Access Toolbar, which removes the tool
- Add Separator, which adds a separator line to the right of the tool

Status Bar

The **Status** bar is located along the bottom of the Revit window (Figure 3.14). When a tool is used, the left side of the status bar provides tips or hints on what to do. When highlighting an element or component, the status bar displays the name of the family and type.

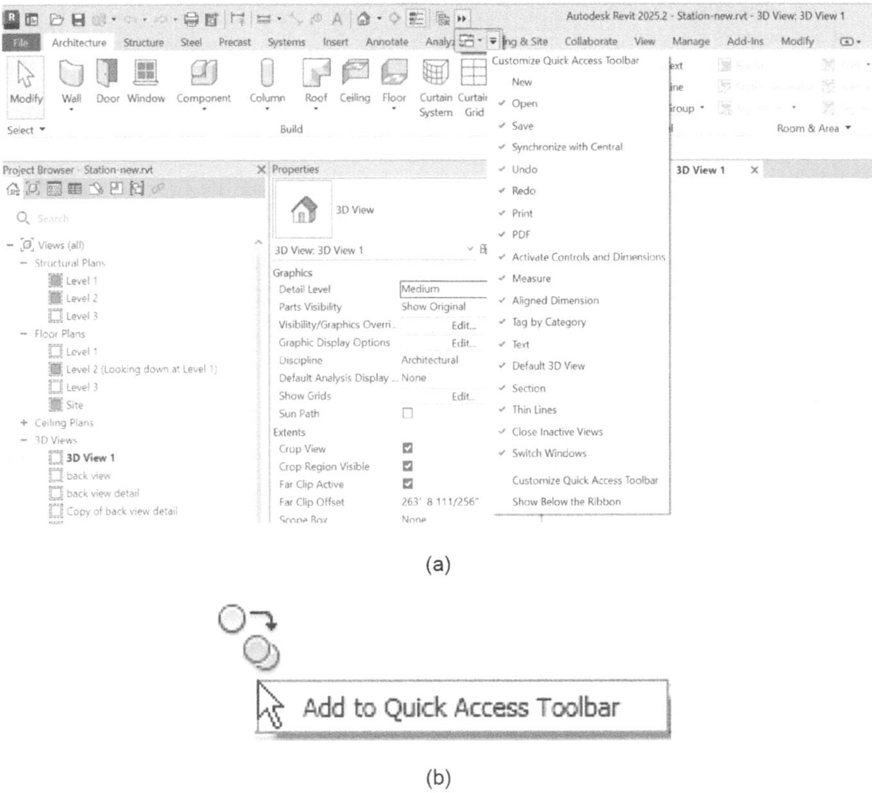

FIGURE 3.13 (a) Customize the **Quick Access** toolbar and (b) Add a **Quick Access** toolbar.

FIGURE 3.14 **Status** bar.

FIGURE 3.15 **Option** bar.

Options Bar

The **Options** bar is located below the ribbon (Figure 3.15). Its contents change depending on the current tool or selected element.

To move the **Options** bar to the bottom of the Revit window (above the status bar), right-click the **Options** bar and click **Dock at bottom**.

Properties Palette

The **Properties** palette is a modeless dialog where you can view and modify the parameters that define the properties of elements in Revit (see Figure 3.16).

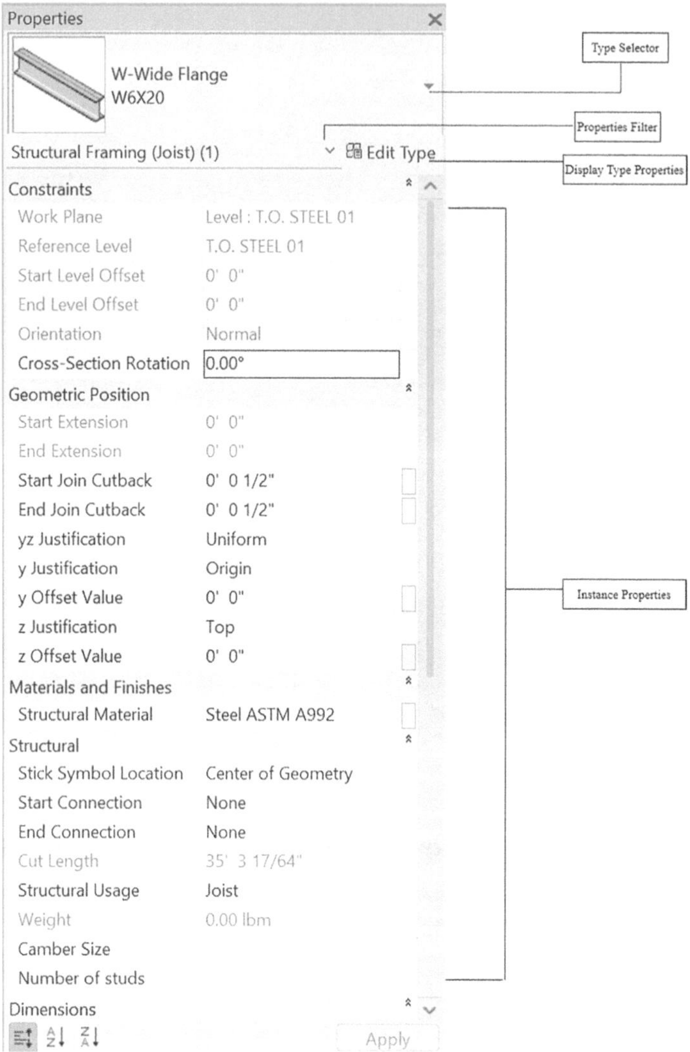

FIGURE 3.16 Properties palette.

Opening the Properties Palette

On starting Revit for the first time, the **Properties** palette is open and docked above the **Project Browser** on the left side of the drawing area. If the **Properties** palette is subsequently closed, it can be reopened using any of the following methods:

- Click **Modify** tab ➤ **Properties** panel ➤ (**Properties**).
- Click **View** tab ➤ **Windows** panel ➤ **User Interface** drop-down ➤ Properties.
- Right-click in the drawing area and click **Properties**.

You can dock the palette to either side of the Revit window and resize it horizontally. You can resize it both horizontally and vertically when it is undocked. The display and location of the palette will persist from one Revit session to the next for the same user.

Typically, the **Properties** palette is kept open during a Revit session to assist in the following:

- You can select the type of element that will be placed in the drawing area or change the type of elements already placed.
- Viewing and modifying the properties of the element one is placing or of elements selected in the drawing area.
- Viewing and modifying the properties of the active view.
- Accessing the type properties that apply to all instances of an element type.

In most cases, the **Properties** palette displays both user-editable and read-only (shaded) instance properties. A property may be read only because its value is computed or given automatically by the software or because it hinges on the setting of another property. For instance, a column's analytical model properties are only editable if the value of its virtual model property **Enable Analytical model** is checked.

BIM IN EDUCATION

Building information modeling is a comprehensive information management and analysis technology that is becoming increasingly essential for academic education. AEC schools have implemented a variety of pedagogical methods for introducing BIM into their curricula. They range from using BIM in architectural studio, sustainable design, and construction management to civil engineering (Önür, 2009; Sharag-Eldin and Nawari, 2010, Barison and Santos, 2010; Sacks and Barak, 2010; Wong et al., 2011). For instance, Önür and Sharag-Eldın described how BIM is integrated into the architectural curriculum. Sacks and Barak (2010) introduced BIM as an integral part of freshman-year civil engineering education.

Several academic institutions have integrated BIM into their curricula, using different approaches; however, there is no commonly agreed-on methodology for teaching BIM in AEC programs (Barison and Santos, 2010). Most schools offer BIM in only one or two courses. Many courses limit their coverage to a short period (one to two weeks) (Becerik-Gerber et al., 2011). The BIM course is limited to a single discipline in 90% of the cases (Barison and Santos, 2010). The majority of schools introduce BIM at a basic level by teaching a specific software tool, limiting their perspective on BIM to viewing it simply as another CAD productivity-enhancing tool for creating 2D and 3D drawings (Sacks and Pikas, 2013). However, BIM by nature goes far beyond digital drafting (Eastman et al., 2011). A comprehensive literature review of the subject can be found in the work of Barison and Santos (2010) and Sacks and Pikas (2013). Their main findings indicate that schools wishing to implement BIM in their curriculum are likely to face many difficulties. The greatest challenge these schools face is promoting integration between different areas of the curriculum using BIM and finding programs from other departments or units that are willing to promote collaboration. As a result, Sacks and Pikas (2013) proposed a framework for academia in which they compiled 39 BIM topics classified in three main areas of competence (processes, technology, and application) and the level of achievement expected by the construction industry for each BIM topic for three levels of education (first degree, master's degree, and work experience). This framework could be utilized by educators to plan and evaluate BIM content for their curricula.

Because BIM differs extensively from traditional CAD, it does require new ways of thinking and teaching. For example, BIM facilitates collaboration and teamwork across disciplines, which must be incorporated in teaching BIM courses. Furthermore, BIM provides rich visualization of building elements and parametric modeling of behavior, which can enhance students' learning experience in virtual construction, such as understanding how building elements fit together just as they must on a physical site (Eastman et al., 2011).

BIM for Students of Structural Engineering and Architecture

With recent technological advancements, engineers and architects have smarter tools to create and analyze artistically efficient structural forms and demonstrate how load combinations affect the stability and behavior of a structure. Specifically, BIM has the potential to provide solutions to the issues related to the conceptual linking and integration between architectural and structural engineering principles and to advance different types of structural knowledge-sharing objectives without compromising their distinct requirements. BIM is a process that fundamentally changes the role of computation in structural design by creating a database of the building objects to be used for all aspects of the structure, from design to construction, operation, and maintenance. Based on this collaborative environment, a new framework is proposed to advance structural design education. This framework is referred to as the structure and architecture synergy framework (SAS framework). The framework explores structural design as an art while emphasizing engineering principles and thereby provides an enhanced understanding of the influence structure can play in creating form and defining spatial order and composition.

EXERCISES

3.1. In reference to BIM basic concepts, which statement is *incorrect*?
   a. BIM is a computable representation of a facility and its related lifecycle information.
   b. BIM includes the entire life cycle from inception onward.
   c. BIM is shared information and supports interoperability and collaboration.
   d. BIM is 3D molding for a purely visualization process.

3.2. BIM model-based workflow in structural design is characterized by _____?
   a. A single source of data from which all drawings and reports can be derived.
   b. A model that itself coordinates all drawings.

c. Design coherency, clash detection, and reduction in errors and omission.
d. All of the above.

3.3. List the major benefits of BIM in contrast to traditional CAD systems.
3.4. Define the following terms within the context of the BIM environment:
a. Building information model
b. Objects
c. Parameters
3.5. Revit as a platform for building information modeling classifies elements by
a. Categories, families, libraries, and types
b. Families, libraries, objects, and types
c. Categories, families, types, and instances
d. Families, libraries, types, and instances
3.6. What is meant by *project* in Revit?
3.7. Explain the term *family* in the Revit platform?
3.8. What kinds of families exist in Revit?
3.9. Give two examples of families in Revit and their respective categories.
3.10. Explain the concept of data sharing and collaboration in a typical BIM environment.
3.11. What is the primary advantage of recent technological advancements in structural Design according to the passage?
A. Reducing construction costs significantly.
B. Simplifying the maintenance process for buildings.
C. Providing smarter tools to analyze artistic and structural efficiency.
D. Eliminating the need for structural engineers.
3.12. What does the SAS framework aim to achieve in structural design education?
A. Minimize the use of architectural elements in structural Design.
B. Emphasize the role of computation in replacing engineering principles.
C. Explore structural Design as an art while emphasizing engineering principles.
D. Focus on operational maintenance of structures over Design.

4 Modeling Elements

STRUCTURAL ELEMENTS

Modeling activities may be grouped into several broad levels that reflect the detailed sequence in which structural designs are encountered. At the core of these design levels is the process of moving from approximations to progressively more precise information. According to the American Institute of Architects (AIA) Contract Document G202-2013 (AIA, 2013), *Building Information Modeling Protocol Form*, levels of development (LOD) can be classified into levels 1–5 (see Table 4.1). They define and illustrate the characteristics of model elements of different building systems at different phases of building information modeling (BIM) project development. The BIM details increase from a space management model to a major systems model, to a simulation model, and finally to the most detailed virtual building model. As the design progresses, there will be increased specificity in the elements required in the BIM.

Level 1 covers design issues dealing with the structural fabric, design intent, and contextual programmatic dictates. It describes the overall morphology of the structures. Level 2 focuses on specific structural strategies in relation to the exact shape of the structural entity itself and how that shape is formed by specific structural materials and components. Level 3 deals with particular element sizes, shapes, patterns, and other related properties. Level 4 specifies more details about the element's relationship to other elements, including fabrication and construction details. In Level 5, the model details are finalized and checked with the actual construction and assembly of the building.

This chapter delves into creating structural element drafting and model content. These component-building skills are necessary for proper structural design. They include geometry, member nodes, material properties, element cross sections, external supports, nodal restraints, and the type of analysis.

PHYSICAL AND ANALYTICAL MODELS

In the Revit platform, the analytical model of the structure is a simplified three-dimensional (3D) representation of the full physical description of a structural model; all structural elements are connected to each other continuously. It is also known as the "stick model." The analytical model consists of structural elements, geometry, material properties, nodal restraints, external supports, and loads that together form an analytical model for structural design. Revit creates the analytical model automatically while the user creates the physical model, which can be exported to analysis and design applications. Figure 4.1a represents the physical model, and Figure 4.1b represents the corresponding analytical model.

DOI: 10.1201/9781003598435-4

TABLE 4.1
LOD Definition

Level	Definition
100	The Model Element may be graphically represented in the Model with a symbol or other generic representation, but does not satisfy the requirements for LOD 200. Information related to the Model Element (i.e., cost per square foot, tonnage of HVAC, etc.) can be derived from other Model Elements.
200	The Model Element is graphically represented within the Model as a generic system, object, or assembly with approximate quantities, size, shape, location, and orientation. Non-graphic information may also be attached to the Model Element.
300	The Model Element is graphically represented within the Model as a specific system, object or assembly in terms of quantity, size, shape, location, and orientation. Non-graphic information may also be attached to the Model Element.
350	The Model Element is graphically represented within the Model as a specific system, object, or assembly in terms of quantity, size, shape, orientation, and interfaces with other building systems. Non-graphic information may also be attached to the Model Element.
400	The Model Element is graphically represented within the Model as a specific system, object or assembly in terms of size, shape, location, quantity, and orientation with detailing, fabrication, assembly, and installation information. Non-graphic information may also be attached to the Model Element.
500	The Model Element is a field verified representation in terms of size, shape, location, quantity, and orientation. Non-graphic information may also be attached to the Model Elements.

Source: AIA Document E202 (2008, pp. 4–6).

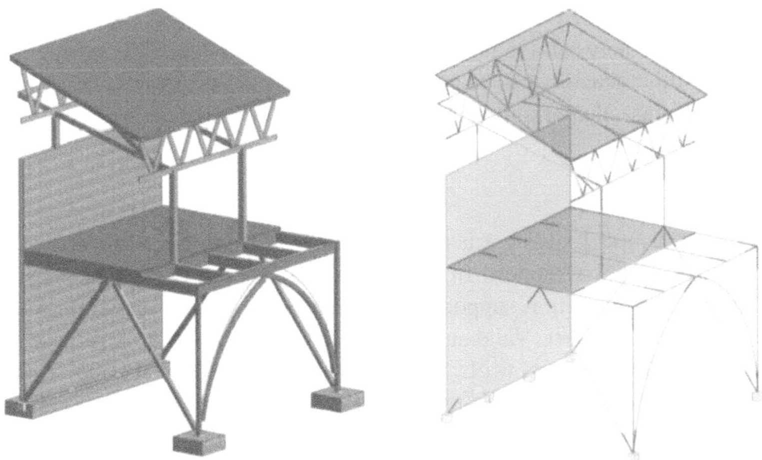

(a) (b)

FIGURE 4.1 Relationship between (a) physical and (b) analytical models.

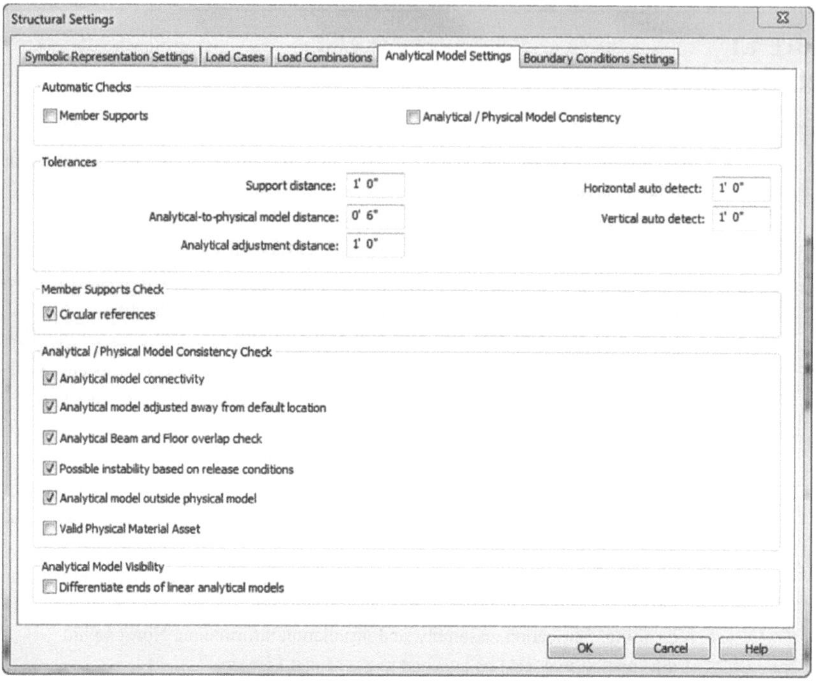

FIGURE 4.2 Analytical Model Setting tab in the Structural Settings dialog box.

The analytical model can be controlled from the **Analytical Model Settings** tab under the **Structural Settings** dialog box (see Figure 4.2).

One of the important settings for the analytical model is the Automatic Checks. When enabled, they provide warnings when members are not supported or an inconsistency is found between the physical and analytical models. This setting is useful when most of the structure has been modeled and you want to know if changes made to the model cause elements to become unsupported or violate consistency rules.

MODELING RULES

In the physical model, every structural object (beam, column, truss, etc.) must be supported with a point support. In other words, a supporting member must have a point of intersection with a supported member. For instance, a column must have at least one point support. Valid supports include another column, isolated or continuous foundations, beams, walls, floors, or ramps. On the other hand, a structural floor must have one of the following supports: at least three point supports, one line support and a point support not located on the line, two line supports that are not colinear, or one surface support. Permissible supports for structural floors include columns, continuous or isolated foundations, beams, or walls.

A structural wall must have at least two point supports or one line support. Allowed supports for the wall include columns, continuous or isolated foundations, beams, floors, or ramps. Similar to a structural wall, a structural beam requires at least two point supports, or one point support located at one end that must have its release condition set to fixed, or one surface support. Valid supports that can be used to support beams include structural columns, continuous or isolated foundations, beams, or walls.

The modeling rule for a structural brace element requires only two point supports. These supports can be one of the following: structural columns, continuous or isolated foundations, beams, floors, walls, or ramps.

MODEL INTEGRATION

Revit integrates the physical model with the independent, editable analytical model. This integration enables one to link the structural model bidirectionally to any structural analysis application, such as SAP2000, RISA, Fastrack Building Designer, Autodesk Robot™ Structural Analysis, and other programs. Please note that the analytical model generated in Revit may have discrepancies that are not suitable for direct integration with some structural analysis and design software. You can use Revit to automatically adjust the analytical model, reducing or eliminating these discrepancies for both existing and newly created elements. Also, utilize Revit to disassociate or reassociate continuity of the analytical model. Furthermore, some structural configurations are not suitable for direct integration with analysis and design software. Thus, adaptive adjustment is required before a structural model is sent to the analysis and design software.

SPATIAL ORDER: GRID LINES

The spatial order of a building structure is a set of patterns of various complexities organized one within another. Every aspect of a building structure—from the plan in its most abstract sense to the slightest physical detail—appears in a pattern, and all of the patterns relate to one another dimensionally to create a seamless continuum of spatial order. The instrument used to establish this spatial order is the concept of "grid lines." They normally help identify the relationship between structural fabric, the intent of the design, and contextual programmatic orders. They are the essential modeling lines that assist in organizing various structural and architectural elements in plan or elevation views.

To create a grid line, invoke the **Grid** tool from the **Structure** tab: **Structure** tab ➤ **Datum** panel ➤ **Grid**. The **Modify | Place Grid** tab will be displayed (Figure 4.3). Any of the options displayed can be used to draw a grid line in the desired plan or elevation views. Grid patterns can be rectangular or curved based on the order requirements (see Figure 4.4).

Grid lines can be modified and customized after they are created. You can change the color, line weight, and line type of the entire grid line or part of the grid line as well as the symbols at the end of the grid line.

To control the visibility of grid lines, you need to invoke the **Scope Box** tool from the **View** tab. The **Scope Box** tool helps in defining the boundary limit for the visibility of these grid lines in various views (Figure 4.5).

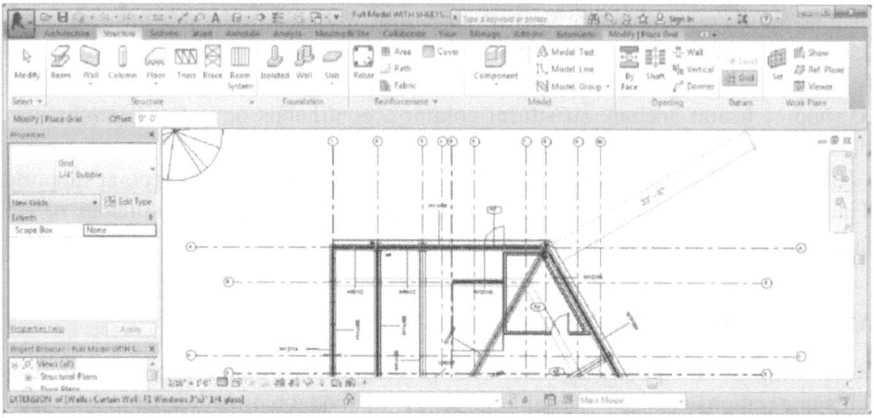

(a)

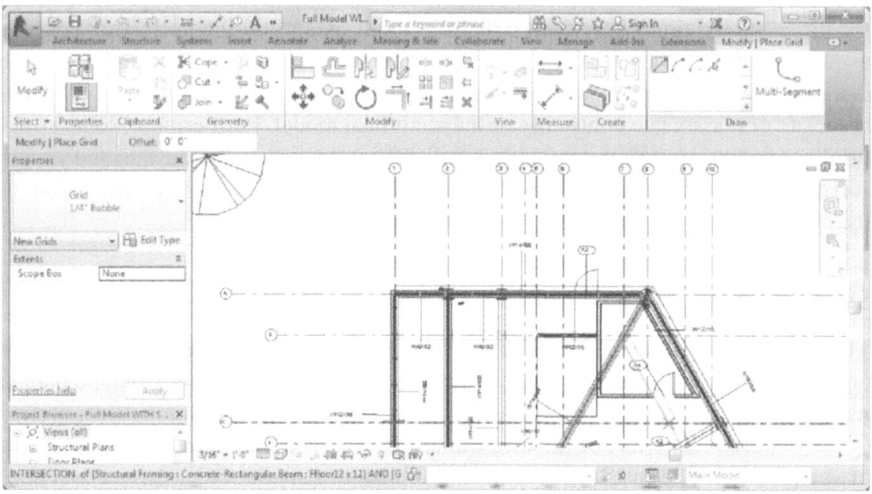

(b)

FIGURE 4.3 Options for placing or modifying grid lines: (a) ribbon showing grid lines icon; (b) sketching tools for grid lines.

LEVELS

Levels define the horizontal plans for hosting various level-hosted objects, such as walls, slabs, columns, beams, and ceilings. Levels are created using the **Level** tool by sketching the required level lines in elevation or section views (see Figure 4.6). All created levels display by default associated label names and elevations.

In Revit, there are two types of levels: story level and nonstory level. Story levels normally have corresponding plan views; nonstory levels do not hold plan views. However, nonstory levels can act as hosts for placing other objects and details, such as

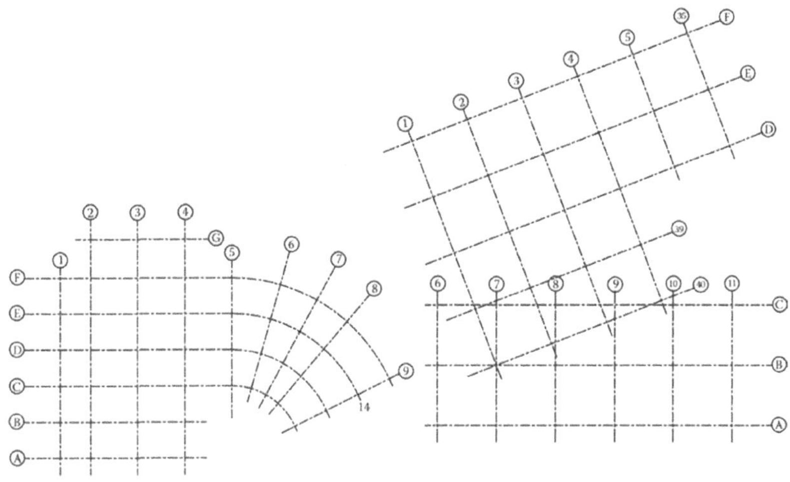

FIGURE 4.4 Grid lines examples.

the top of walls, foundations, and slabs. To distinguish between these levels, different colors are used to display the level bubble for each type of level.

Similar to other datum objects like grid lines and reference lines, the visibility of levels can be controlled by using the **Scope Box** tool. It allows specifying the views in which the level objects will display.

COLUMNS

In Revit software, there are two kinds of columns that can be modeled: architectural columns and structural columns. Structural columns are load-bearing elements and are part of the resisting analytical model of the structure. Architectural columns are non-load-bearing elements and are present primarily for the aesthetic and overall ambience of the building model.

In terms of modeling, structural columns connect with other structural members, such as beams, trusses, and floors, whereas architectural columns functionally cannot connect to other structural objects. Furthermore, structural columns are displayed in the **Graphical Columns Schedule**, whereas architectural columns are not.

Each of these column types has some predefined families that are shipped with the software. These families include hot-rolled steel columns, wood columns, and reinforced concrete columns. They are available in both metric and imperial systems of units. Figures 4.7 and 4.8 show how to load a column family using the **Insert** tab.

To model a column, it is recommended to first establish grid lines and then place columns at the intersection of these grid lines (see Figure 4.9). Tagging each column can also be achieved while modeling in columns by selecting the **Tag on Placement** tool.

It is also critical to specify the various options related to the alignment constraints and annotation of columns when invoking the **Structural Column** tool to place a column in plan, elevation, or 3D views. These options will display in the **Option** bar (Figure 4.10).

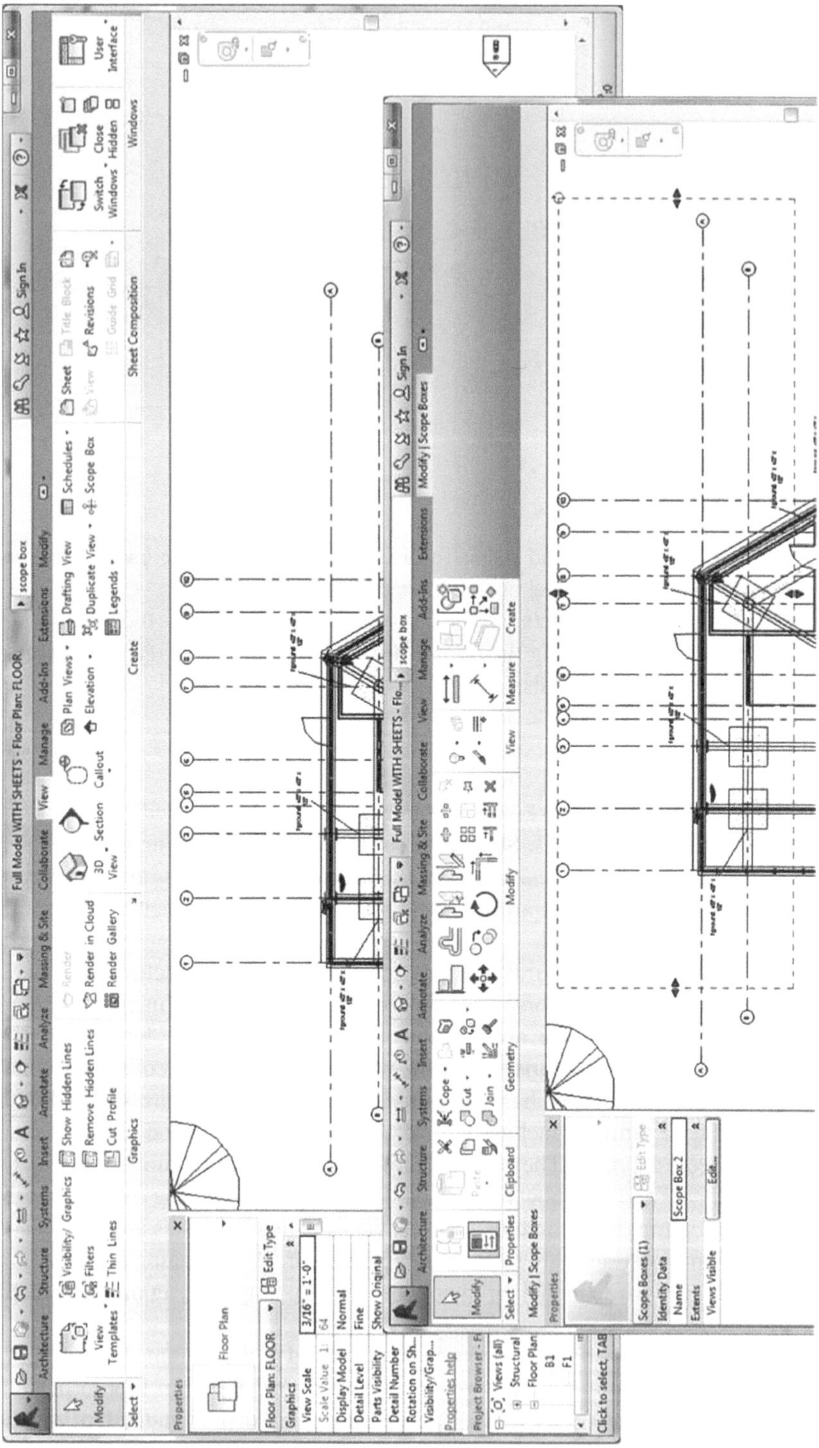

FIGURE 4.5 Scope Box tool.

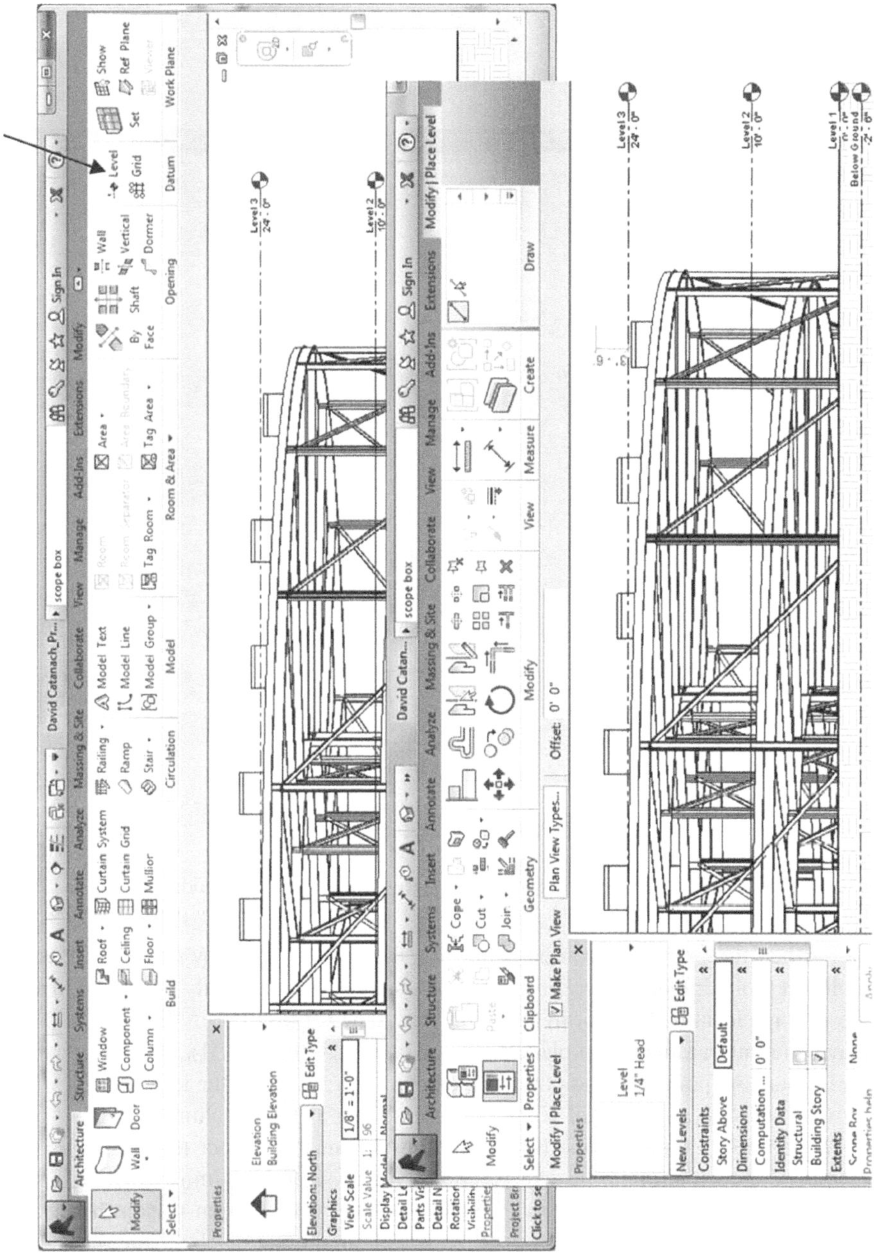

FIGURE 4.6 The tool.

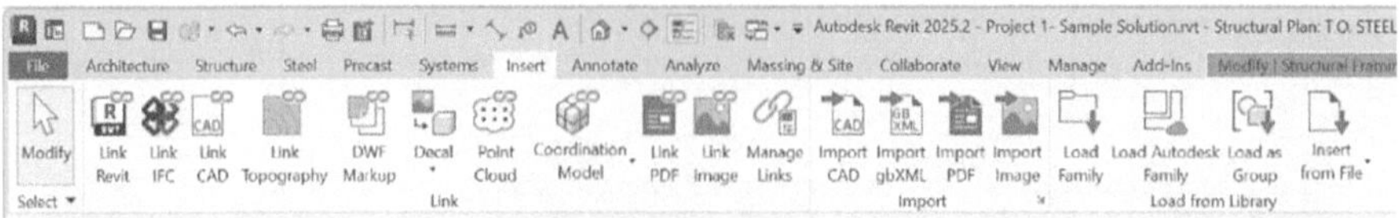

FIGURE 4.7 Loading a family into a Revit project.

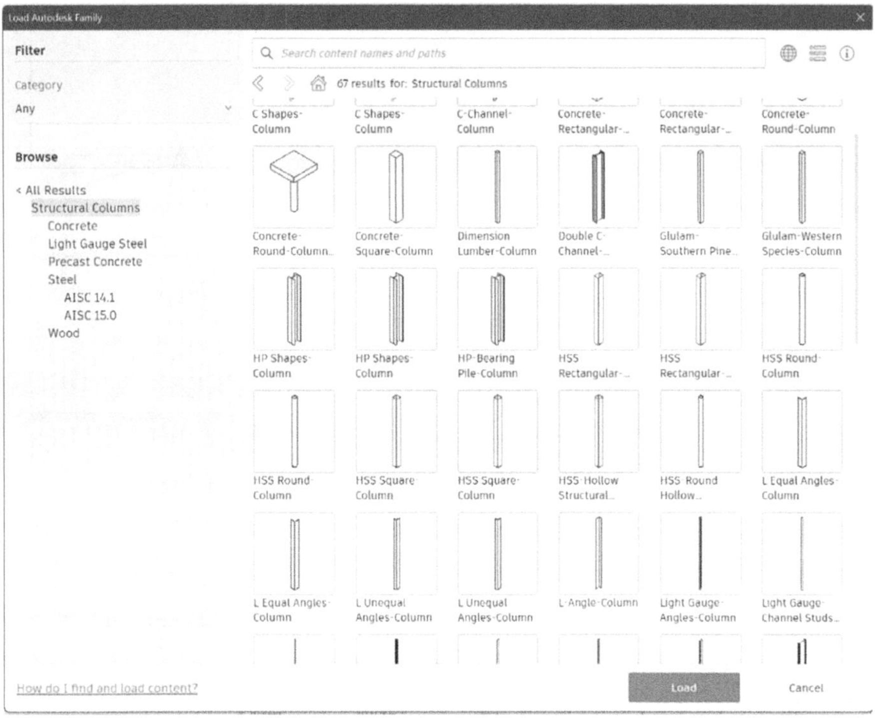

FIGURE 4.8 Loading a column family into a Revit project from Autodesk Library.

Furthermore, columns can be defined as inclined columns by using the **Slanted Column** tool. Unlike vertical columns, inclined columns can only be placed in a 3D view or an elevation view or in a section view. The **Option** bar in this case will display different parameters to place a slanted column (Figure 4.11). Define the values for the starting point and end point of the column by selecting the level and the offset for the **1st Click** and the **2nd Click** when modeling a sloped column.

Sometimes, columns must be cut to adjust them to connect to other structural members. Cuts and openings can be created in structural columns by invoking the **Structure ➤ Opening** tool. Once the face is selected, you can sketch the desired opening profile. Changes can be made to any created openings by using the options in the **Modify ➤ Structural Opening Cut** tab.

For structural analysis purposes, it is important to define the top and bottom release conditions of a column. This can be achieved by setting the parameters for

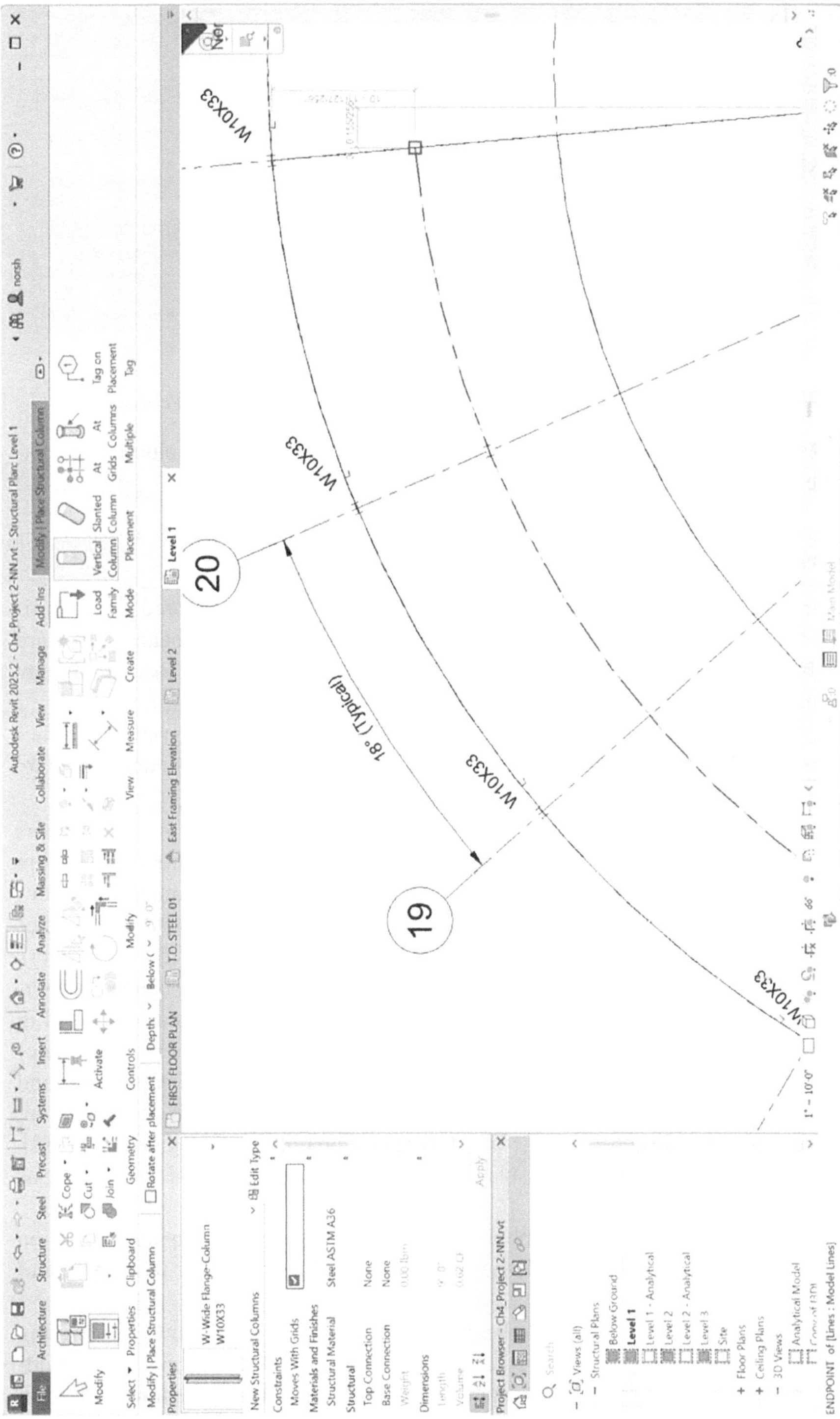

FIGURE 4.9 Columns placed at the grid line intersections in a plan view.

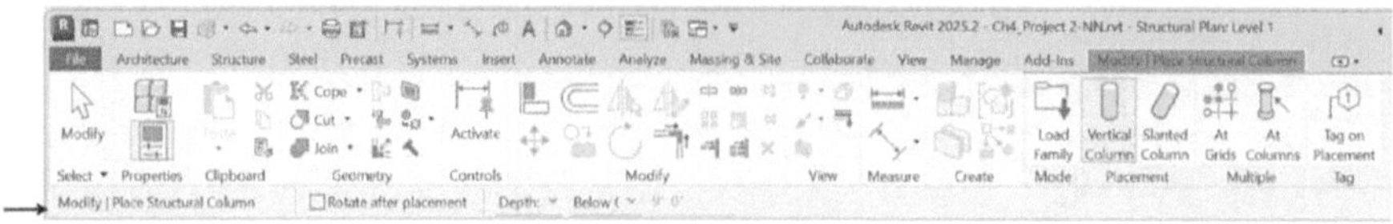

FIGURE 4.10 The **Option** bar displaying a column's parameters.

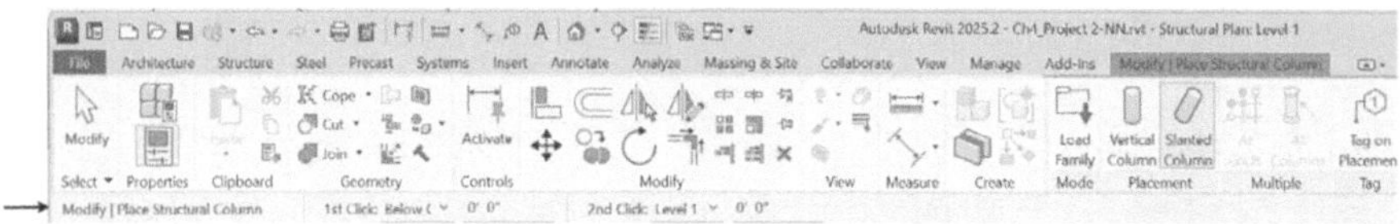

FIGURE 4.11 The **Option** bar showing parameters for inclined structural column.

Top Release and **Bottom Release** in the instance properties of the column or invoking the **Boundary Conditions** tool from the **Analysis** tab.

BEAMS

The structural beams that are available in the software include the predefined families of hot-rolled steel beams, wood beams, and reinforced concrete beams. They are available in both metric and imperial systems of units. They can be loaded into any project using the **Insert** tab, similar to loading structural column families.

It is recommended to model in beams after finishing placing gridlines, structural columns, and walls. A beam can be modeled by invoking the **Structure** tab ➤ **Structure** panel ➤ **Beam** in plan, elevation, or 3D views (Figure 4.12).

It is also recommended to place beams after creating grid lines because beams snap easily to grids. However, structural beams can be added without an existing grid line. A number of options are available to create a structural beam in plan view. These include (see Figure 4.13) (1) **Sketch** individual beams, (2) **Select** grid lines that lie between structural elements, and (3) **Create** a chain of beams (beam system).

After creating a structural beam, it is necessary to define its structural properties by setting up the necessary parameters in the **Properties** palette as displayed in Figure 4.14. For instance, for the beam selected in Figure 4.14, the structural material is Steel ASTM (American Society for Testing and Materials) A992; its start and end connections with the columns are of type moment connection (fixed connection, moment frame); its structural usage is girder (i.e., a beam connecting two columns). The ends of the beam will now display the moment frame symbol in a plan view as shown in Figure 4.14.

Additional connection type symbols can be loaded and assigned using connection symbol families in the **Symbolic Representation Settings** tab under the **Structural Settings** dialog (see Figure 4.14).

In addition to modeling a single beam at a time, the software allows for defining a beam system (a group of similar parallel beams) in a selected area of the building. This beam system can be created by invoking the **Beam System** tool from

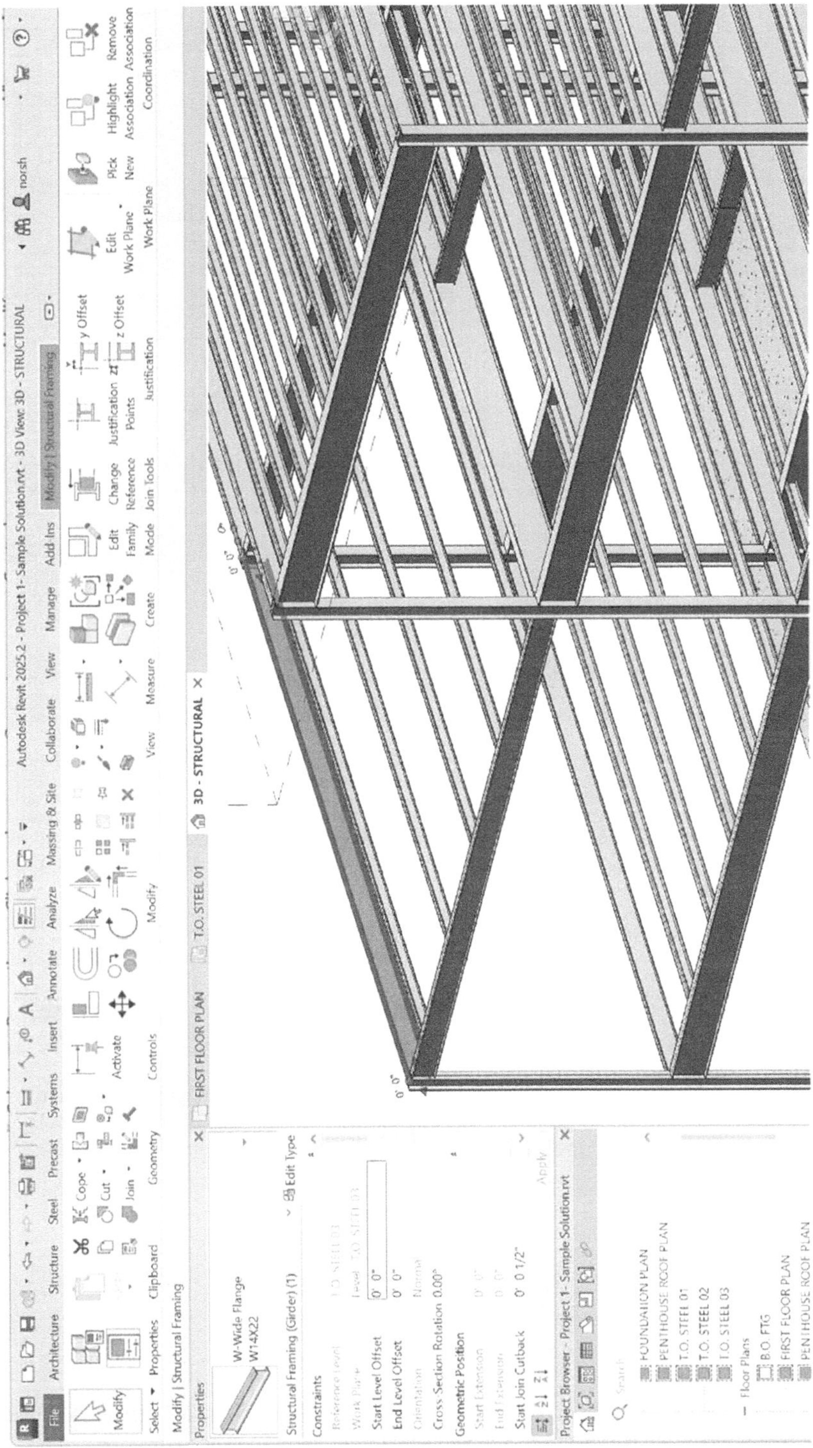

FIGURE 4.12 Modeling structural beams.

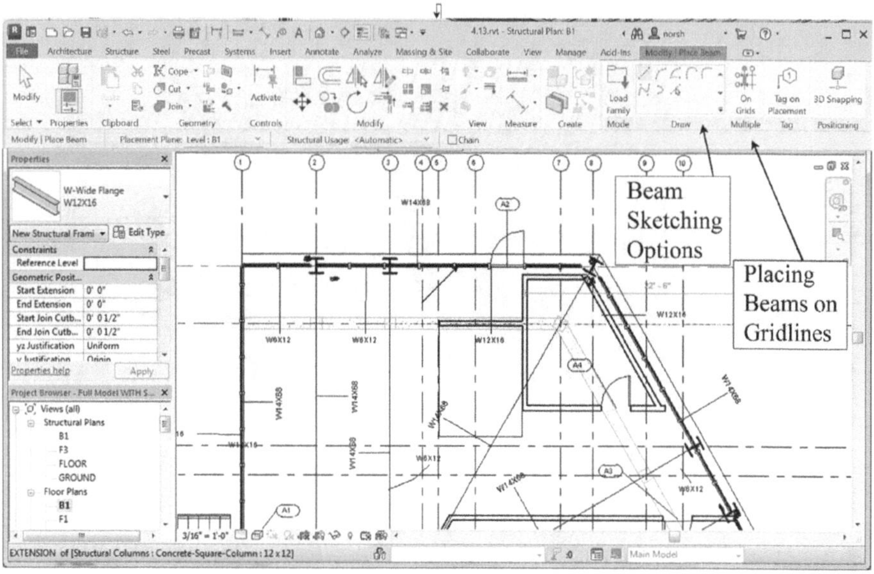

FIGURE 4.13 Structural beam placing options.

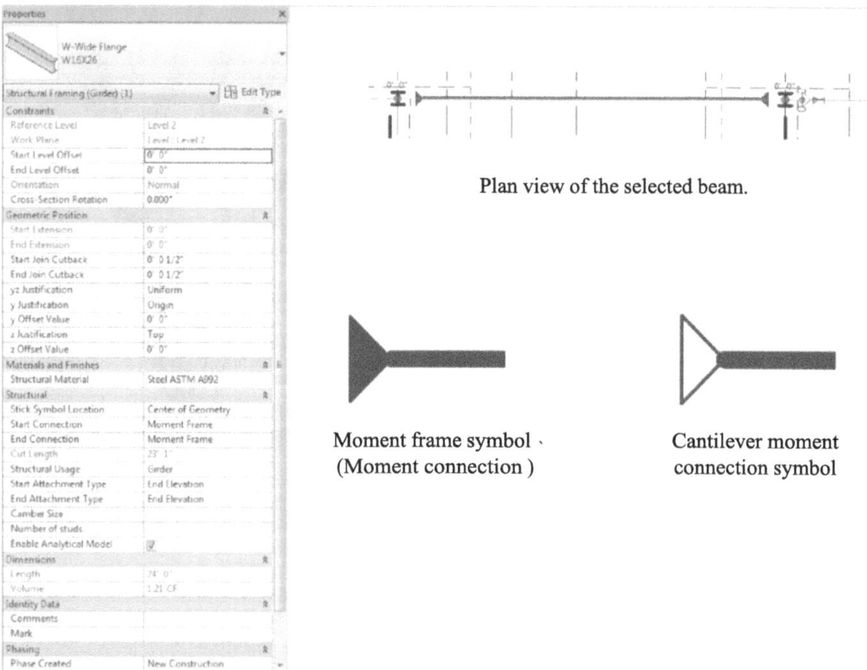

(a) (b)

FIGURE 4.14 Structural beam **Properties** palette. (a) Parameters for structural steel W-Wide flange beam. (b) Explanation of moment and cantilever moment connection symbols.

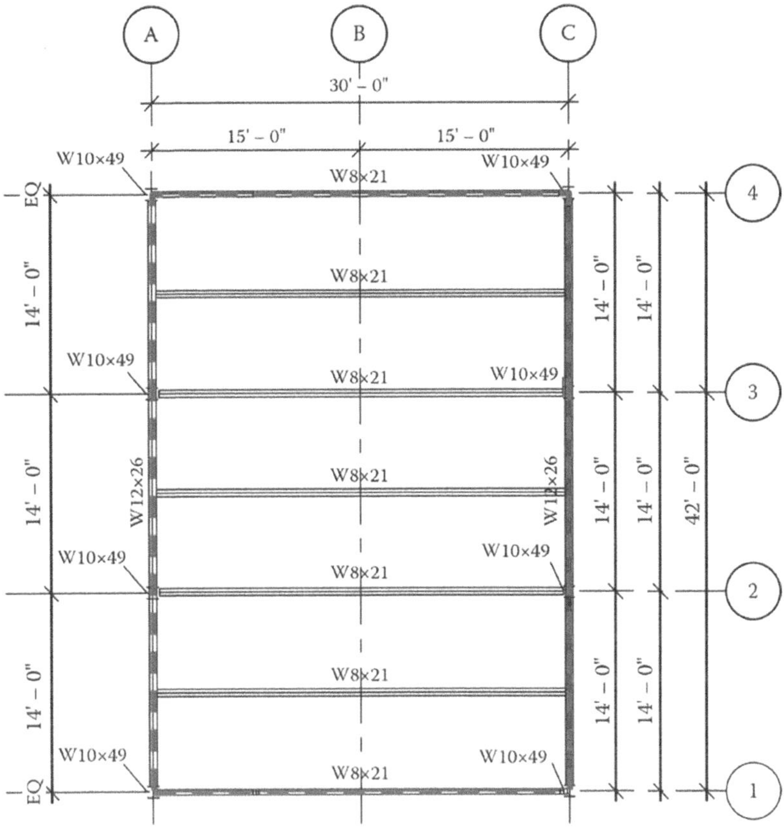

FIGURE 4.15 A beam system of steel beam W8X12 placed inside a rectangular area.

the **Structure** tab. Once the **Beam System** tool is invoked, the **Modify ➤ Place Structural Beam System** tool will be displayed. At this stage, a number of options will be available to create the beam system. You can use an existing area bounded by girders or structural walls to automatically establish the beam system for the area (Figure 4.15). Alternatively, one can sketch a new area by defining its boundary lines.

WALLS

Like columns, walls can be defined as structural or architectural walls. In essence, all wall types within the basic wall family have an instance property called **Structural Usage**, which can have one of the following values (Figure 4.16): (1) shear, (2) bearing, (3) nonbearing, and (4) structural combined. The first option sets up the wall as a structural shear wall with a rigid planar surface that inherently resists lateral forces. The second selection specifies the wall as a bearing wall that supports vertical load in addition to self-weight. All architectural walls have a nonbearing default value for the **Structural Usage** property, which defines the walls as space dividers (partitions) with no support capacity for additional vertical or lateral load beyond their own weight. The last option is normally specified if the wall has more than

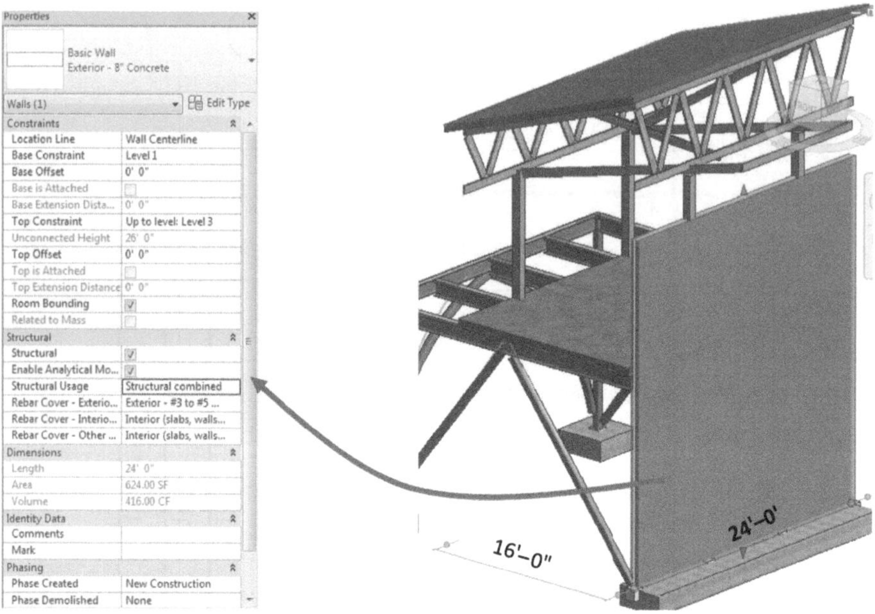

FIGURE 4.16 **Structural Usage** property of walls.

one purpose (i.e., the wall can function as a shear wall, a bearing wall, and a space divider at the same time). Structural walls can be created using one of two methods:

- By invoking the **Structure** tab ➤ **Structure** panel ➤ **Wall** drop-down ▢ **Wall: Structural**
- By invoking the **Architecture** tab ➤ **Build** panel ➤ **Wall** drop-down ▢ **Wall: Structural**

Walls can be placed in plans, elevations, or 3D views. Similar to columns and beams, a number of sketching options are available to establish structural walls. Again, grid lines will provide great assistance in modeling structural walls.

To create the wall correctly, it is good practice to define the following parameters on the **Option** bar before you start sketching:

- **Depth:** Select a level for the wall's bottom constraint or enter a value for the default setting of **Unconnected**. Or, if you want the wall to extend upward from the base constraint, select **Height**.
- **Location Line:** Select which vertical plane of the wall you want to align with the cursor as you draw or with the line or face you will select in the drawing area.
- **Chain:** Select this option to draw a series of wall segments connected at end points.
- **Offset:** Optionally, enter a distance to specify how far the wall's location line will be offset from the cursor position or from a selected line or face (as described in the next step).

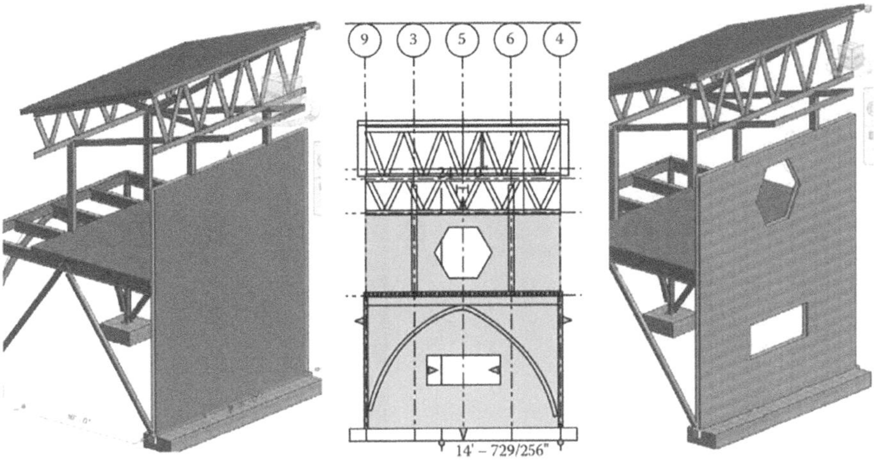

FIGURE 4.17 Modifying the shape of a structural wall.

One can always modify the appearance of structural walls through their properties after placement. However, modifying the shape of the wall or adding openings does require editing its elevation profile. To edit a wall's elevation profile, the view must be parallel and can be either a section or an elevation view (Figure 4.17).

TRUSSES

Modeling a structural truss element can be accomplished by using the truss tool when invoking the **Structure** tab ➤ **Structure** panel ➤ **Truss**. This tool allows users to create the truss according to the layout and other parameters specified in the truss family type selected. The following structural truss families are available to load into any project:

- **Fink Truss** with and without cambered bottom **Chord Truss**
- **Howe Flat Truss**, **Howe Gabled Truss-6 Panel l**, **8 Panel**, **10 Panel**, and **12 Panel**
- **Howe Gabled Truss-6 Panel-JT**
- **Pratt Flat Truss**, **Pratt Gabled Truss-6 Panel 8 Panel 10 Panel**, and **12 Panel**
- **Scissors Truss-4 Panel 6 Panel**, and **8 Panel**
- **Scissors Truss-6 Panel-JT**
- **Simple Fink** or **W Truss**
- **Warren Truss-5 Panel 7 Panel 9 Panel**, and **11 Panel**
- **Warren With Cambered Top Chord Truss-6 Panel 10 Panel**, and **14 Panel**

All types within a truss family share the same layout. Individual types specify other parameters, such as the structural framing families to be used for modeling chords and web members. To use the **Truss** tool, select a truss family type and then specify the truss start point and end point in the drawing area. Notice in case of placing a truss between columns, the extra flange at the column location will not be necessary in such cases (Figure 4.18). Such flanges at the end of the truss must be selected and deleted.

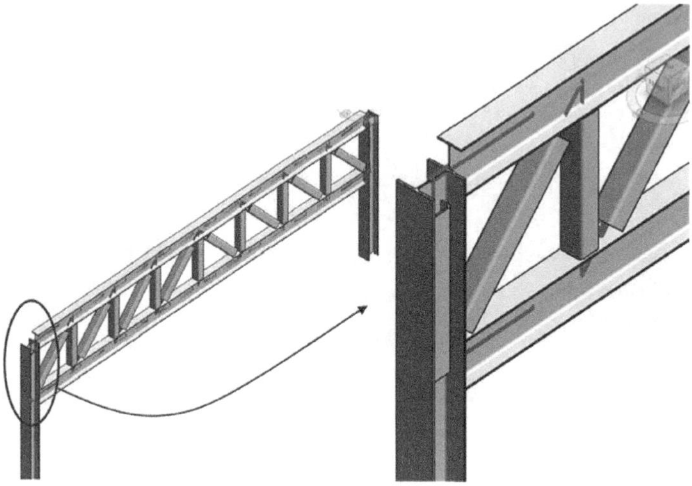

FIGURE 4.18 End truss flanges at the connection with a structural column.

CUSTOMIZING THE TRUSS ELEMENT

A structural truss element can be customized in a number of ways. Each of the truss families mentioned can be modified by changing member materials and sizes.

For example, some of the truss members can be changed into an **HSS-Hollow Structural Section**; others can be selected as a **Wide Flange** section (Figure 4.19).

The shape of the truss can be also modified to match the roof or floor profile. By attaching a truss to a roof or structural floor, a truss will conform its chords to that element. This can be attained by selecting a truss object and then invoking **Modify | Structural Trusses** tab ➤ **Modify Truss** panel ➤ (**Attach Top/Bottom**).

On the **Options** bar, select **Attach Trusses: Top or Bottom** for the appropriate truss chord being attached. After that, select the structural roof or floor (see Figure 4.20).

Please note that not all truss families can be modified to correctly attach to a roof or structural floor. For the truss chord to adapt to the shape of the corresponding roof or structural floor, the layout family's chord sketch lines must coincide with the top reference plane. The structural floor or roof profile defines the transformation of the family's reference plane, not the shape of the chord.

To drastically modify the truss and have the maximum flexibility, you can remove the truss family from a project and leave its chord and web members in place. This can be accomplished by selecting the truss and then invoking **Modify | Structural Trusses** tab ➤ **Modify Truss** panel ➤ **Remove Truss Family**. As a result, the truss family drops from the selection, leaving its individual framing elements in place. These elements can now be modified or removed as desired.

FLOORS

Modeling floors is available under both **Structural** and **Architectural** tabs. The structural floor in a structural model represents a floor slab or a floor deck.

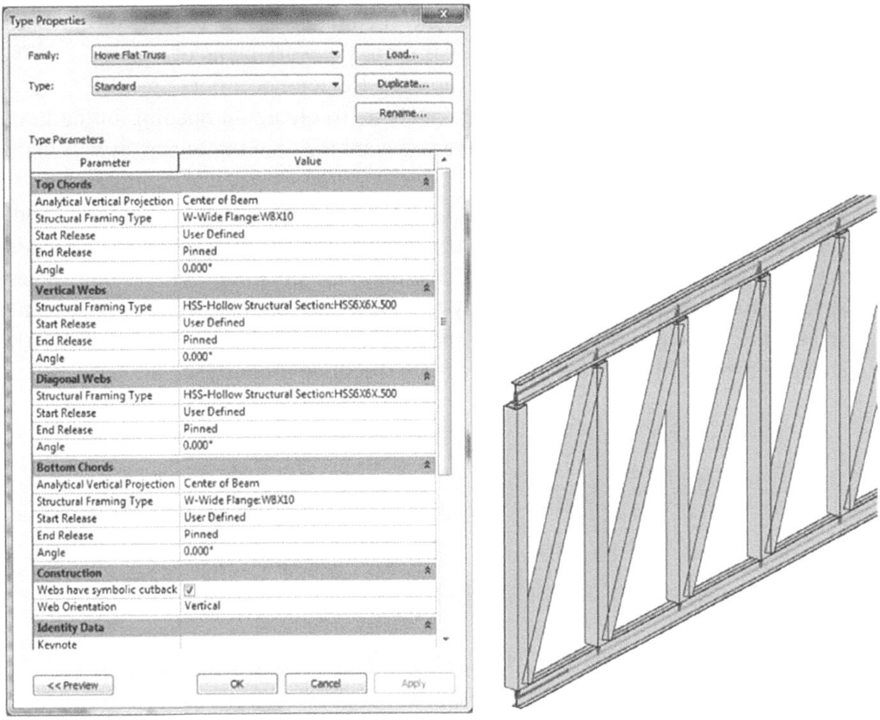

FIGURE 4.19 Truss customization.

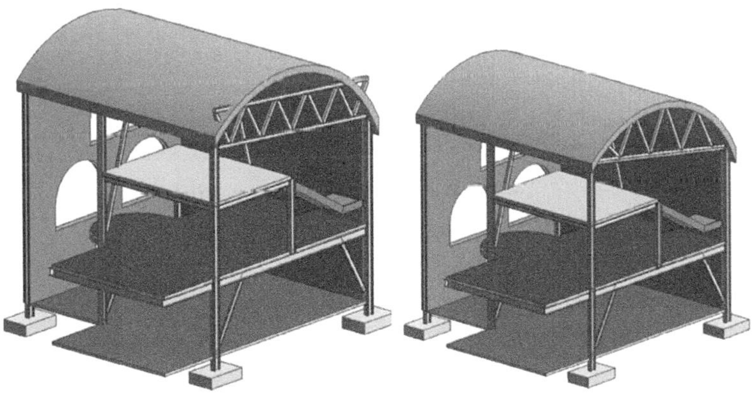

FIGURE 4.20 Modifying a truss profile to match the roof form.

Various parameters can be specified for these floors, including thickness, material, and structural composition.

One creates structural floors by invoking **Structure** tab ➤ **Structure** panel ➤ **Floor** drop-down ➤ **Floor: Structural**. Two methods are available for sketching, either by the selecting boundary walls or by using the sketching tool box (see Figure 4.21). Typically, a floor is sketched in a plan view, although you can use a 3D view if the work plane of the 3D view is set to the work plane of a plan view. The floor boundary must be a closed loop (profile). Remember, floors are offset downward from the level on which they are sketched. To create an opening in the floor, you can sketch another closed loop where you want the opening to appear within the original floor closed polygon.

Slopes can be defined for any structural floor after creating its boundary. To add a slope, select the floor and edit its profile, then choose the **Slope Arrow** tool in the **Draw** panel to assign a slope to the selected floor. On clicking on the **Slope Arrow** tool, the list box on its right will display two sketching options: **Line** or **Pick Line.** To sketch the slope arrow, you need to specify the start and end points of the arrow line (see Figure 4.22) and then give values for the slope properties.

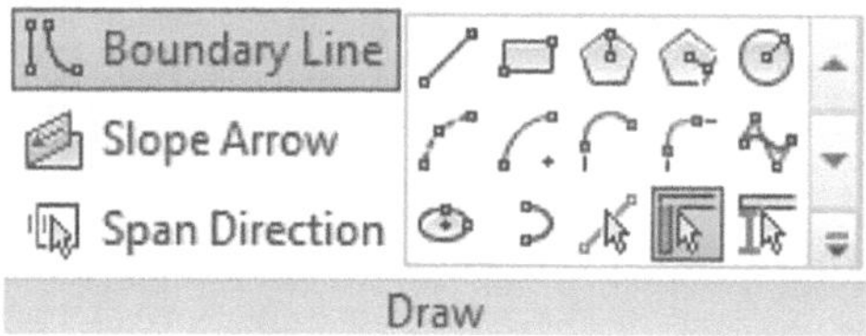

FIGURE 4.21 **Sketching** tool box for a structural floor.

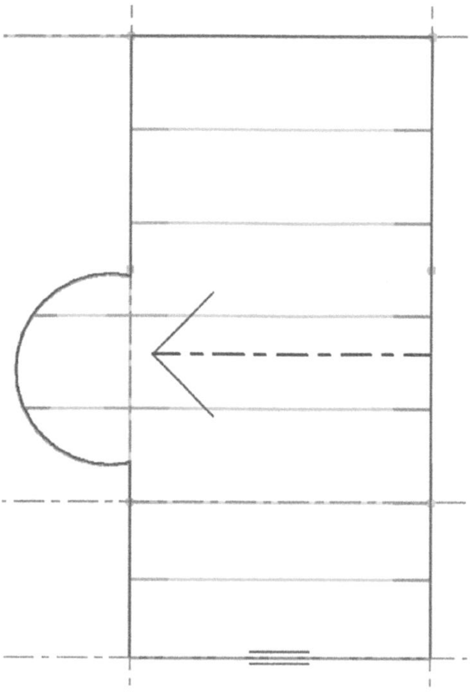

FIGURE 4.22 Defining a slope for a structural floor.

Note: In general, you define a slope using the following schema shown below (Figure 4.23).

Similar to the slope of the floor, cantilevers can be added to the structural floor either before or after sketching the floor. The **Slab Cantilever** property permits the sketch of the structural floor to be both constrained to its supports and its edge to be extended beyond them. Cantilevers are created by specifying the offset parameters that represent the concrete and steel deck of a structural floor. This can be achieved by selecting the **Modify | Create Floor Boundary** tab ➤ **Draw** panel ➤ **Boundary Line** and then click (**Pick Supports**). Then, on the **Properties** palette, under **Other**, enter offset values for both **Concrete Cantilever** and **Steel Cantilever**. Finally, apply and finish the edit mode.

Note: The cantilever offset is applied in relation to the direction that the slab edge was created. This is applicable to line sketching, **Pick Supports**, or **Pick Walls**

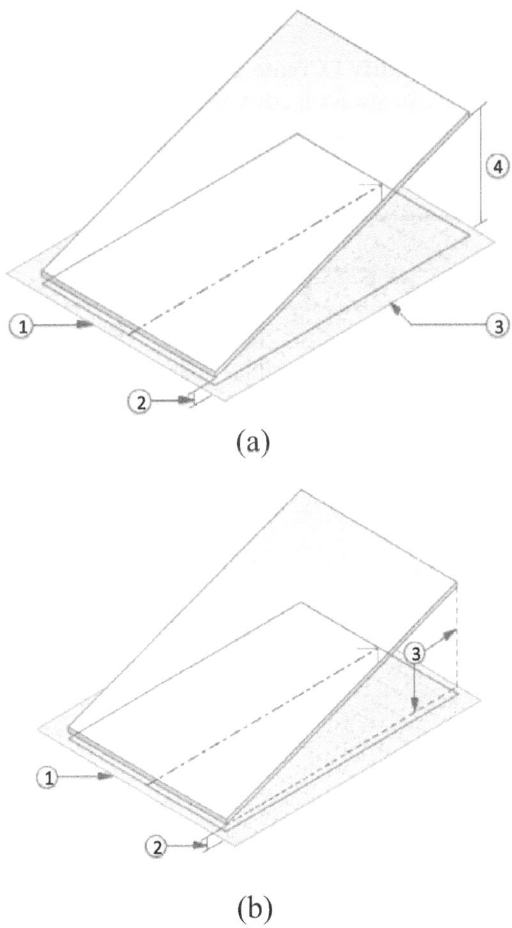

FIGURE 4.23 (a) Defining a slope for structural floor using heights of sides. (b) Defining floor slope using rise/run.

structural floor creation. If the cantilever appears to be inside the structural floor, enter sketch mode, select the edge, and adjust the cantilever values to a negative number. The section view in Figure 4.24 shows a concrete structural floor with a metal deck. The cantilevered concrete edge extends beyond the supporting beam.

Also, one can modify or add slab edges by invoking **Structure** tab ➤ **Structure** panel ➤ **Floor** drop-down ➤ (**Floor: Slab Edge**), then selecting horizontal edges of floors. As you select edges, Revit treats this as one continuous slab edge (Figure 4.25).

Structural floors can be also modified to add drop panels to reinforce floors at column locations. This is achieved by sketching a second smaller structural floor. The easiest way to model drop panels is to open a plan view with a structural floor over a column. Then, invoke **Structure** tab ➤ **Structure** panel ➤ **Floor** drop-down ➤ **Floor: Structural**. After that, click the **Create Floor Boundary** tab **Draw** panel **Boundary Line** and use the sketching tools to complete a sketch of the drop panel (see Figure 4.26). Please note that the sketch for the drop panel must form a closed loop or boundary condition.

When finished, click **Modify | Create Floor Boundary** tab ➤ **Mode** panel ➤ **Finish Edit Mode**. One can always modify the properties of the created drop panel by adjusting its constraint parameters on the **Properties** palette so that it is at the desired elevation in the model.

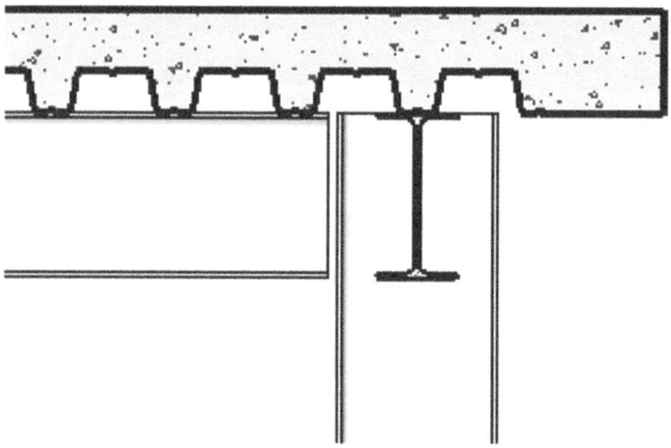

FIGURE 4.24 Cantilever definition of a structural floor of concrete on a metal deck.

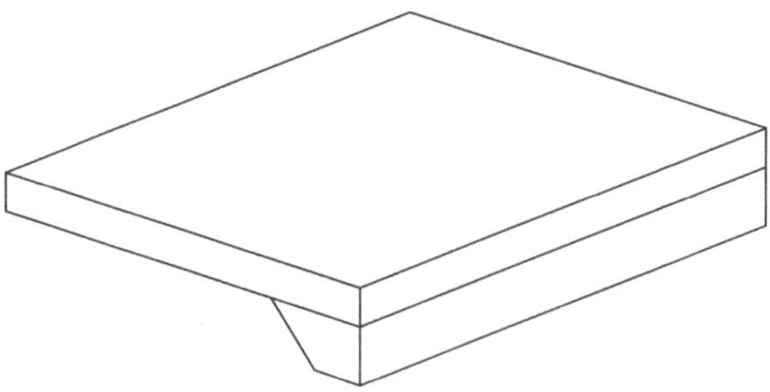

FIGURE 4.25 Modeling slab edges.

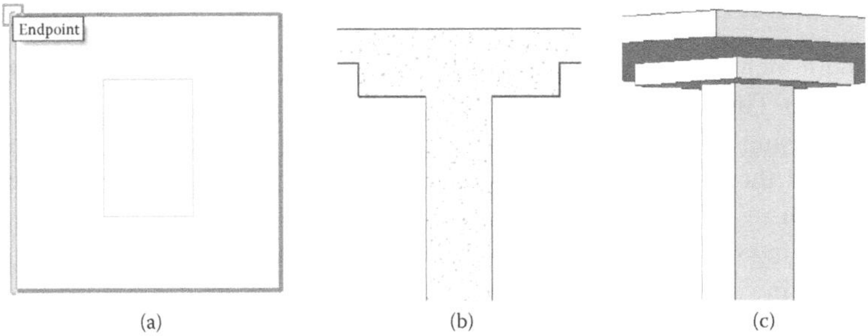

FIGURE 4.26 Modeling floor drop panels at column locations: (a) plan view; (b) elevation view; (c) 3D view.

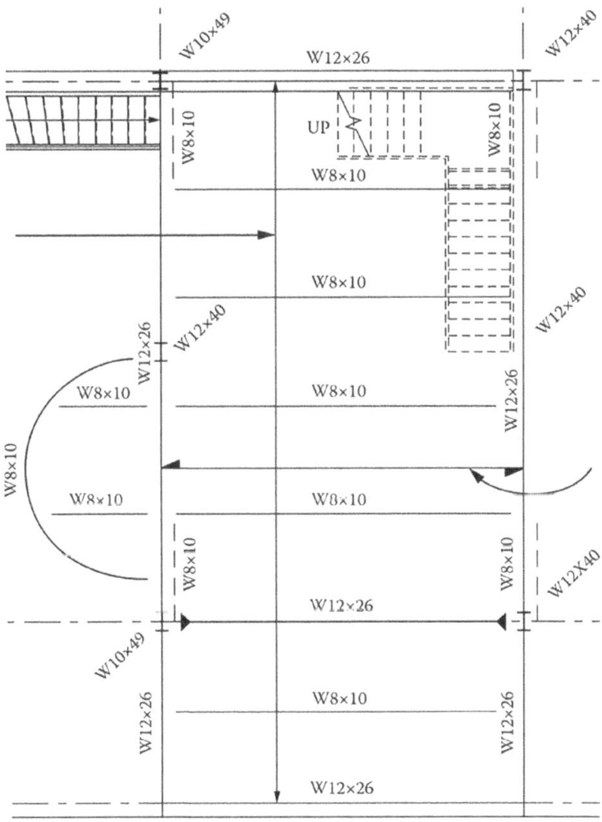

FIGURE 4.27 Structural floor **Span Direction** component.

Another important aspect of structural floor modeling is the span direction component. This span direction symbol is normally placed in the plan view of the structural floor. The span direction component is used to change the orientation of the steel deck in the plan. Deck span direction is designated by the direction of the filled half-arrows (see Figure 4.27).

STEEL CONNECTIONS

In the main menu items of Revit user interface, there is a new Tab for "**Steel**" (see Figure 4.28a). This tab includes all the tools you need for structural steel connections.

To start using the Steel Connection tool, one needs to click on the small arrow at the corner of the first panel to activate the structural steel connection settings (see Figure 4.28b).

In the setting screen, one needs to select all the connections needed for the project. In the following sections, the modeling of main steel connections will be explained.

Base plate connection is used assist to spread the columns load and transfers it to the concrete foundation. One begins by selecting the column and then invokes the "Connection" tool from the "Steel" tab as shown in Figure 4.28c and then select from the **Type Selector** dropdown "Baseplate connection". One needs to finish this task by pressing the Enter key. The base plate connection will be depicted as illustrated in Figure 4.28d.

One can edit the properties of the base plate connection just like any other Revit objects. You select the connection and then in the **Properties Inspector**, click on "**Edit Type**". Next, click on "**Edit**" to modify parameters (see Figure 4.28e). Now you can modify any of the parameters defining the base plate connection such as material, thickness, anchor bolt size and type, and anchor bolt arrangements.

(a)

(b)

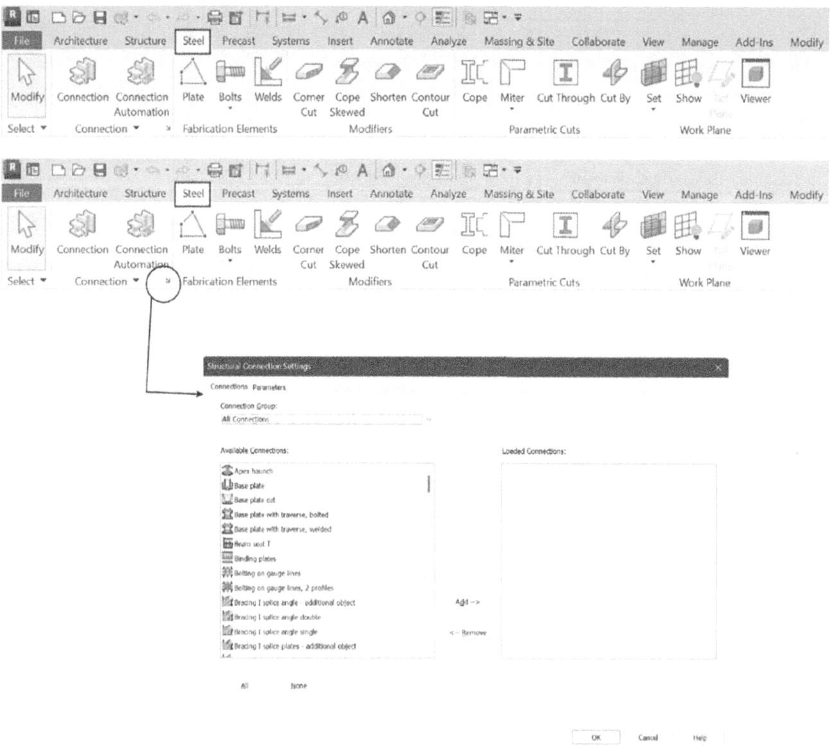

FIGURE 4.28 (a) The New Steel tab. (b) Activating the structural steel connections settings.

(Continued)

(c)

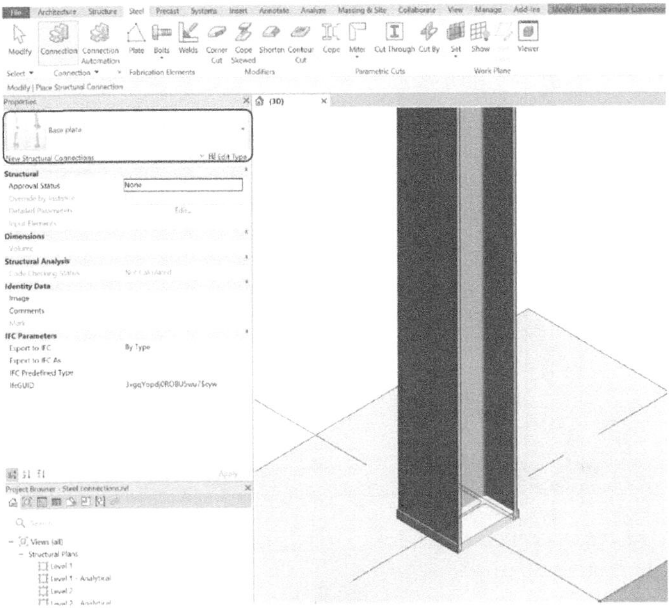

(d)

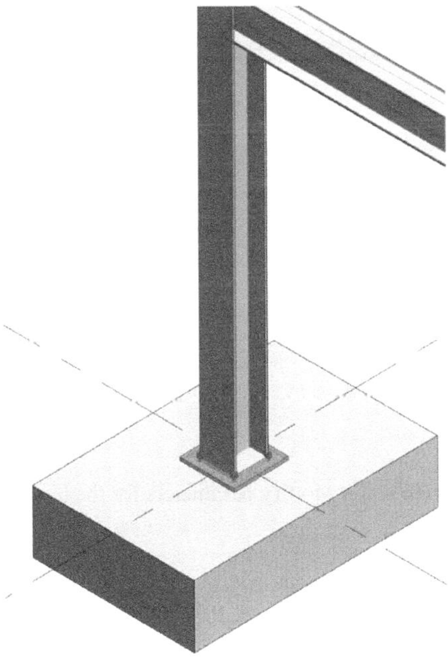

FIGURE 4.28 (*Continued*) (c) Choosing the type of connector. (d) A base plate connection.

(*Continued*)

(e)

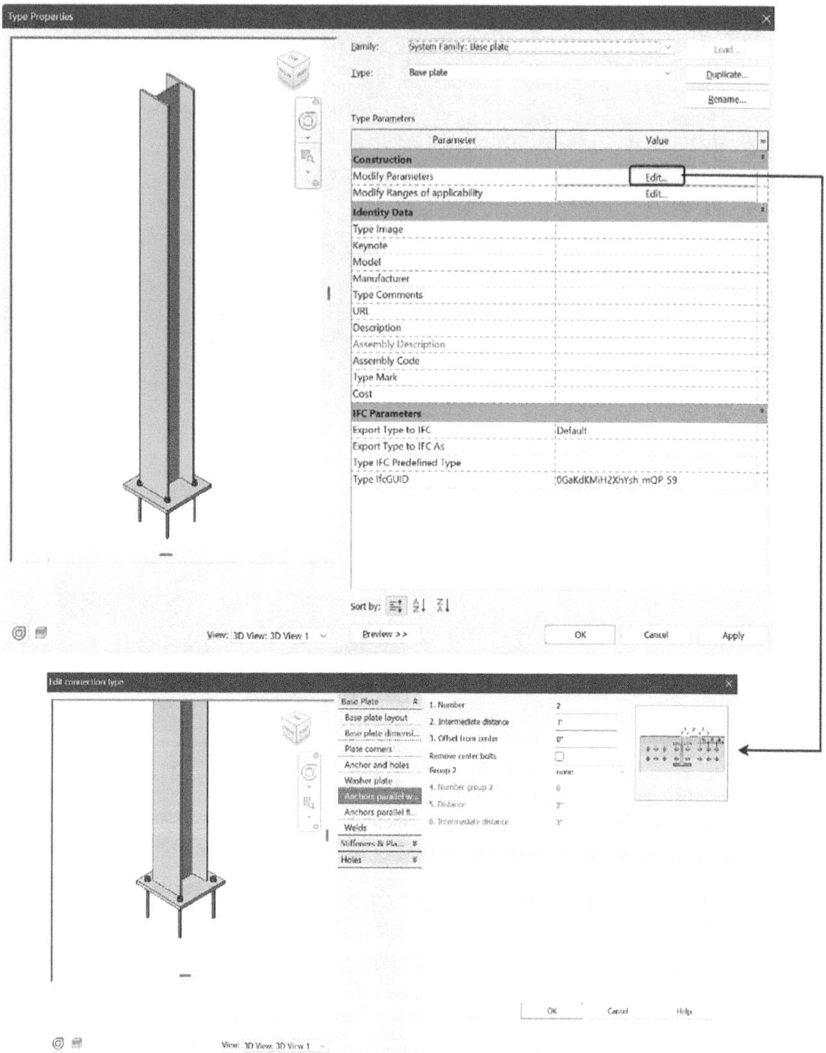

FIGURE 4.28 (*Continued*) (e) Modify parameters for the base plate connections.

(*Continued*)

Framing angle or clip angle is an important connection type in steel construction. It connects beams to girders. The modeling of these types of connections starts by selecting the two members that need to be connected and then invoke the **Connection** tool and choosing clip angle (Figure 4.28f).

After the clip angle is placed correctly between the two members, one can select it and modify its parameters to suit the design forces. In Figure 4.28g the number of bolts are reduced to 4 to satisfy the structural needs.

Tube Connection with sandwich plate is used to connect brace elements to columns. The modelling approach is similar to the other type of connections. One needs

(f)

(g)

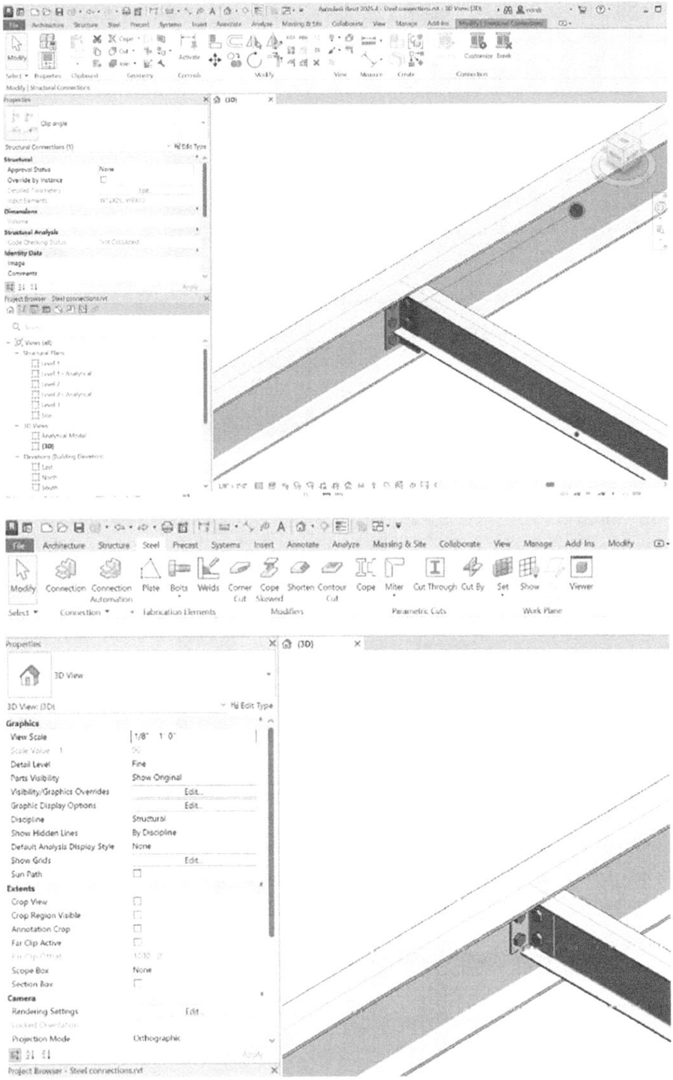

FIGURE 4.28 (*Continued*) (f) Clip angle connection. (g) Modifying the clip angle properties. (*Continued*)

to invoke the **Connection** tool → **Type Selector** drop-down → Tube connection with sandwich plate. Then select the column and the brace element, and press Enter to place the connection (see Figure 4.28h). After the tube connection with sandwich plate is placed, one can alter its properties by selecting "**Edit Type**" from the **Properties** pallet as depicted in Figure 4.28i.

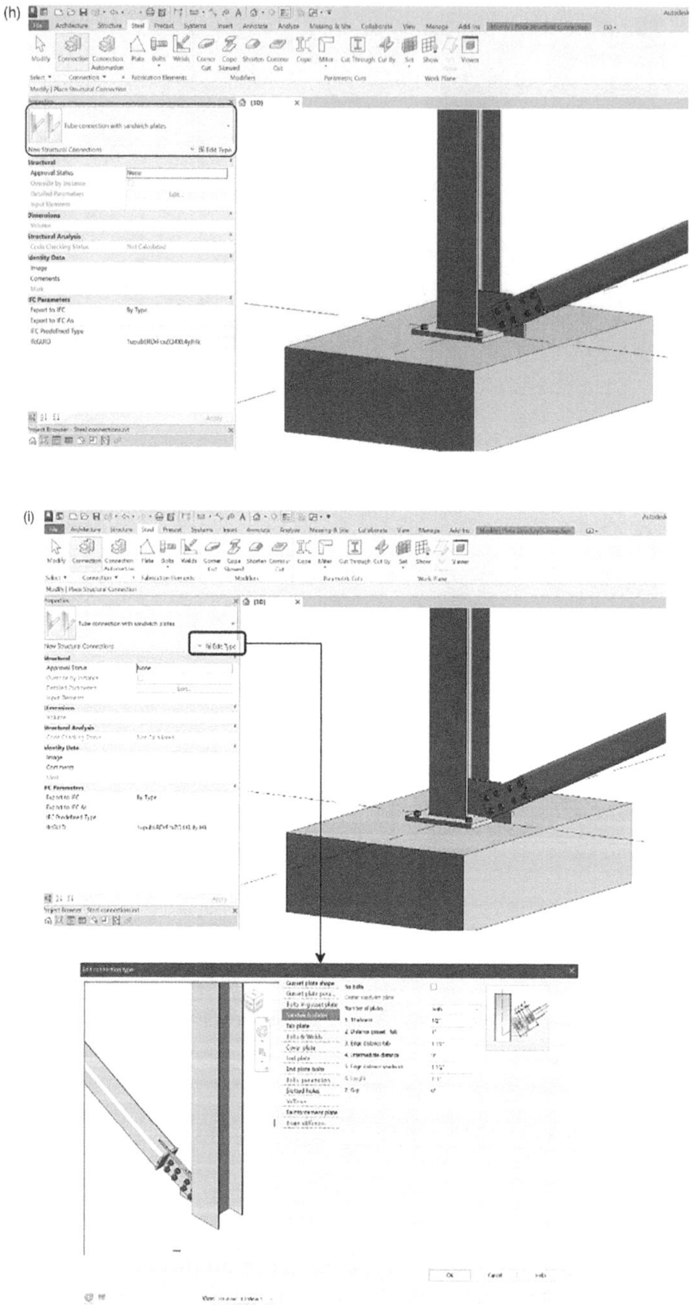

FIGURE 4.28 (*Continued*) (h) The modeling of tube connection with sandwich plate. (i) Editing the tube connection with sandwich plate properties.

(Continued)

(j)

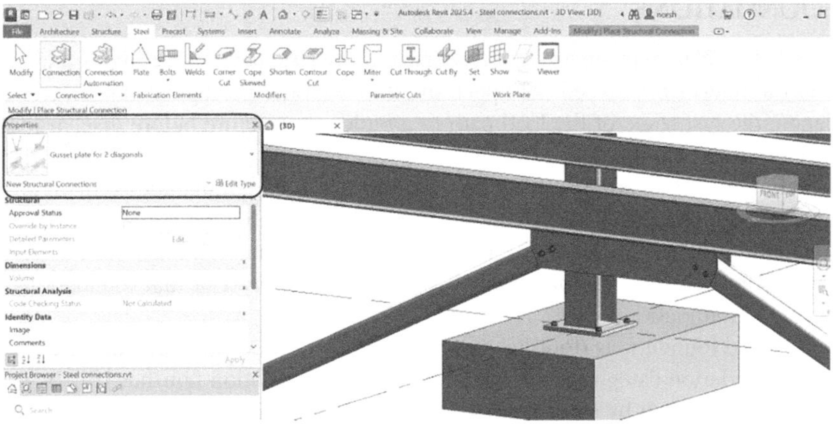

(k)

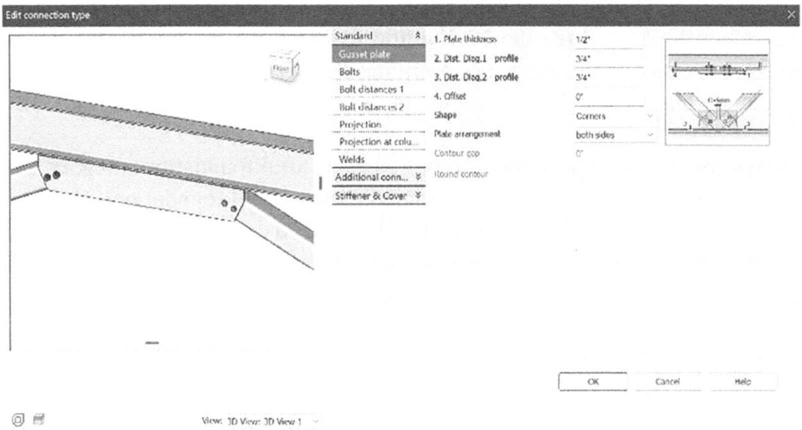

FIGURE 4.28 (*Continued*) (j) Modeling of a Gusset plate connector for two diagonals. (k) Adjusting the parameters of the gusset plate connector.

The last connection type that will be described in this section is the guest plate to connect two diagonals. This is a typical type of connection for braced frames such as chevron brace. One begins again from the **Steel** tab, invoke **Connection** tool → **Type Selector** drop-down → Guest plate for 2 diagonals. Then select the two diagonals and the beam to place the guest plate connector (see Figure 4.28j)

You can modify the properties of the gusset plate connector by clicking on "**Edit Type**" from the Properties pallet and then click on **Edit** to modify the parameters of the connection (Figure 4.28k). Depending on the design loads on the connection, the gusset plate material, thickness and the bots types, diameter and their arrangement can be changed using this screen (Figure 4.28k).

FOUNDATIONS

Foundation elements are an important part of the structural modeling as they represent the interface between the upper structure and the geological subsurface. They denote the last stage of the load path in a building structure before discharging the applied loads into the ground.

Revit allows modeling of both shallow and deep foundations in a project. Specifically, you can create isolated footing, wall, or slab foundations, as well as pile foundations. These cover different physical foundation conditions, such as isolated footing for columns, retaining walls, stepped footing, raft or mat foundations, and deep pile foundations.

Isolated foundations (footings) are stand-alone families that are part of the structural foundation category. You can load isolated foundation families if they are not present in a project by selecting **Modify | Place Isolated Foundation** tab ➤ **Model** panel ➤ **Load Family**. You can create an isolated footing instance in a project by invoking the **Structure** tab ➤ **Foundation** panel ➤ **Isolated**. After that, you must select an isolated footing type from the **Properties** palette, the **Type Selector** drop-down. The **Isolated** foundation can be placed in either a plan or a 3D view (Figure 4.28).

Wall foundations are also members of the structural foundation category, but they are hosted only by walls. Similar to isolated footing, wall foundations can be placed along structural walls in either a plan or a 3D view. Also, notice that wall foundations are constrained to the walls that they support and change with them. In other words, there is a relationship between the foundation and the host element (wall). As a result, whenever there is a modification in the form or location of the wall, the associated foundation also changes.

You create a wall foundation by invoking **Structure** tab ➤ **Foundation** panel ➤ **Wall**, then selecting a wall foundation type from the **Type Selector** drop-down. Currently, there are both retaining and bearing wall foundation types available for selection. After selecting the type, you need to identify the wall receiving the wall foundation type (see Figure 4.29). The wall foundation is then created at the bottom of the selected wall.

The **Structural Usage** (either **Retaining** or **Bearing**) of a wall foundation can be adjusted at any time using one the following methods: The first method is by modifying the wall foundation to an appropriate wall foundation type. Alternatively, you can change its structural usage by setting its **Structural Usage** parameter in the **Properties** palette for the type. This parameter can be set to one of the following values: **Retaining** or **Bearing**.

If **Retaining** is chosen, you may need to specify values for **Toe Length**, **Heel Length**, and **Foundation Thickness**. Toe and heel length define the width of the foundation. On the other hand, if **Bearing** is selected, you would need to specify values for **Width** and **Foundation Thickness**.

In Revit, you can also define **Matt** foundation by invoking the **Foundation Slab** tool from the **Foundation** panel under the **Structural** tab (**Structure** tab ➤ **Foundation** panel ➤ **Slab** drop-down ➤ **Structural Foundation: Slab**). Foundation slabs can also be used to model structural floors on a grade, which do

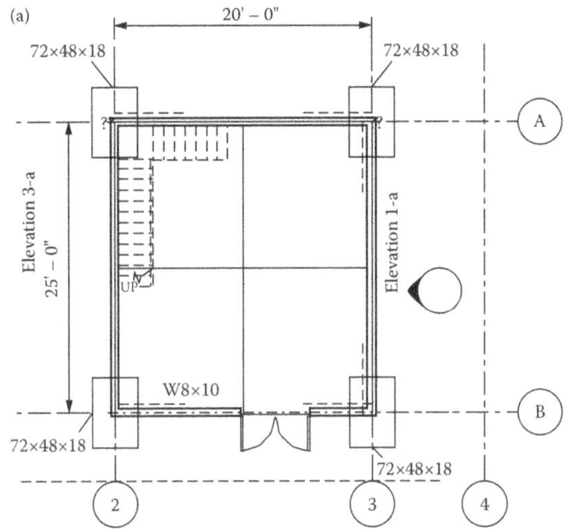

(a)

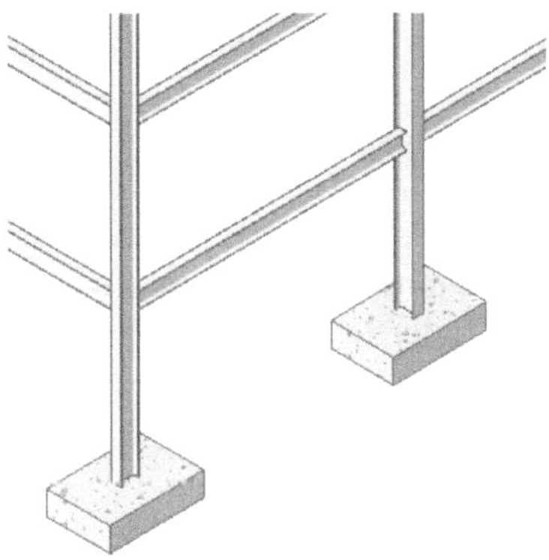

(b)

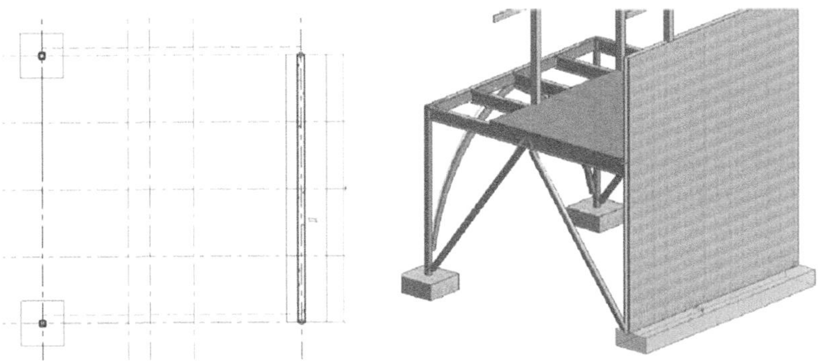

(b)

FIGURE 4.29 (a) Modeling of a rectangular isolated footing in (a) plan and (b) 3D views. (b) Modeling a wall foundation.

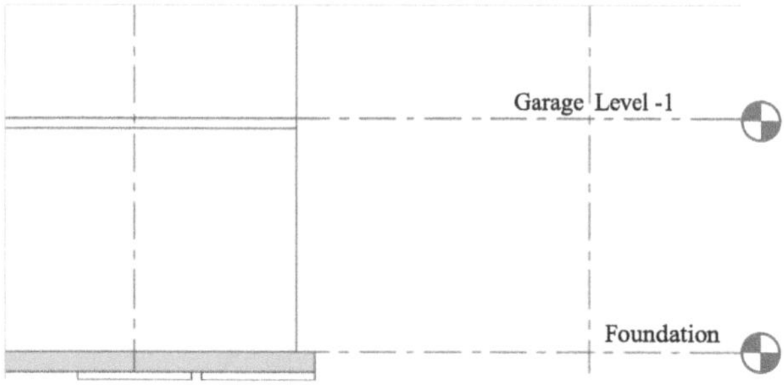

FIGURE 4.30 Matt foundation below the foundation level.

not require external support from other structural elements. **Foundation Slabs** can provide a structural support solution in case of modeling complex foundation shapes that cannot be created using **Isolated** or **Wall Foundation** tools.

Foundation slabs are normally placed below the level at which they are drawn. For example, if you add a foundation slab in level 1, it will be placed below level 1 and will not be visible in the level 1 plan view. To view the foundation slab in a plan view, you need to create a new level below level 1 (see Figure 4.30). Furthermore, once you create a new level below level 1, you will be able to see the foundation slab as an underlay (displayed in halftone) in level 1.

An important setting for the foundation slab is the value of the instance parameter: **Analyze as** under the **Structural Analysis** group in the **Instance Properties** palette. You can specify the **Analyze as** parameter of a foundation slab: either **Foundation** or **Slab on Grade**. If set to **Foundation**, the slab will provide support for superstructure. If set to **Slab on Grade**, the slab will only support itself. The **Foundation Slab** type also has an additional, read-only parameter, **Elevation at Bottom**. The **Elevation at Bottom** parameter is used for tagging the **Bottom of the Foundation** elevation.

FAMILIES'

In Revit, all building elements (beams, columns, walls, roofs, windows, doors, etc.) that are used to assemble a building model, and all the dimensions, callouts, fixtures, tags, and detail components that are used to document a BIM model are created with families. A family in the Revit context is simply a subclass of a category and represents a group of elements with a common set of properties, called parameters, and a related graphical representation. Different elements belonging to a family may have different values for some or all of their parameters, but the set of parameters (their names and meanings) is the same. These variations within the family are called family types or types. For example, the **Structural Column** category includes families and family types that can be used to create different columns, such as steel wide-flange columns, concrete columns, and **HSS-Hollow Structural Section** columns (see Figure 4.31).

By utilizing predefined families and creating new ones in Revit, you can easily add both standard and custom objects to building models. Families also provide a

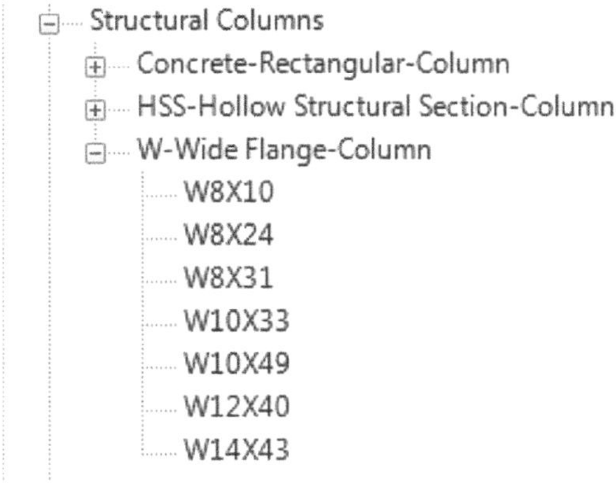

FIGURE 4.31 **Structural Columns** category and families.

level of customization and control over objects that are similar in use and behavior, allowing designers to have more freedom and simply make design modifications and manage projects more efficiently.

There are three kinds of families in Revit Architecture: system families, loadable families, and in-place families. Most elements created in a project are system families or loadable families. Loadable families can be combined to create nested and shared families. Nonstandard or custom elements are created using in-place families.

System families create basic building elements, such as walls, roofs, ceilings, floors, and other elements that would be assembled on a construction site. System settings, which affect the project environment, including types for levels, grids, drawing sheets, and viewports, are also system families. System families are predefined in Revit, and they are not loaded into a project from external files or saved in locations external to the project. If the system family type cannot be found in a project, you can create a new type by changing the properties of an existing type by duplicating (copying) a family type and changing its properties.

Because system families are predefined, they are the least customizable of the three kinds of families, but they include more intelligent behavior than the other standard loadable families and in-place families. For example, a wall created in a project automatically resizes to accommodate doors and windows placed in it. There is no need to cut openings in the wall for the windows and doors before placing them. Examples of system families include **Ceilings**, **Curtain Systems**, **Curtain Wall Mullions**, **Detail Items**, **Floors**, **Fluids**, **Model Text**, **Railings**, **Ramps**, **Roofs**, **Site (Pad)**, **Stairs**, **Structural Columns**, **Structural Foundations**, **Structural Framing**, and **Walls**.

On the other hand, loadable families are families used to create both building objects and some annotation elements. Loadable families create the building components that would usually be purchased, delivered, and installed in and around a building, such as composite sections and members, conceptual masses, windows, doors, fixtures, furniture, and plantings. They also include some annotation elements that are routinely customized, such as symbols and title blocks. Because they are highly customizable in nature, loadable families are the families that one most commonly creates and modifies in Revit. Unlike system families, loadable families are created in external RFA files and imported, or loaded, into a BIM project.

One can share a loadable family with many projects or families by loading instances of families in other families to create new families. By nesting existing families inside other families, you can save appreciable modeling time. A loadable family example, includes **Annotations**, **Balusters**, **Casework**, **Columns**, **Curtain Panel by Pattern**, **Curtain Wall Panels**, **Detail Components**, **Doors**, **Electrical Components**, **Entourage**, **Furniture**, **Furniture System**, **Lighting Fixtures**, **Mass**, **Mechanical** components, **Plumbing** components, and **Structural** components.

In-place families are unique elements that are created when there is a need to create a distinctive component that is specific to the current project, for example, modeling a unique or unusual geometry for a nonstandard roof shape. Unlike system families and loadable families, however, you cannot duplicate in-place family types to create multiple types. Although it may seem easier to create all families as in-place elements, the best practice is to use them only when necessary because in-place families can increase file size and degrade software performance.

Unlike system families, which are predefined, loadable and in-place families are always created in the **Family Editor**. However, system families may contain loadable families that are modifiable in the **Family Editor**.

The design environment for creating families is called **Family Editor**. The **Family Editor** is a graphical editing mode in Revit that lets you create and modify families. Once the **Family Editor** is launched, you have to select a template from the drop-down menu to start modeling a family. The template normally includes multiple preconfigured views, such as plans and elevations. The **Family Editor** interface has the same appearance as the project environment in Revit, but it depicts different tools (Figure 4.32). The family topic is a large topic that may deserve a full book to cover all the details. The following is an example and workflow strategies that would be helpful in understanding families and their potential.

Example 4.1: Creating a Loadable Family

For the best results, family creation should be approached in a systematic manner. For instance, the following steps are essential for successful family modeling:

- Before beginning family creation, plan the family by sketching it and define the necessary views and parameters.
- Create a new family file (RFA) using the correct family template.
- Define subcategories for the family to support various views of the model.
- Model the family skeleton and components.
- Make sure to clearly identify the origin (the insertion point) of the family when used in a project.
- Start laying out reference planes and reference lines to assist in drawing skeleton geometry.
- Define linear and nonlinear dimensions to postulate parametric relationships.
- Label dimensions to create type or instance parameters of the family.

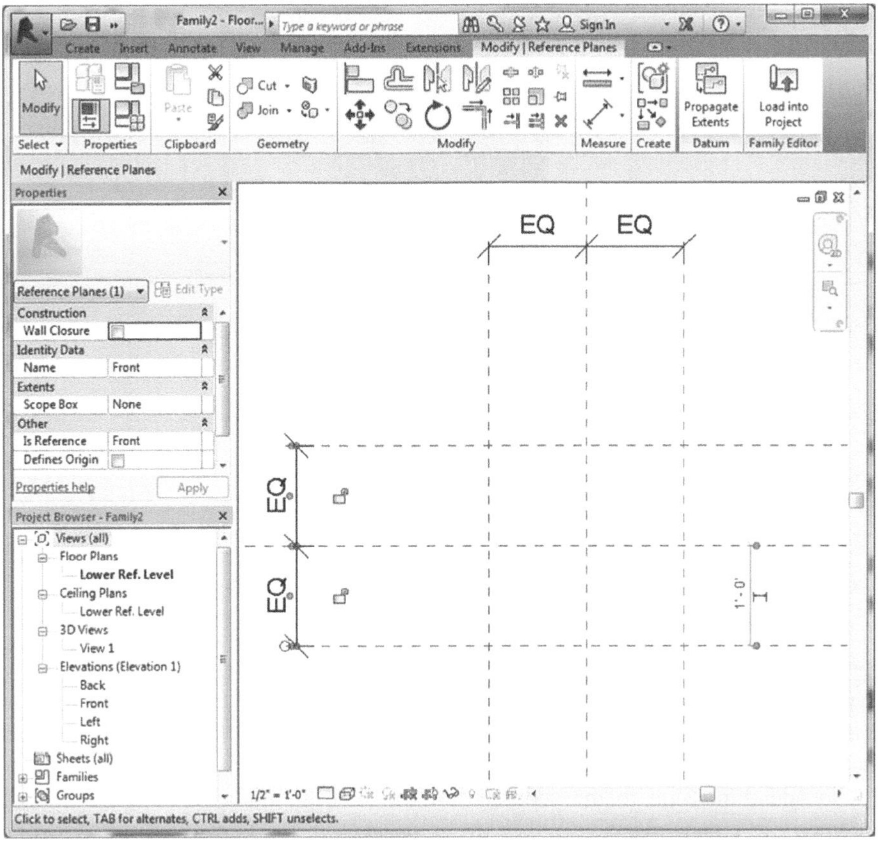

FIGURE 4.32 Family Editor.

- Add family type variations by defining different values for the parameters.
- Finally, test, or flex, the family by trying different values for each parameter defined.

In this example, an L-shaped concrete column family is created (Figure 4.33). Invoke the **Family editor** by selecting **new Family**. Then, choose the **Structural columns** template (see Figure 4.32).

The new family opens in the **Family editor**. For most families, two or more dashed green lines display. These are reference planes, or the working planes that are meant to help when creating the family skeleton. In the plan view, these reference planes define the boundaries of the column. The next step is to save the file with a meaningful name.

The L column has a cross section of with a fillet curve as shown in Figure 4.34. Starting with the floor plan "Lower Ref. Level", an extrusion will be created by launching the **extrusion** tool in the **Family editor**: the **create** tab **Forms extrusion**. Then from the **draw** panel, select line to draw the shape shown in Figure 4.34.

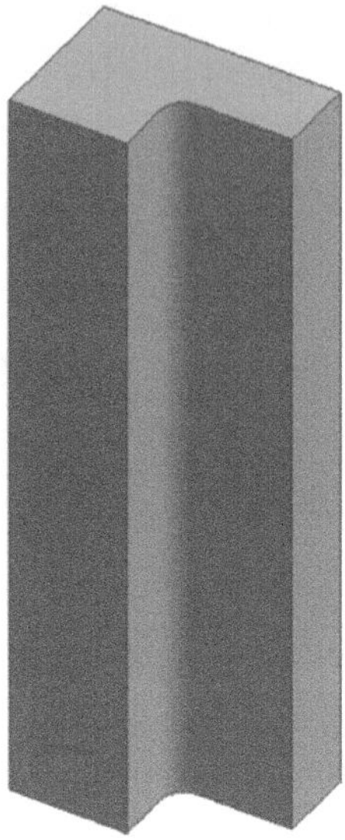

FIGURE 4.33 L-section concrete column family.

Next, you define family parameters. The parameters that you define at this stage usually control the size (length, width, height) of the element and let you add family types. The best workflow for defining geometric properties is to insert dimensions for the critical part of the geometry in the plan and elevations. In this case, the width and depth are the main parameters defining the cross section of the column.

To add dimension to the geometry of the family, you need to define family geometry reference planes as either strong or weak in the **Family editor**. A strong reference has the highest priority for dimensioning and snapping. Thus, as when creating the family, temporary dimensions snap to any strong references in the family. When selecting the family in the project, temporary dimensions appear at the strong references.

A weak reference has the lowest priority for dimensioning. When placing the family into the project and adding dimension to it, you may need to press **tab** to select a weak reference because any strong references highlight first.

To define parameters for the geometry, you need to label the dimensions to create parameters. For instance, the dimensions in Figure 4.30 have been labeled with **depth** and **width** parameters to define the cross section of the column.

Now, you can test or flex the parameters that have been applied to the family framework. To flex the family, the parameter values are changed, making certain that the reference planes to which one applied the parameter change accordingly. Flexing is a way to test the integrity of the parametric relationships. Flex early and frequently when creating families to ensure the stability of the families. To flex the family, launch the **Family properties** panel: **create** tab ➤ **Family properties** panel ➤ **types**. The **Family types** dialog displays (Figure 4.35). Although no any family types are defined yet, the dialog lists the parameters that have been created.

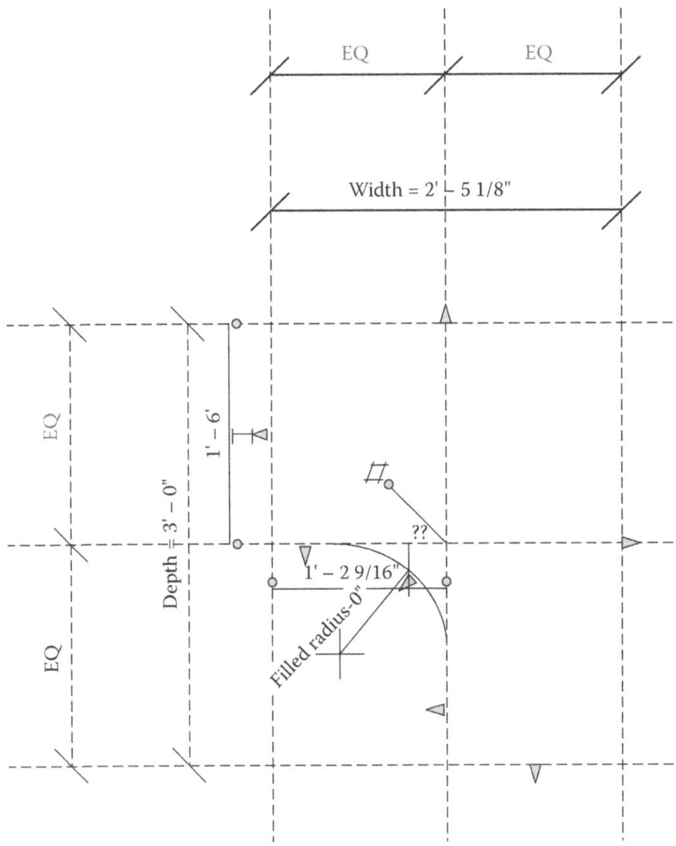

FIGURE 4.34 Plan view of the L-shaped column.

Using the **Family types** tool, many types (sizes) for a family can be created. Each family type has a set of properties (parameters) that includes the labeled dimensions and their values. It is possible also to add values for standard parameters of the family (such as **Material**, **Model**, **Manufacturer**, **type Mark**, and others). To create family types from the **Family types** tool, click on **new** and enter the family name. In the **Family types** dialog, enter the values for the type parameters for each named type. In this example, the following types are created: **l.3 × 2 concrete** and **l.2 × 1.5 concrete** columns.

Parameters can depend also on other parameters using the formula. This can be achieved by modifying the **Formula** column in the **Family types** tool (Figure 4.35). A simple example would be a **depth** parameter set to equal twice the width of an object. In practice, formulas can be used in various ways, varying from simple to complicated applications. Typical examples include embedding design relationships, defining angular associations, and relating a number of instances to a variable length (e.g., changing the diagonals in an open web joist as the length increases).

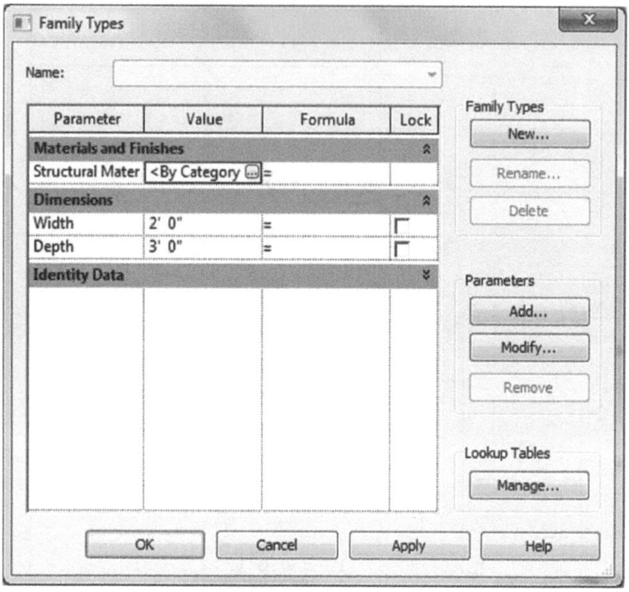

FIGURE 4.35 **Family** type tool.

Furthermore, you can add conditional statements in formulas to define actions in a family that depend on the state of other parameters. For example, the statement: = **IF (Depth > 10', "Too Deep", "Normal")** specifies the condition that if the depth is greater than 10 ft, then it is too deep; otherwise, it is normal. With conditional statements, the software enters values for a parameter based on whether a specified condition is satisfied.

TESTING A FAMILY IN A PROJECT

A final step in testing a family is to load it in at least one project and create elements with the family types to ensure it works correctly. It is recommended to select a test project that contains any geometry with which the family must interact.

To load the family in a test project, you can do either of the following: In the family, select **Create** tab ➤ **Family Editor** panel ➤ **Load into Project**. Or, from the test project, select **Insert** tab ➤ **Load** from **Library** panel ➤ **Load Family**, navigate to the location of the family, select it, and click **Open**.

In the project, click the **Home** tab and then click the appropriate tool to begin creating an element from one of the new family types (Figure 4.36).

LOADS

Before carrying out any structural analysis, it is required to define the loads that the structural model must support. The magnitude and the direction of these forces may

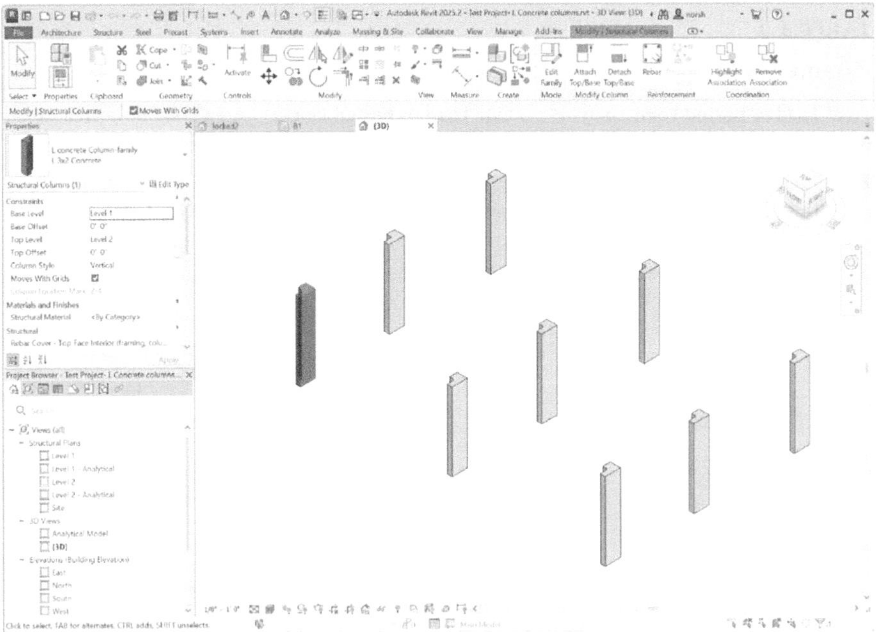

FIGURE 4.36 Test project for the **L-Concrete** column family.

vary depending on various conditions. These loads can then be grouped into different load cases and combined according to the building code specifications. Using Revit, one can apply point, line, and area loads. Each of these three loads represents a family that contains instance and type parameters. Loads can be applied either by sketching or by using host elements, such as floors and walls. These loads can be edited before or after placing them.

You can add loads by invoking the **Loads** tool from the **Analyze** tab: **Analyze** tab ➤ **Loads** panel. Then, select one of the options displayed (**Point Load**, **Line Load**, **Area Load**, **hosted Point Load**, **hosted Line Load**, **hosted Area Load**). Parameters that need to be defined for each of these loads include (1) load case, (2) x component of the force F_x, (3) y component of the force F_y, (4) z component of the force F_z, (5) x component of the moment M_x, (6) y component of the moment M_y, and (7) z component of the moment M_z. Table 4.2 summarizes the main input parameters for these loads.

Load modeling depends on the type of coordinate system. Revit utilizes several coordinate systems for loads. These coordinate systems include project coordinate system, current work plane, and host work plane. The project coordinate system appears in the view when you click **Analyze** tab ➤ **Loads Panel**. Text is also displayed under the coordinate system to indicate whether the load is defined in terms of project, work plane, or host work plane coordinates.

The work plane is the current plane of object placement. When the current work plane is used to orient loads, loads will be placed perpendicular to the current work plane.

TABLE 4.2
Parameters for Defining Structural Loads

Point Load or hosted Point Load (kips or KN)	Line Load or hosted Line Load (kip/ft or KN/m)	Area Load or hosted Area Load (Ksf or KN/m²)
Place Loads tab ➤ **Loads** panel **Hosted Point Load**	**Place Loads** tab ➤ **Loads** panel **Line Load.**	**Place Loads** tab ➤ **Loads** panel **Area Load.**

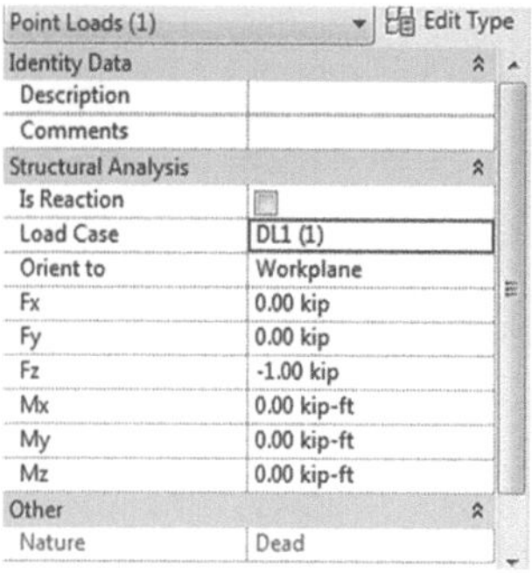

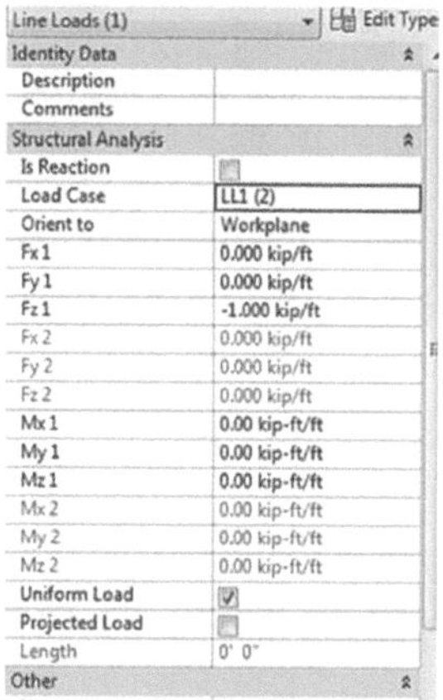

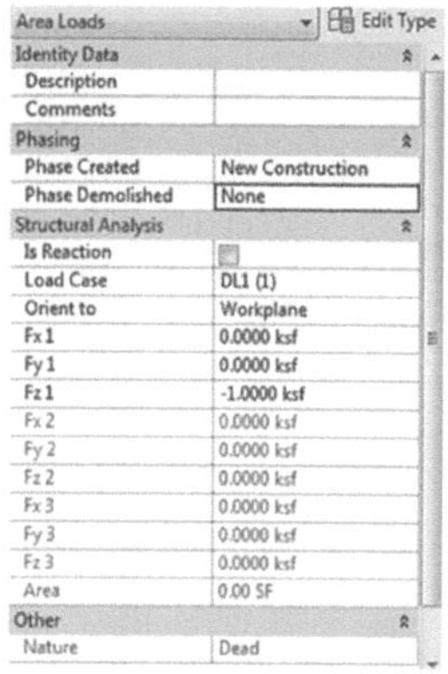

TABLE 4.3
Predefined Load Cases in Revit

Name	Case Number	Nature	Category
DL1	1	Dead	Dead Loads
LL1	2	Live	Live Loads
WIND1	3	Wind	Wind Loads
SNOW1	4	Snow	Snow Loads
LR1	5	Roof Live	Roof Live Loads
ACC1	6	Accidental	Accidental Loads
TEMP1	7	Temperature	Temperature Loads
SEIS1	8	Seismic	Seismic Loads

The host work plane is the plane in which the element chosen to host a load resides.

The load cases that are already defined by Revit, including their nature and category, are listed in Table 4.3. You can also apply additional load cases and load combinations. To add a load case, simply invoke the **Load** case tool from the **Analyze** tab: **Analyze** tab ➤ **Loads** panel ➤ **Load Cases**. You can also utilize the **Structural Setting** dialog box from the **Manage** tab under the **Settings** panel to add load cases.

Similarly, load combinations can be created by invoking the **Load Combinations** tool from the **Analyze** tab: **Analyze** tab ➤ **Loads** panel ➤ **Load Combinations**.

BOUNDARY CONDITIONS

The support conditions of a structural model are generally defined through the boundary conditions settings. They specify the support conditions of a structural element by its surrounding environment. These boundary condition objects are used to communicate structural engineering assumptions about support conditions to analysis software programs. You define boundary conditions in Revit by activating the **Analyze** tab and then invoke the **Boundary Conditions** tool from the **Boundary Conditions** panel (**Analyze** tab ➤ **Boundary Conditions** panel ➤ **Boundary Conditions**) in a 3D view of the analytical model.

Depending on the type of structural element (point, line, or area), boundary conditions can be set to **Fixed**, **Pinned**, **Roller**, or any other user-defined boundary condition. In case of user-defined boundary conditions, specify the various translational and rotational constraints listed (see Figure 4.37).

ADDITIONAL ANALYTICAL MODEL TOOLS

Under the **Analyze** tab, there are analytical model tools that can be used to adjust the analytical model or check its consistency (Figure 4.38). For example, some structural configurations are not suitable for direct integration with analysis and design software. Adaptive adjustment is required before a structural model is input into the analysis and design software. For this reason, the geometry of the structural member analytical model may also be adjusted in relation to those elements to which it joins. The additional tools include adjusting nodes, linear members, and planar or curved panels.

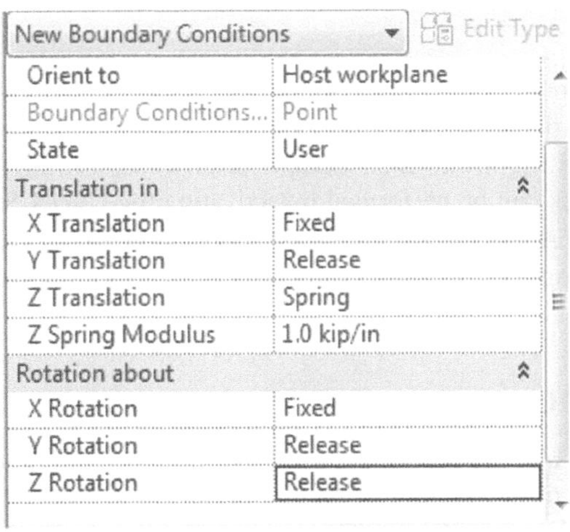

FIGURE 4.37 Setting values for user-defined boundary condition.

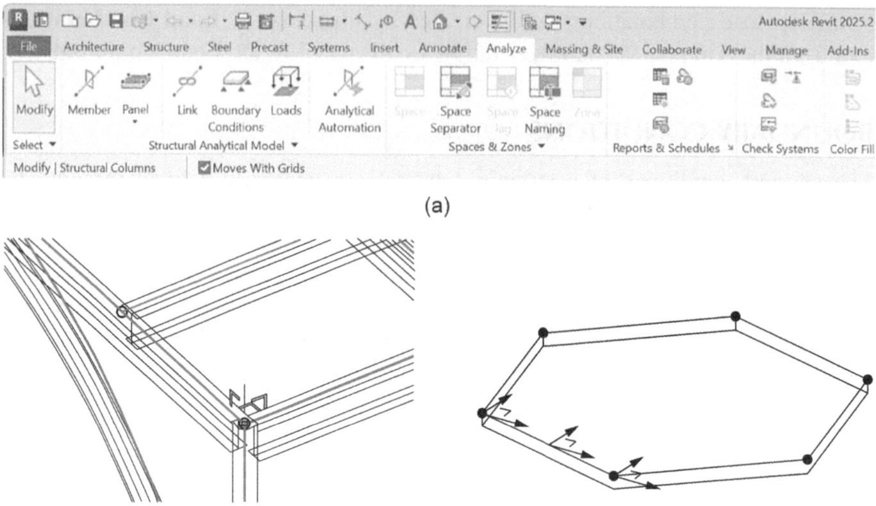

(a)

(b) (c)

FIGURE 4.38 (a) Analytical model tools. (b) Directly manipulating nodes and edges of the analytical model: (b) adjusting a linear element (beam); (c) modifying a surface element (slab).

ADJUSTING THE ANALYTICAL MODEL

One can manually modify the analytical model by invoking the **Member or Panel** tool from the **Analyze** tab as follows: **Analyze** tab ➤ **Analytical Model Tools** panel ➤ **Analytical Adjust**. The drawing area will display an edit mode in which non-analytical elements are disabled (grayed out). Linear and surface analytical model elements can then be directly manipulated (see Figure 4.38).

Check Supports

The **Check Supports** tool is a useful tool for analytical consistency and provides warnings in the early stages of design about the stability of the structure. This gives designers greater insight into their designs prior to submitting them for complete analysis. The check can be performed by activating the **Check Supports** tool from the **Analyze** tab: **Analyze** tab ➤ **Analytical Model Tools** panel ➤ **Check Supports**. A dialog box displays results that need to be reviewed and necessary actions taken to make changes (Figure 4.39).

CONSISTENCY CHECKS

The **Consistency Checks** tool is used to verify the analytical and physical model consistency. To activate the tool invoke **Analyze** tab ➤ **Analytical Model Tools** panel ➤ **Consistency Checks**. Then, review the warnings results displayed and make the appropriate changes to your model.

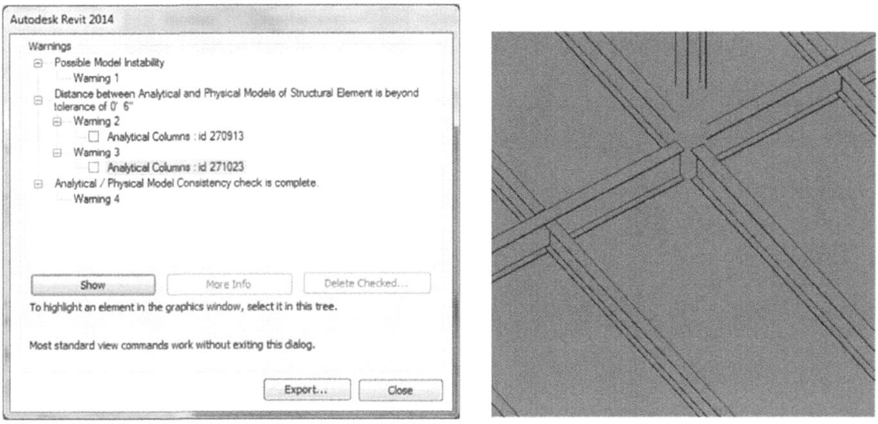

FIGURE 4.39 Example warning: unsupported elements.

ADDITIONAL ANALYTICAL MODEL TOOLS

Analytical Automation

With the new Revit platform, one can automatically generate an analytical model from the physical model—and vice versa, the physical model can also be created directly from the analytical model (see Figure 4.38a). To use this tool, invoke the **Analytical Automation** tool as follows: Analyze tab ➤ Structural Analytical Model panel ➤ **Analytical Automation.** Then select the physical model or part of it for the creation of the analytical counterpart model and click **Run** (Figure 4.41).

It is important to mention that the analytical automation tool is limited to major structural elements. This includes, for instance,

- Structural Columns: Vertical and Slanted Columns.
- Structural Framings: Straight line beams and Arc curved beams.
- Floors
- Foundation Slabs
- Walls: Straight line walls.

The properties carried over from physical members to their analytical counterparts are:

Material, Structural Role, Cross-section: in the form of Family and Type, Cross-section, and Rotation (Figure 4.40).

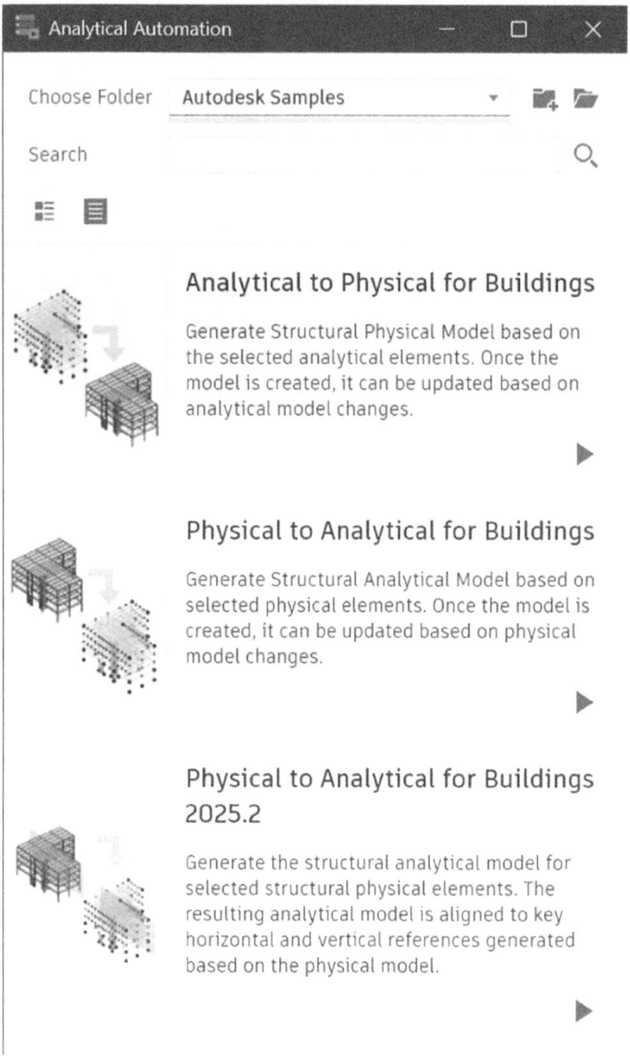

FIGURE 4.40 Structural analytical automation.

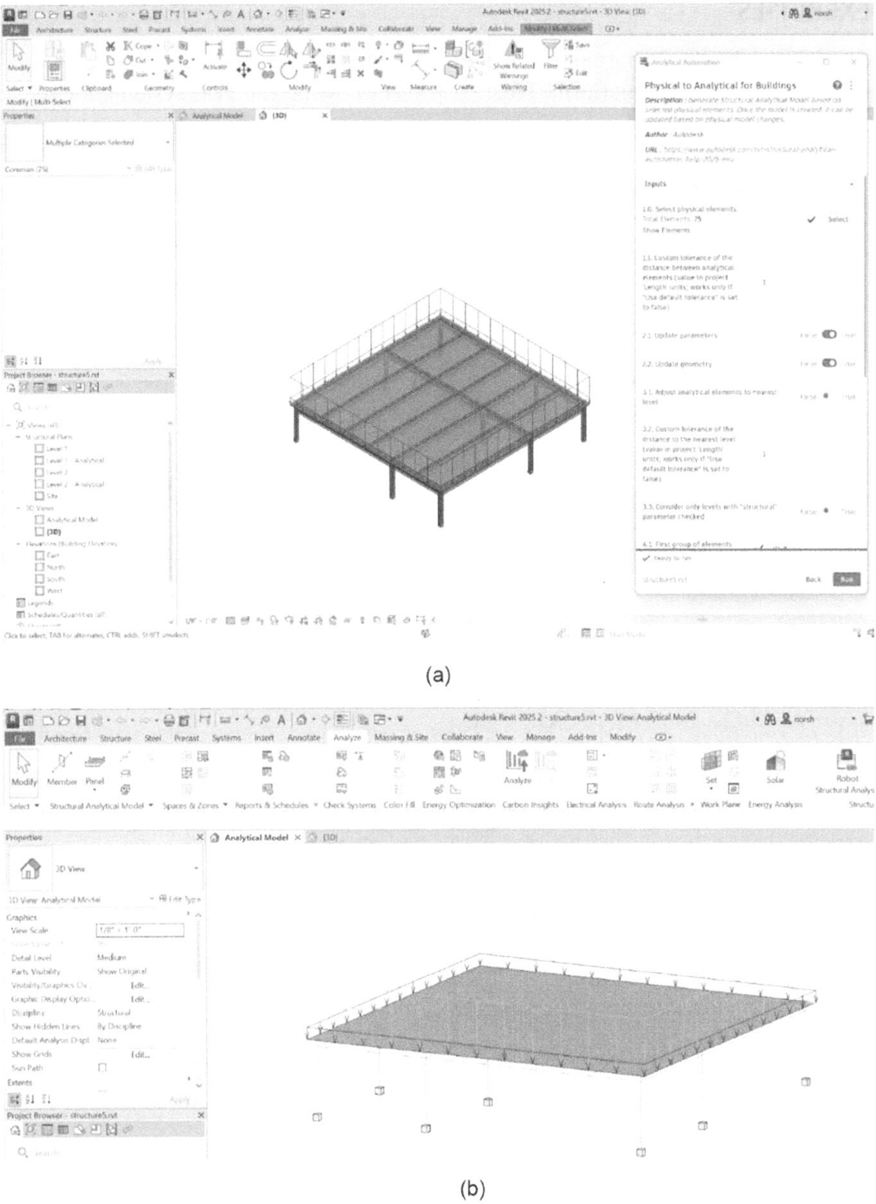

FIGURE 4.41 (a) Creating the analytical model from the physical model. (b) The analytical model produced from the physical model.

EXERCISES

4.1. Describe the importance of grid lines in structural modeling.

4.2. Using the BIM platform, create a project hosting the grid systems for the structural floor plans given below.

   a. Assume grids at 20 ft.
   b. Assume grids at 20 ft.
   c. Assume R1 = 20 ft, R2 = 50 ft, and R3 = 80 ft.
   d. Assume equal circular arcs.

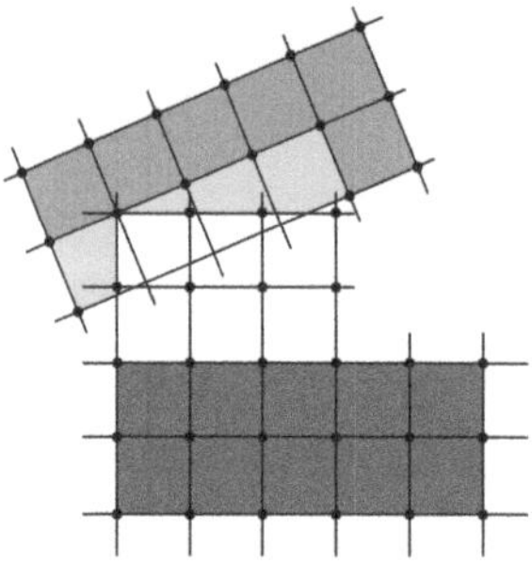
(a)

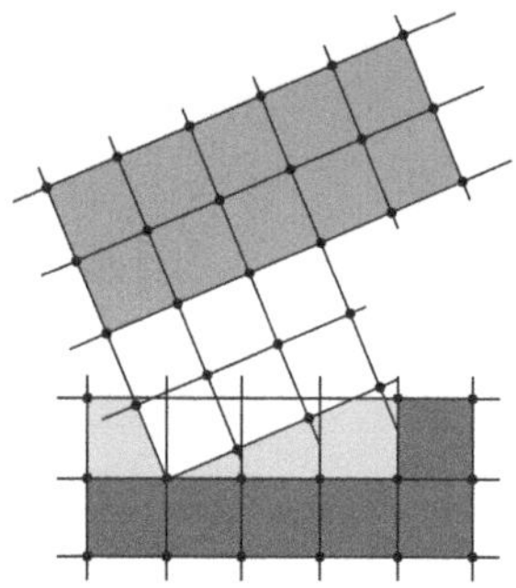
(b)

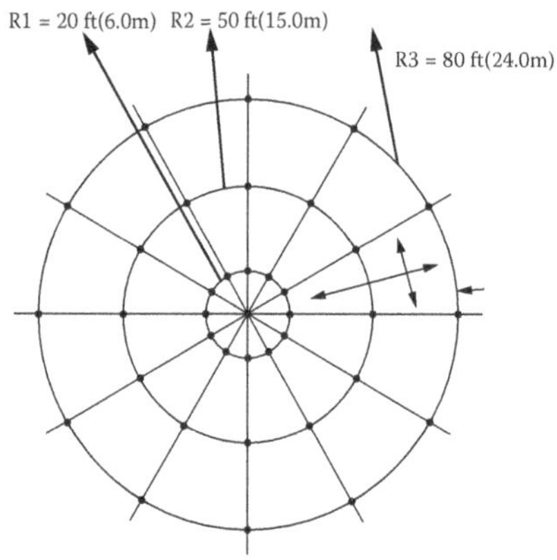

(c)

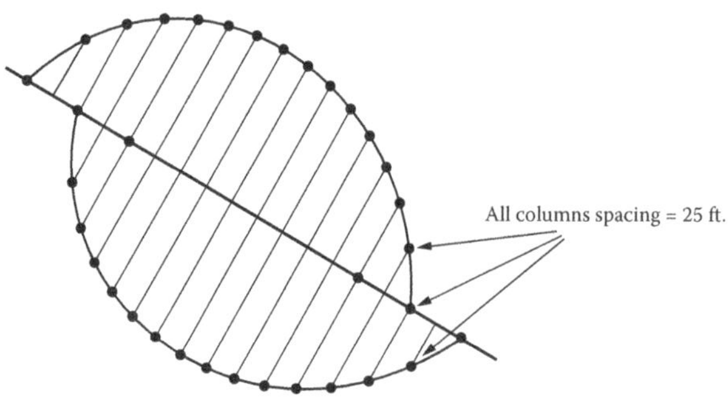

(d)

4.3. How would you control the visibility of gridlines in a **Work Plane**?
4.4. What is meant by structural model bidirectionality?
4.5. Describe the relationship between the physical and analytical models.
4.6. What is the purpose of levels that do not hold plan views?
4.7. How would you define the top and bottom release conditions of a structural column for structural analysis purposes?
4.8. How would you create openings in a structural column?
4.9. Describe how you would model fixed-end beams in Revit?
4.10. Model a beam system for the floor area shown below.

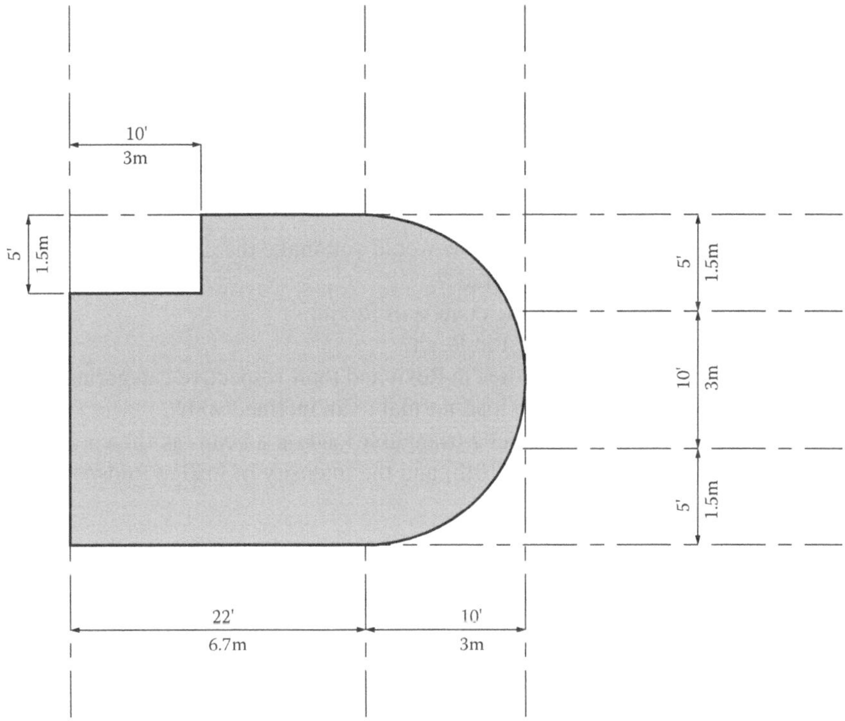

Floor plan

4.11. In modeling a structural foundation, explain the importance of the **Structural Usage** parameter.
4.12. Describe how you would model a combined footing.
4.13. Create a Revit project showing the modeling of a cantilever or strap footing. Assume both columns are 14×14 in. (350×350 mm). Footing supporting the left column is 6×12 feet (2×4 m) and the footing on the right is 6×6 ft (2×2 m). The strap beam is 14×20 in. (350×500 mm) and 10 ft (3 m) long.

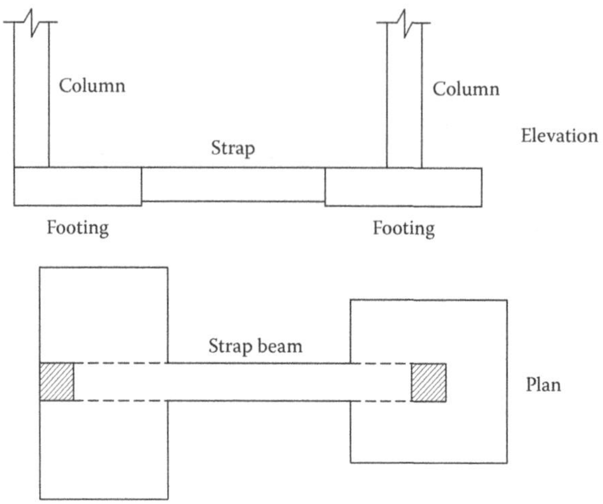

Cantilever footing

4.14. When modeling foundations, how would you make the distinction between foundation slab and slab on grade?
4.15. Define the term *family* in the context of Revit.
4.16. What kinds of families exist in Revit?
4.17. Give two examples of families in Revit and their respective categories.
4.18. How can you apply an area load normal to an inclined wall?
4.19. Describe how you can model a triangular load on a beam as shown below. Assume the span to be 20ft (6m) and the intensity of load at midspan is 5 kips (22 kN).

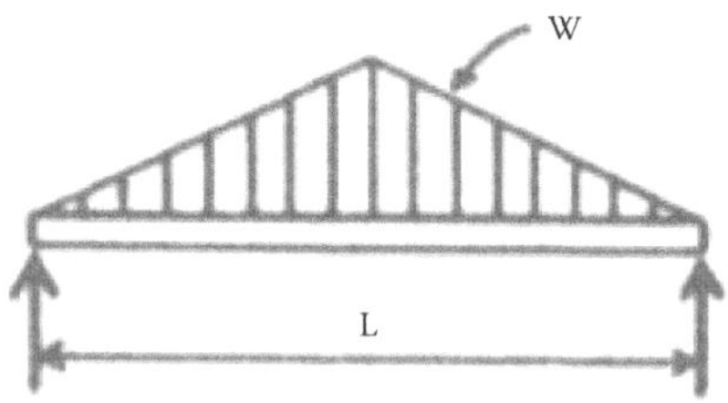

Triangular load on a simple beam

4.20. A torsional loading on a beam is required to be modeled in Revit. Describe the steps that can be undertaken to model a beam loaded with torsion.

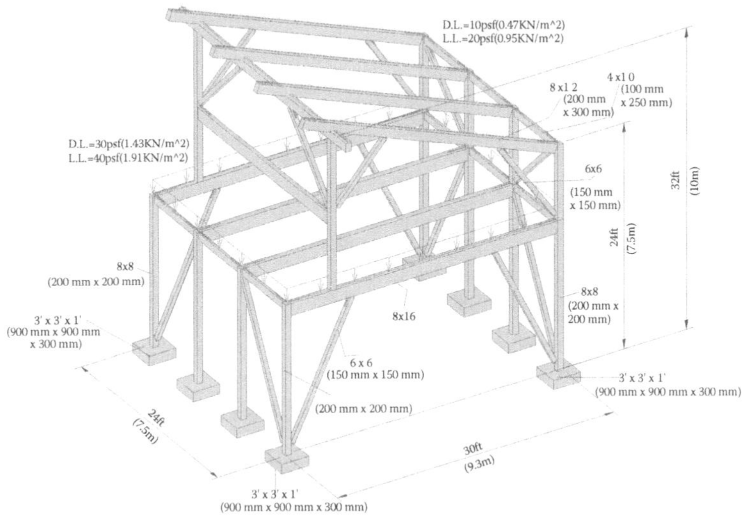

4.21. How would you apply the boundary conditions to the structure shown below?

4.22. Which tool can be used to check the support stability of a structural model?

4.23. Model all the elements shown in the models below and perform consistency checks for the model. Apply a dead load of 10 psf and a live load of 20 psf for the roof. Also, apply a dead load of 30 psf and a live load of 40 psf for the floor.

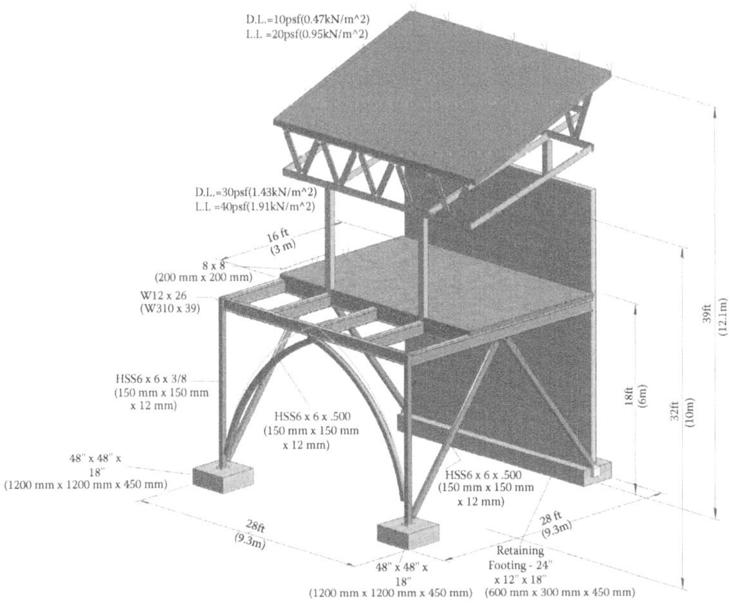

4.24. Repeat Exercise 4.20 for the buildoid shown below.

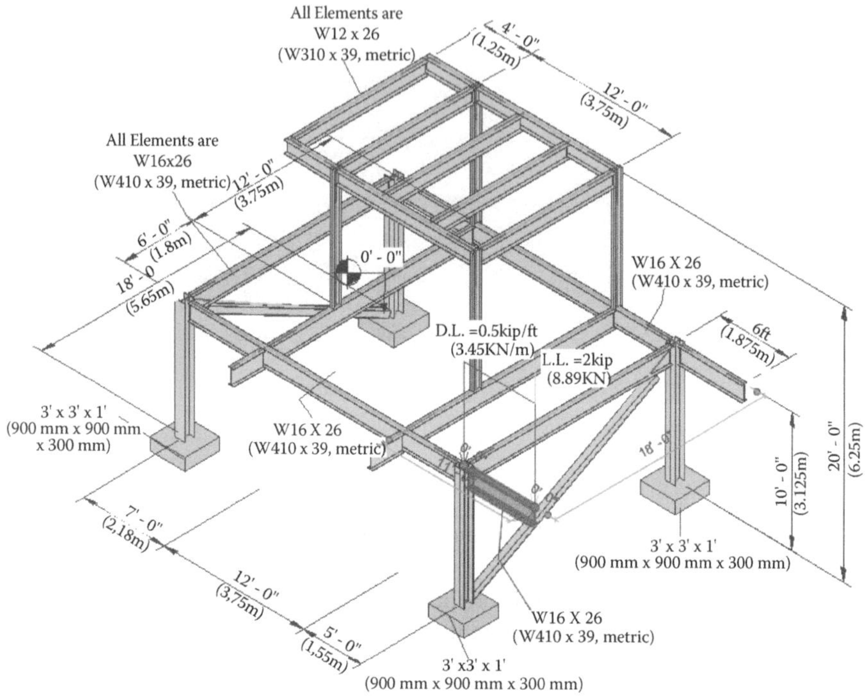

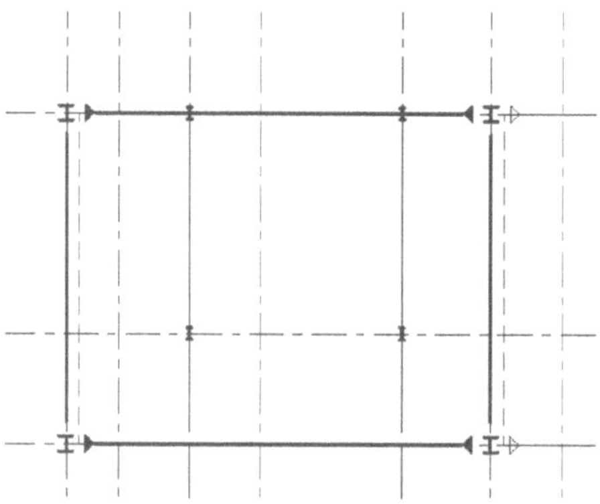

Structural framing floor plan

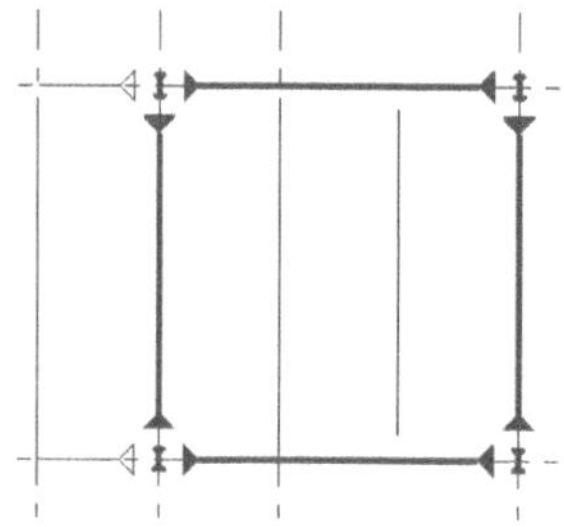

Structural framing roof plan

4.25. Using Revit, create a new family by altering the steel W-Wide flange beam family by adding double-sided wood nailers as shown below.

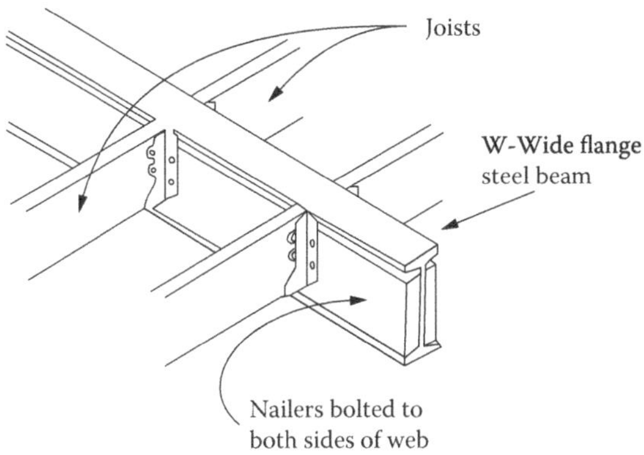

4.26. Create a BIM model for the structure shown below. Model all the elements shown in the given plans, elevations, and 3D views and perform consistency checks for the model.

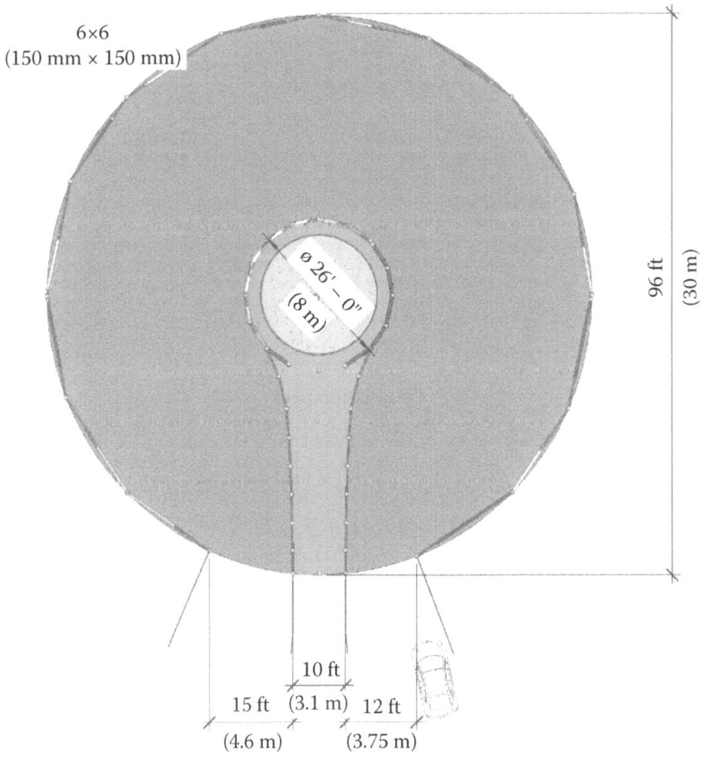

Plan view

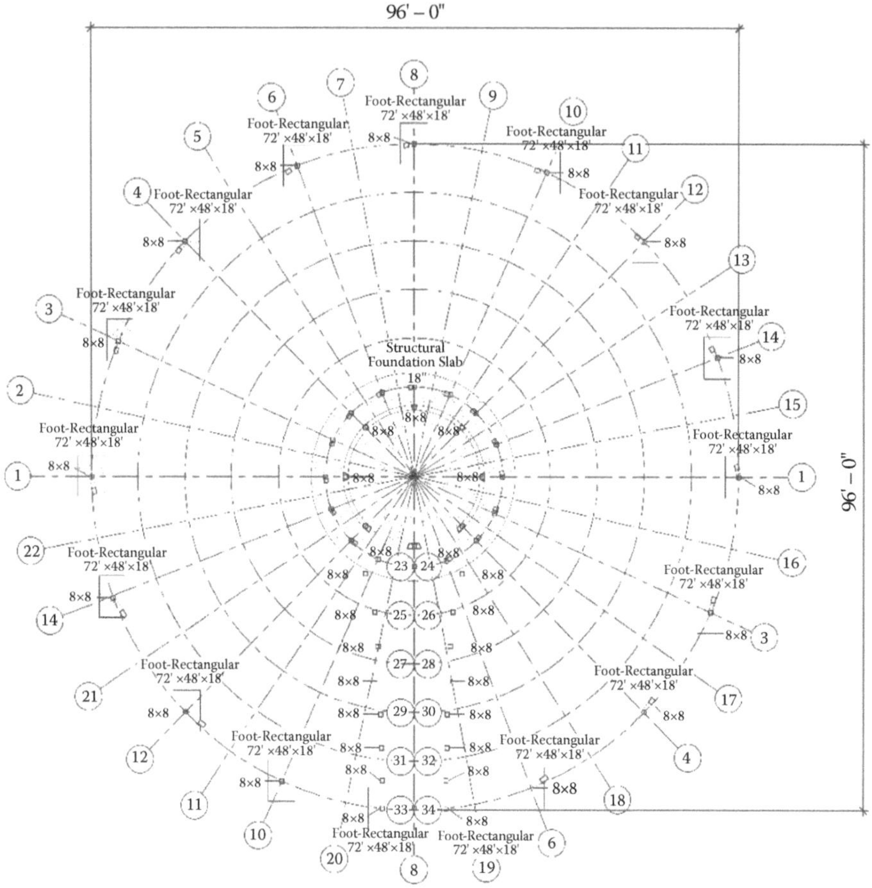

Foundation plan

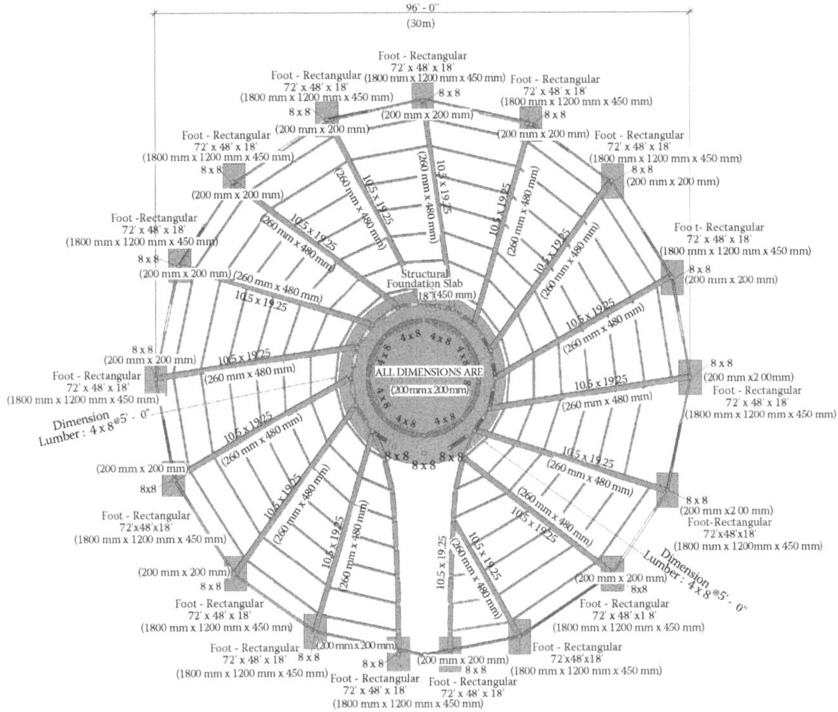

Framing plan

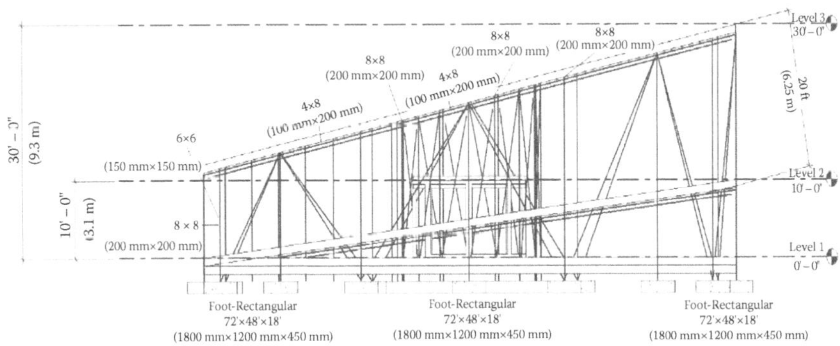

Framing Elevation

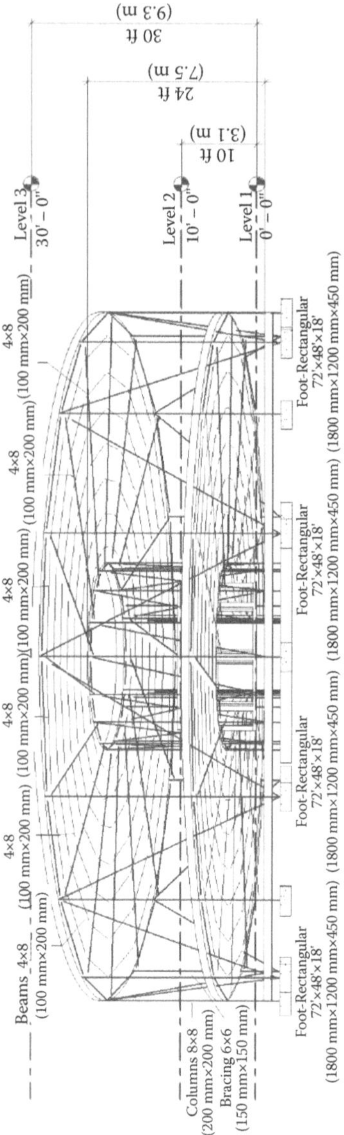

3D View

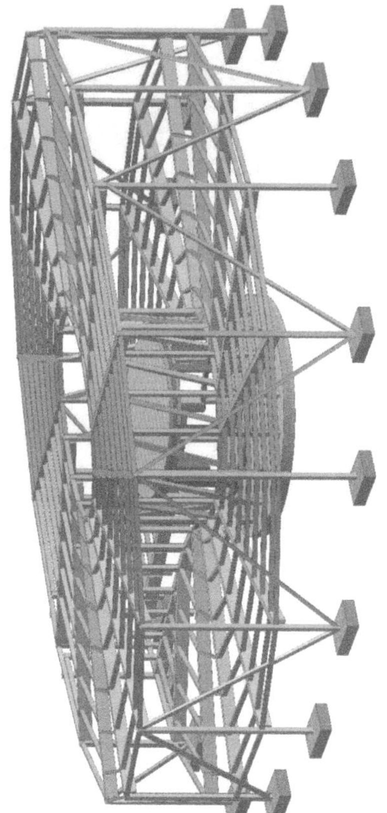
3D view

4.27. Referring to the swimming pool Cabana details given in figures below, create a BIM model that includes:

- The grid system along with columns and walls. Add Gridlines with appropriate designations (number in NS and letter on the EW directions).
- Add foundation elements.
- Add columns and beams.
- Add beams and beam systems as shown on the given framing plans.
- Add structural trusses.
- Add a roof element
- Various Structural 3D views.

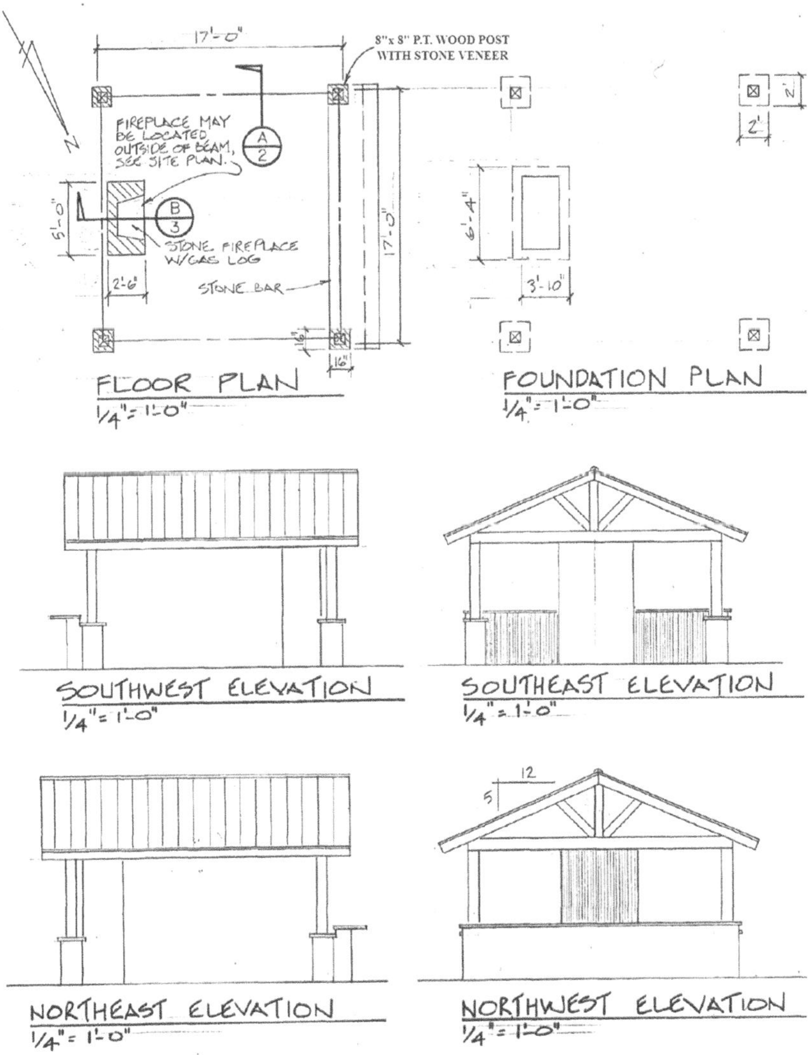
17'-0"
8" x 8" P.T. WOOD POST WITH STONE VENEER
FIREPLACE MAY BE LOCATED OUTSIDE OF BEAM, SEE SITE PLAN.
A
2
B
3
5'-0"
STONE FIREPLACE W/GAS LOG
2'-6"
STONE BAR
17'-0"
16"
N
FLOOR PLAN
1/4" = 1'-0"
2'
2'
6'-4"
3'-10"
FOUNDATION PLAN
1/4" = 1'-0"
SOUTHWEST ELEVATION
1/4" = 1'-0"
SOUTHEAST ELEVATION
1/4" = 1'-0"
12
5
NORTHEAST ELEVATION
1/4" = 1'-0"
NORTHWEST ELEVATION
1/4" = 1'-0"

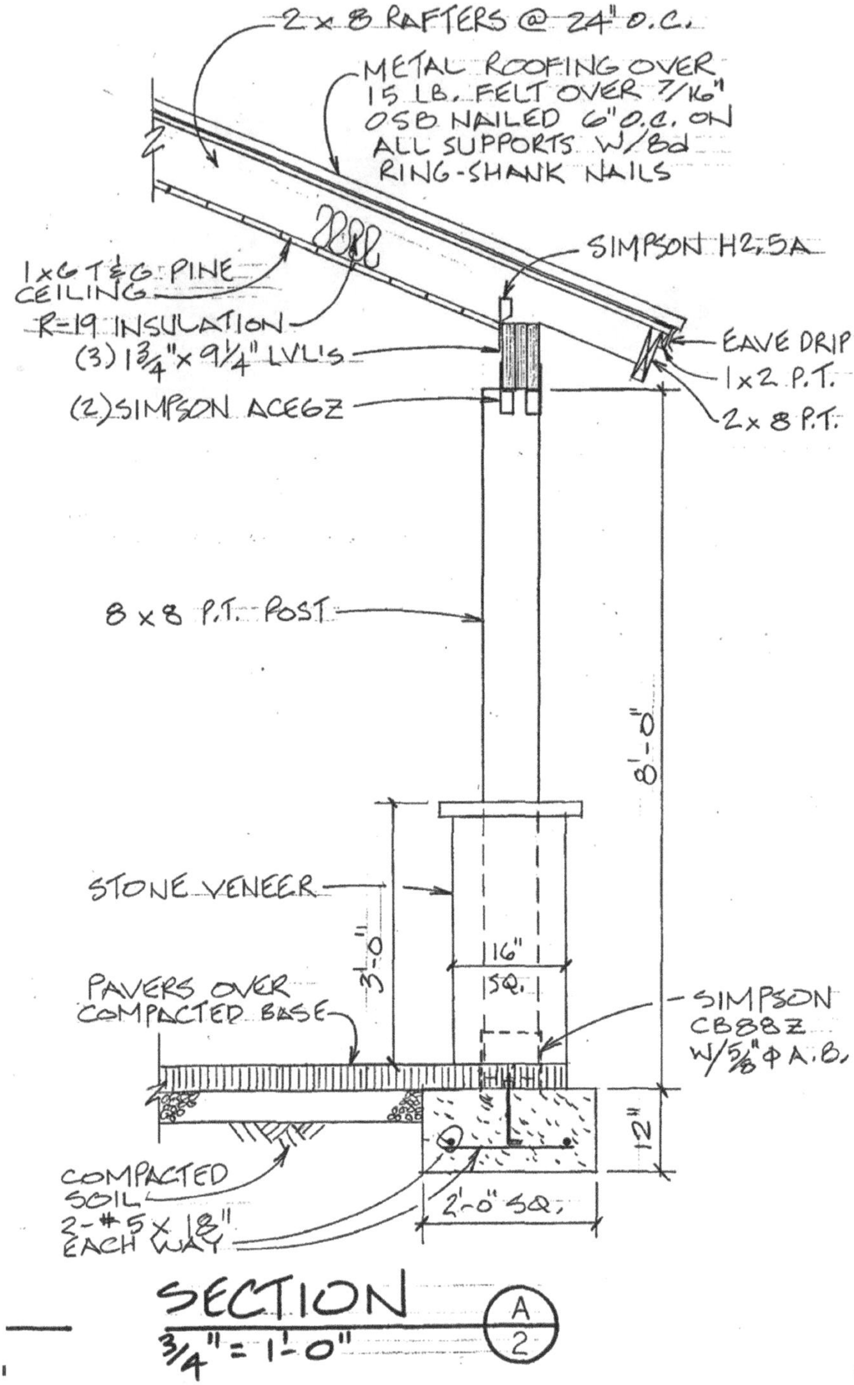
2 x 8 RAFTERS @ 24" O.C.
METAL ROOFING OVER 15 LB. FELT OVER 7/16" OSB NAILED 6" O.C. ON ALL SUPPORTS W/8d RING-SHANK NAILS
SIMPSON H2.5A
1x6 T&G PINE CEILING
R=19 INSULATION
(3) 1 3/4" x 9 1/4" LVL'S
(2) SIMPSON ACE6Z
EAVE DRIP
1x2 P.T.
2x8 P.T.
8 x 8 P.T. POST
8'-0"
STONE VENEER
3'-0"
16" SQ.
PAVERS OVER COMPACTED BASE
SIMPSON CB88Z W/5/8" Φ A.B.
12"
COMPACTED SOIL
2-#5 x 18" EACH WAY
2'-0" SQ.
SECTION
3/4" = 1'-0"
A
2

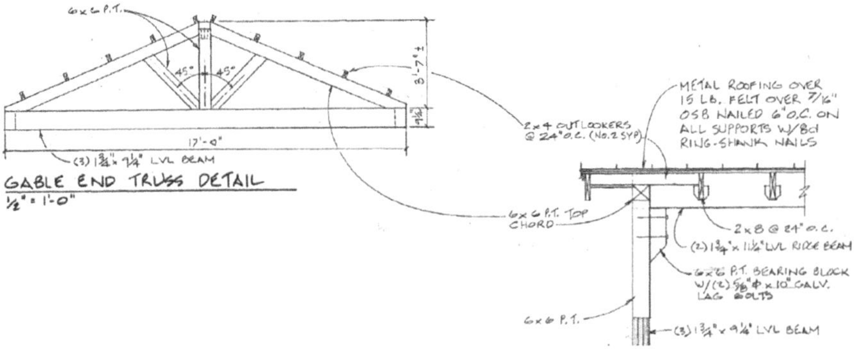
6x6 P.T.
45°
45°
3'-7" ±
9¼"
17'-0"
(3) 1¾" x 7¼" LVL BEAM
GABLE END TRUSS DETAIL
½" = 1'-0"
2x4 OUTLOOKERS @ 24" O.C. (NO.2 SYP)
METAL ROOFING OVER 15 LB. FELT OVER 7/16" OSB NAILED 6" O.C. ON ALL SUPPORTS W/8d RING-SHANK NAILS
6x6 P.T. TOP CHORD
2x8 @ 24" O.C.
(2) 1¾" x 11¼" LVL RIDGE BEAM
6x6 P.T. BEARING BLOCK W/(2) 5/8" Φ x 10" GALV. LAG BOLTS
6x6 P.T.
(3) 1¾" x 9¼" LVL BEAM

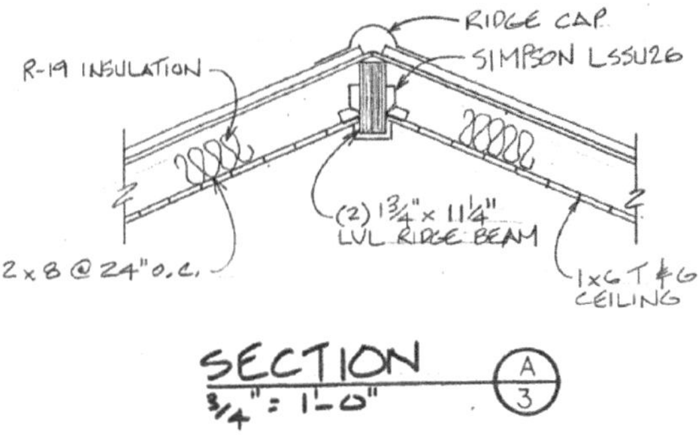
RIDGE CAP
SIMPSON LSSU26
R-19 INSULATION
(2) 1¾" x 11¼" LVL RIDGE BEAM
2x8 @ 24" O.C.
1x6 T&G CEILING
SECTION
¾" = 1'-0"
A
3

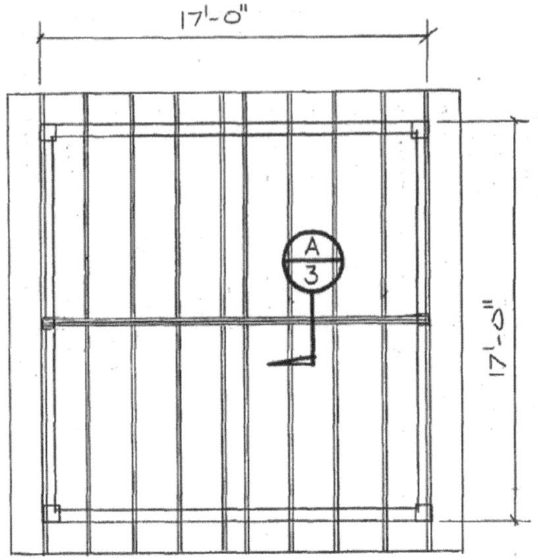

ROOF FRAMING PLAN
1/4" = 1'-0"

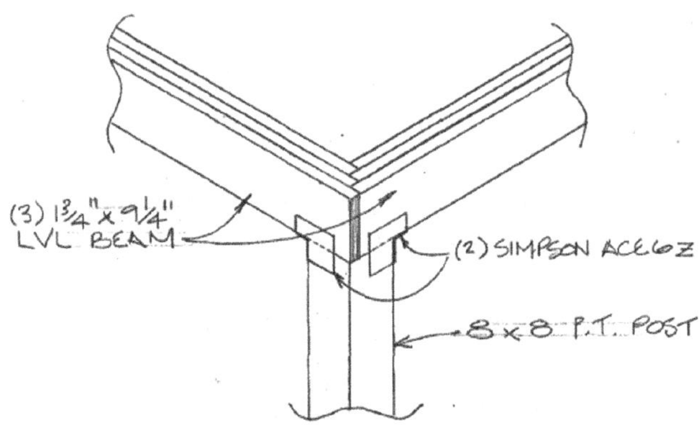

CORNER DETAIL
3/4" = 1'-0"

5 Architectural Elements

INTRODUCTION

In the initial steps of the architectural design process, design projects generally go through sequential phases of design that correlate roughly to the levels of development (LOD) described within the American Institute of Architects (AIA) Contract Document G202-2013 (AIA, 2013), *Building Information Modeling Protocol Form*, referred to in Chapter 4 (see Table 4.1). This chapter delves into the creation of architectural modeling elements that cover primarily levels 1 and 2 from the viewpoint of the architectural design and the spatial component elements of a building. This includes site modeling, conceptual massing, walls, floors, ceilings, roofs, windows, doors, stairs, and furniture.

A particular distinction must be drawn between the geometry types and the information embedded within the geometry. In many cases (as is evident within the Revit interface), modeling elements are based on the element type (i.e., walls, columns, doors, etc.). It is important to note that the elements under the **Architecture** tab are not considered structural or load-bearing unless explicitly defined as such in the **Structural Usage** property. However, these architectural modeling elements can eventually be specified with materials, exterior/interior finishes, and assembly layers that are not associated with the structural model but pertain more to the spatial organization and aesthetic quality of the building (Figure 5.1).

These architectural elements can be specified to a high level of detail and customization as loadable families created by either the designer or the supplier. However, in some cases, especially early in the schematic design phase, these elements cannot yet be defined clearly enough to be modeled with the level of specificity inherent to a building information modeling (BIM) package. Revit provides generic geometric types for these situations, with the option to convert, host, or use them as reference geometry for further development. For example, the **Mass Family** type is used to generate spatial masses and voids for early design analysis without having to specify actual building elements. This is a powerful way to analyze, modify, and iterate broader design ideas, maintaining flexibility while retaining the capability of using the **Mass** object to eventually host more specific architectural elements when the design direction is clarified and becomes more concrete.

SITE MODELING

The **Massing & Site** tools will allow modeling site context for situating a project. This section covers basic techniques for modeling site topography, creating a building pad, and creating landscape site objects using Revit (Figure 5.2).

Site modeling typically begins by establishing the topography, then adding property lines, parking areas, landscape objects, and other site components. The following

DOI: 10.1201/9781003598435-5

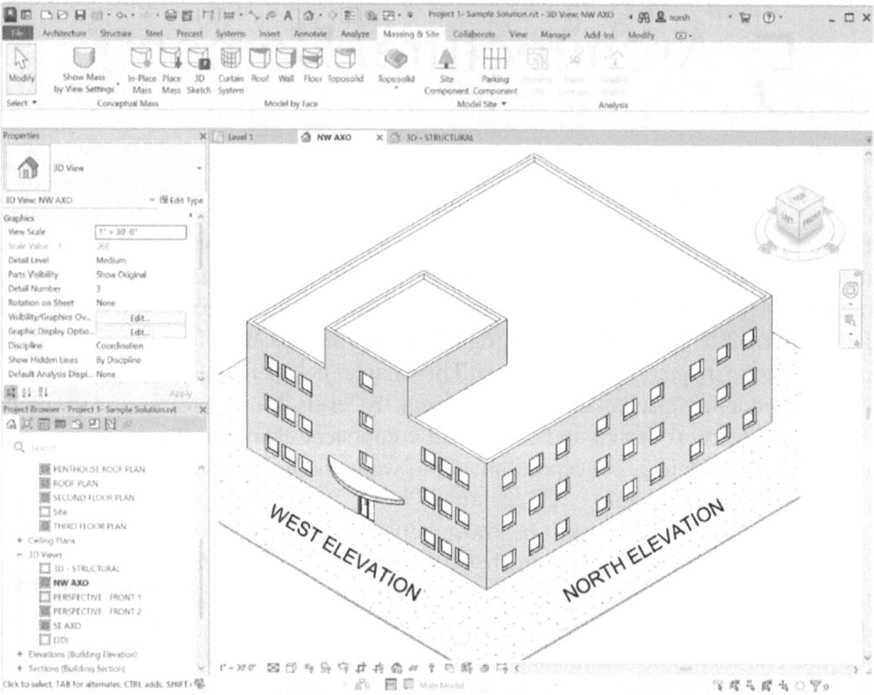

FIGURE 5.1 Massing and site tab.

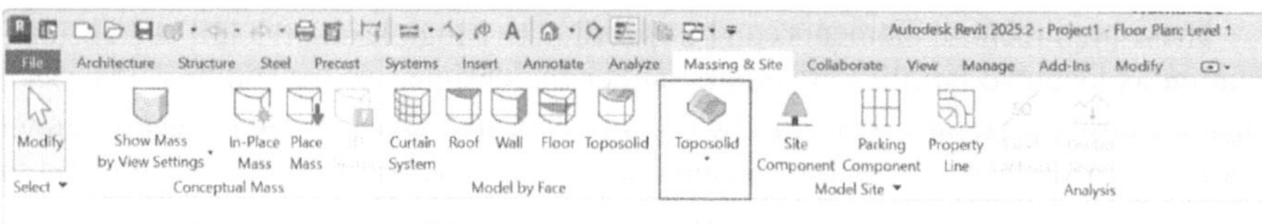

FIGURE 5.2 Toposolid tool.

sections provide a detailed walkthrough of the steps involved in developing a site's topography (Figure 5.2).

Topography

To represent the topography of the site, one begins with creating a **Toposolid** in the Revit model. There are two main ways to create site topography within Revit: by manually placing points or by importing a computer-aided design (CAD) file with topographical data. Before starting, select and make active the **Site** view from the **Project Browser** under the **Floor Plans** category.

The Toposolid tool defines a topographical element using defined elevation points. Folding lines are edges on toposolids. As one adds points to generate a toposolid, **Folding Lines** are automatically created—though they remain hidden by default.

To view these lines, simply enable the "Model Categories/Toposolid" option in the Visibility/Graphics dialog.

Toposolid

The simplest way to model Toposolid is by sketching a boundary and placing elevation points. Place points that will define and interpolate a surface based on their elevation. It is advisable to create some reference geometry beforehand to snap to using detail lines that will help you create an accurate outline.

One begins with the **Site view** to create the boundaries of the toposolid in sketch mode (Figure 5.3). Then, modify elevation lines and points on the boundary sketch. Use the controls on the ribbon to specify different elevations for points on the toposolid. You can use other objects in the model to cut the toposolid element.

One can modify the elevations of the different points on the toposolid by clicking on the **Modify Sub Elements** tool to change the elevation of the points and edges of the boundary. Figure 5.3a shows a plan of the toposolid with contour lines. Figure 5.3b depicts a section view illustrating the elevations of the toposolid.

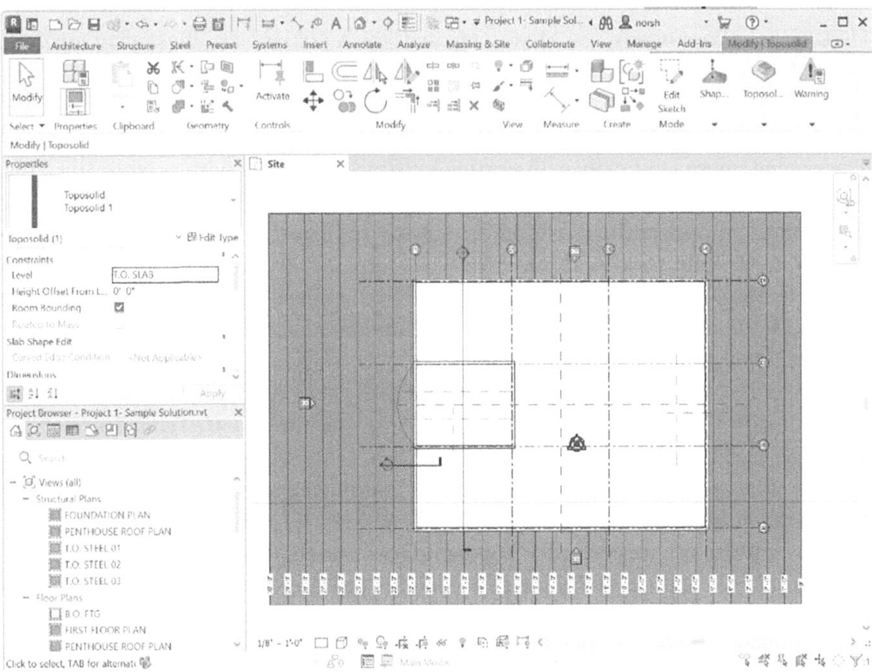

(a)

FIGURE 5.3 (a) Modify toposlid. (b) A section view illustrating toposolid elevation.

(Continued)

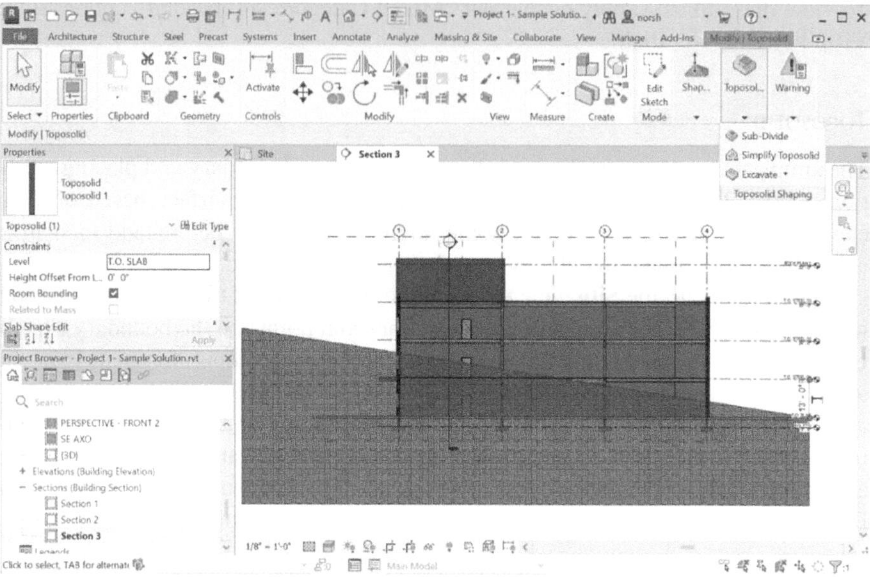

(b)

FIGURE 5.3 (*Continued*) (a) Modify toposlid. (b) A section view illustrating toposolid elevation.

Importing CAD File

The second approach involves importing existing CAD data with contour lines that are correct in elevation, then using those data to generate the topography. To do this, you must import the CAD file first using the **Insert** tab ➤ **Import** panel ➤ **Import CAD** command with the **Site** view active. This will bring up the **Import CAD Formats** dialog box (Figure 5.4), which will allow you to browse to your saved CAD file. Please note the available import options regarding the color, units, and positioning of the imported geometry.

Invoke the **Massing & Site** tab ➤ **Model Site** panel ➤ **ToposSolid** command to enter **Modify | Edit toposolid** mode. This time, use the **Create from Import** and select the CAD file (Figure 5.4).

Note that you may assign a **Material** parameter to the topography in the **Properties** panel.

You may also use the **Simplify Toposolid** tool smooth the curves of the Toposolid. Each point in a topography generates triangulated geometry, increasing computational overhead. To enhance system performance, simplify toposolid that contain a large number of points. Simplifying an element may impact the accuracy and geometry of the contours, depending on the topography's specific conditions. After simplification, review the contours and geometry, adding points as needed to maintain accuracy.

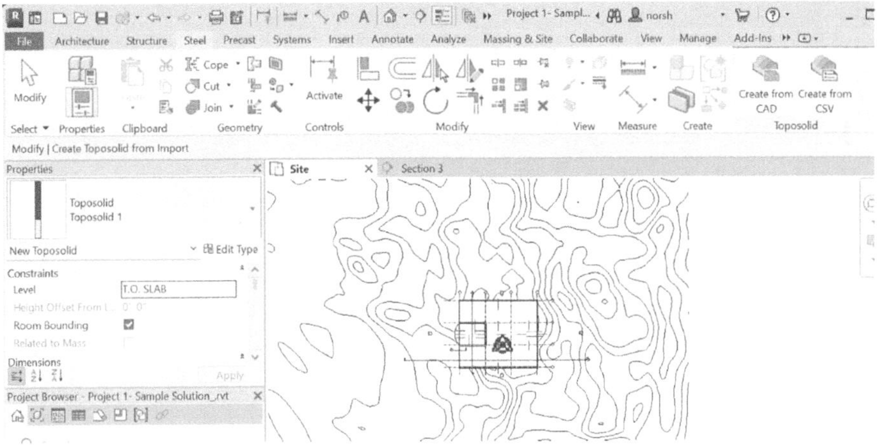

FIGURE 5.4 Imported CAD file to the toposolid.

Create a Building Pad

A building pad in Revit can be defined as like a floor, except that it is used to cut into (or fill) topography to provide a flat surface for the bottom floors of the building to sit on. It is generally associated with the level 1 reference plane (unless you change it), and you specify an offset distance. Please note that you may have multiple building pads at different levels in the same site topography, but they cannot overlap. Floor, roof, and toposolid elements that intersect with toposolids can be used to excavate volume from the host toposolid. For instance, placing a floor element on a sloped toposolid and using the excavate tool allows you to remove the toposolid volume up to the floor's surface. Similarly, overlapping toposolid elements can be used to create excavations for features like building pads or parking lots.

Invoke the **Modify | Toposolid ➤ Toposolid Shaping** panel ➤ **Excavate** command to enter the select mode (Figure 5.5). Then select the intersecting element to use for excavation.

Landscape and Site Objects

Adding site components is identical to adding furniture, doors, or any other hosted objects. Because these generally are trees, shrubbery, or landscape elements such as lamp posts, they are generally inserted onto the site topography.

To place **Site** objects, invoke the **Massing & Site** tab ➤ **Model Site** panel, **Site Component** command. Revit will automatically prompt the placement of the last active component family (Figure 5.6). Choose the **Modify | Site Component Load Family** tool to bring up the **Load Family** dialog box, where you can browse to any saved RFA family files (default location for greenery is in the **Program Files/RVT 2014/Libraries/Planting** folder). These families can then be accessed through the **Family Type** drop-down in the **Properties** panel. Note that you can specify its

FIGURE 5.5 Creating a building pad.

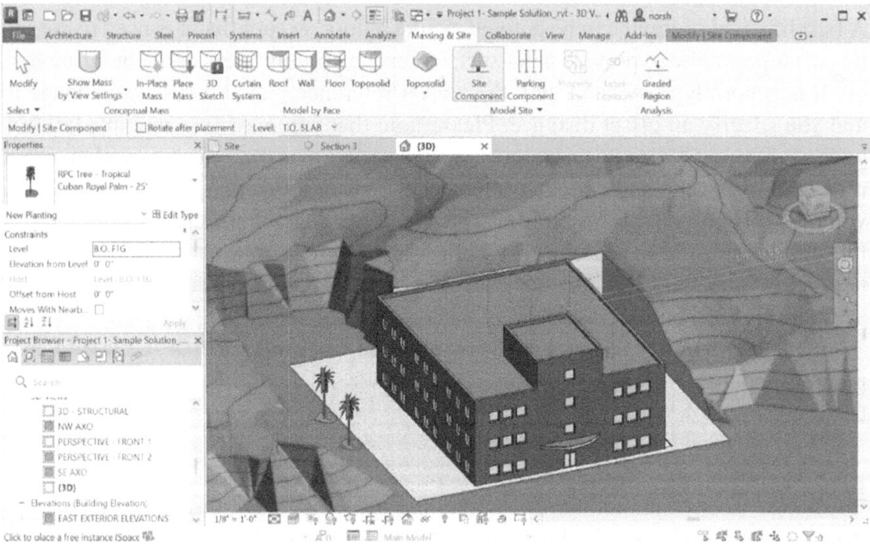

FIGURE 5.6 Modeling site objects.

hosted level as well as offset distance, rotate it after placement, or go into the **Edit Type** properties to change the height parameters for the tree.

Subregions

If for some reason you wish to take your initial toposolid and separate it into smaller entities for editing purposes, you may use the **Sub-Divide** tool to do so under the **Toposolid Shaping** panel.

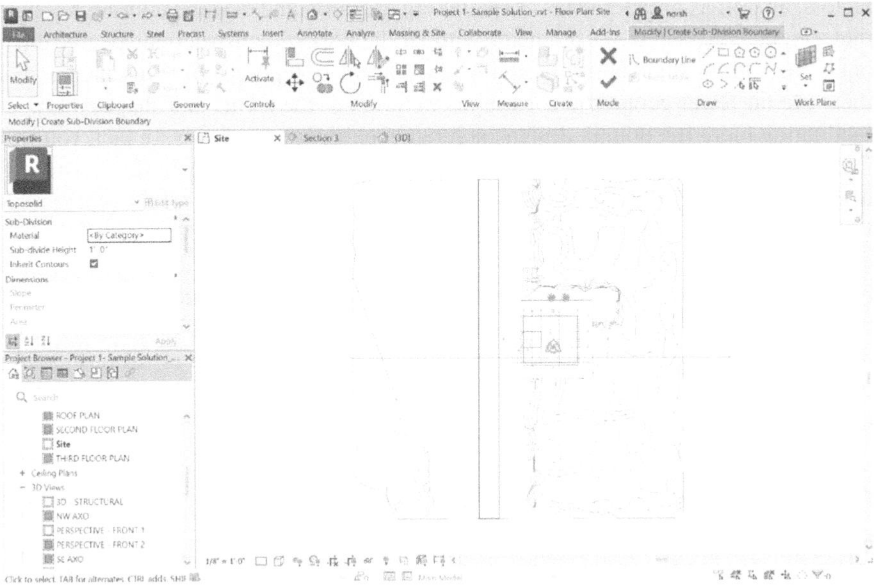

FIGURE 5.7 Creating a subregion on a site.

To create a subregion, select the toposolid and Invoke the **Modify Toposolid** panel ➤ **Sub-Divid** command to enter **Modify | Toposolid** mode. As always, you may use the sketching toolbox to draw an open curve to specify a division line or a closed loop if it is an area within the larger toposurface you wish to split (see Figure 5.7).

After finishing the command, you will be able to select the subregion (use **Tab** to cycle through selections) and apply a separate material parameter to it in the **Properties** panel. Note that the subregion is still considered part of the toposolid, and editing the toposolid will also affect the subregion as well.

GRIDS AND LEVELS

Because grid lines and levels are considered datum objects that serve as reference geometry for the placement and alignment of other modeling objects, functionally their behavior is identical regardless of modeling within a structural or architectural context; both types of elements are hosted and referenced in the same manner. Please refer to Chapter 4 for a detailed description.

CONCEPTUAL DESIGN AND ANALYSIS

Conceptual Mass Modeling

Mass objects are a unique category of three-dimensional (3D) elements within Revit that are used to define geometrically the shape of a building. As implied by the name, they are used in the preliminary stages of the conceptual design process to explore massing qualities of the project in relationship to the site and other environmental factors, as well as broader ideas of spatial organization and design patterns. As such, they can be considered 3D geometry without the normal construction-specific attributes that other Revit objects require or undertake (Figure 5.8). This means that there are far fewer geometric constraints in terms of the forms and shapes allowed, facilitating more freedom in design exploration early in the process. In fact, you will find that to create many of the more complex shapes and surfaces in Revit, you will need to use **Mass Objects** at first to define the geometry because all the other architectural objects have much stricter construction and geometric constraints. However, although early in the modeling process they are considered generic geometry, as the project develops, they can be used as reference geometry to quickly define architectural elements such as walls, floors, and curtain wall systems. In addition, Revit integrates a variety of analysis tools to aid in the quick evaluation of various passive design strategies, such as volumes, floor areas, shading, orientation, as well as energy analysis.

Creating an In-place Mass and Mass Families

There are two means of modeling mass objects in Revit: creating an **In-Place Mass** within the project environment and creating an independent **Mass** family using the Conceptual Design Environment (CDE) or simply the conceptual mass family editor. The **In-Place Mass** is used for unique elements within the project, interfacing with other project geometry, and generally occurs just once. The **Mass** family approach creates an external mass family (loadable family) that can be loaded and placed into the project like a component family, and is useful for repeating elements.

Mass Visibility Settings

Before one starts mass modeling, it is recommended to check the mass visibility settings. By default (unless you are in the CDE), Revit does not show **Mass** objects; you will need to enable their display mode for the project (this will apply to all views). To do this, go to the **Massing & Site** tab ➤ **Conceptual Mass** panel ➤ **Show Mass Zones and Shades** pull-down (Figure 5.9). Here, you will see the following view options: **Show Mass by View Settings**, **Show Mass Form and Floors**, **Show Mass Surface Types**, and **Show Mass Zones and Shades**. Revit will also enable the **Show Mass** mode when you first create an **In-Place Mass** within the project environment. You can also go into the **Visibility/Graphics Overrides** settings under the **Graphics** parameter group of any given view to modify individual view settings (Figure 5.10).

In-place Mass

Within the project environment, invoke the **Massing & Site** tab ➤ **Conceptual Mass** panel ➤ **In-Place Mass** command and you will be prompted to name your mass. You will see the familiar sketching tool box that will allow you to begin drawing curve profiles on the active work plane. Note that as soon as you start sketching, the contextual ribbon offers more options as the **Massing** tool set (Figure 5.11).

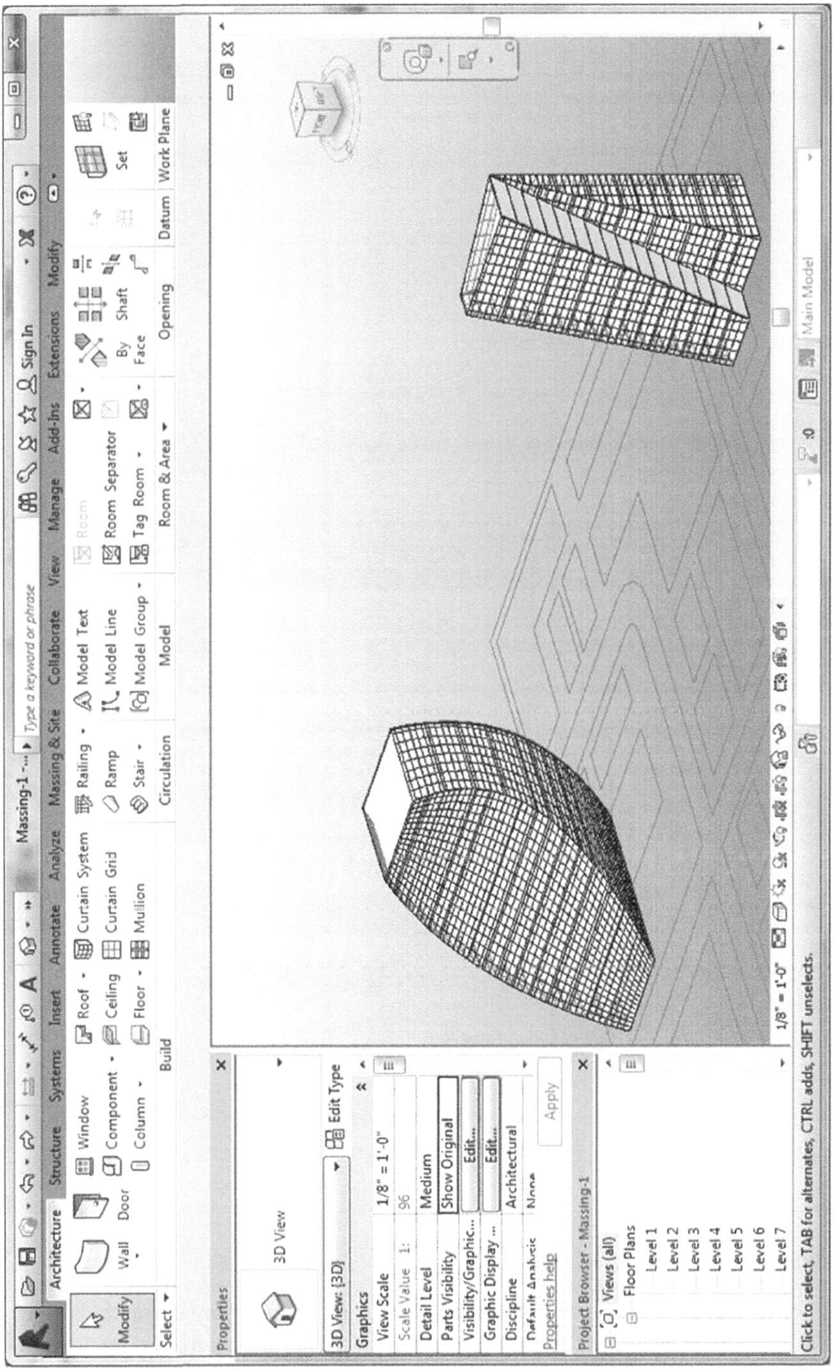

FIGURE 5.8 Conceptual massing model examples.

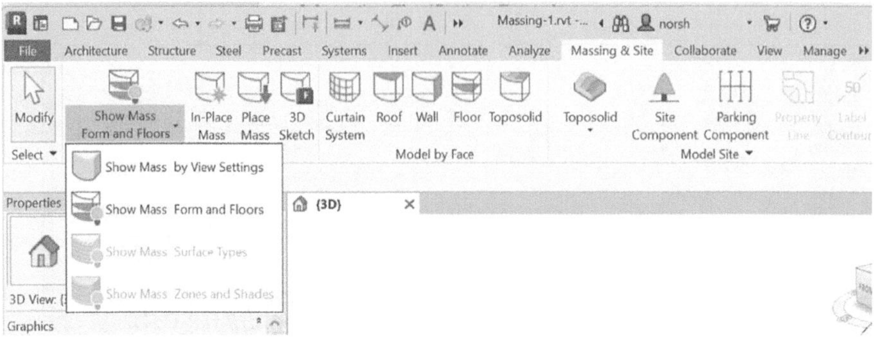

FIGURE 5.9 Show mass by view settings.

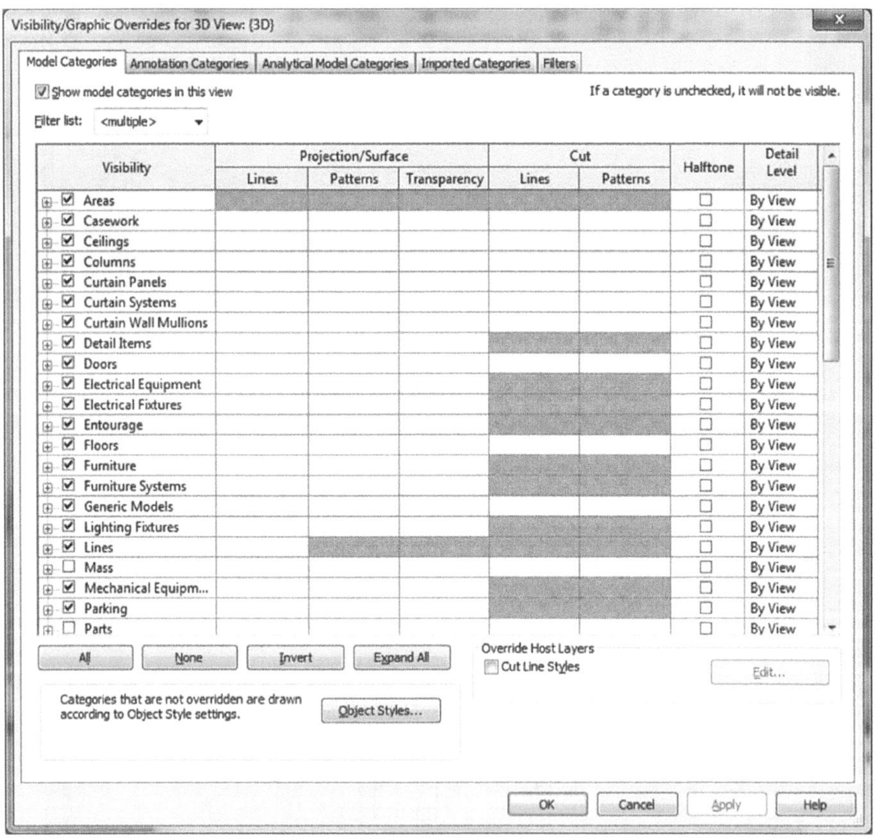

FIGURE 5.10 Visibility/graphics overrides.

When modeling masses, defining work planes and setting them correctly is essential for successful results; at this point, you are generally starting from scratch, so you may need to create reference planes or levels to facilitate your modeling. While you are in the mass creation mode, the **Create** tab ➤ **Work Plane** panel **Set** will allow

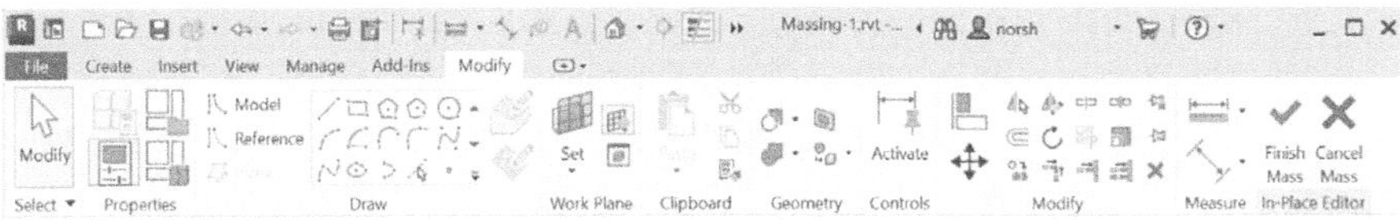

FIGURE 5.11 Massing tool set.

you to set and pick a surface or reference plane as the active work plane. If you are unsure what your current work plane is, use the **Create** tab **Work** transparent square (Figure 5.12). This is particularly helpful when sketching on vertical surfaces or a reference plane that is higher in elevation.

The most common method of creating mass objects is to draw profile shapes using the sketching tool box and then using the **Modify | Lines Form Create Form ➤ Solid Form** command to create the mass. Revit will use your input to determine which of the following types you are trying to create: **Surface** forms, **Extrusions**, **Lofts**, **Revolves**, and **Sweeps**. For example, if you have selected two or more closed profiles on different levels, the software engine will know you are trying to create a **Loft** form. The most common approach is to start with a profile on the ground level, which will create an **Extrusion**. After the extrusion has been created, it will be highlighted in blue, and you may use the 3D form control arrows to further refine and manipulate its shape (Figure 5.13).

If one selects any part of the form, one will enter **Modify | Form** editing mode, which provides a robust set of tools. This is called **Push and Pull Editing**; you can simply use the 3D form control arrows to intuitively modify the vertices, surfaces, points, or edges of the form. Use the **Tab-selection** method to cycle through selection sets. Note that one can also edit the dynamic dimension lengths to enter a specific dimension. To simplify form editing, you can activate the **Modify | Form ➤ Form ➤ Element X-Ray** mode, which will show the underlying geometric framework of the mass with profiles, edges, and vertices that are easier to highlight and select (Figure 5.14).

Of particular note is the **Modify | Form ➤ Form Element ➤ Add Edge** tool, which will allow one to add an edge on an existing surface. The **Modify | Form ➤ Form Element ➤ Add Profile** tool will allow you to duplicate the base profile along the direction of your extrusion or loft path. Also, note that after you have finished editing, selecting the mass object will highlight it and enable the **Modify | Mass** mode with shape handle arrows, allowing you performing basic control of the mass proportions by push/pulling faces. If you need to edit the underlying geometric structure, however, you will still have to double-click the mass to enter the **In-Place Editor** and access the more detailed tool sets.

While creating the mass form, notice that there are two options: **Solid Form** and **Void Form** (see Figure 5.15). You have already created a **Solid Form**, a volumetric representation of the building mass. The **Void Form** is used to a cut from a **Solid Form**, so it can be used to remove or hollow out specific shapes from a **Solid Form** similar to a Boolean subtraction operation. Although **Void Forms** can be highlighted via **Tab-selection**, they are hidden by default and can be harder to manipulate.

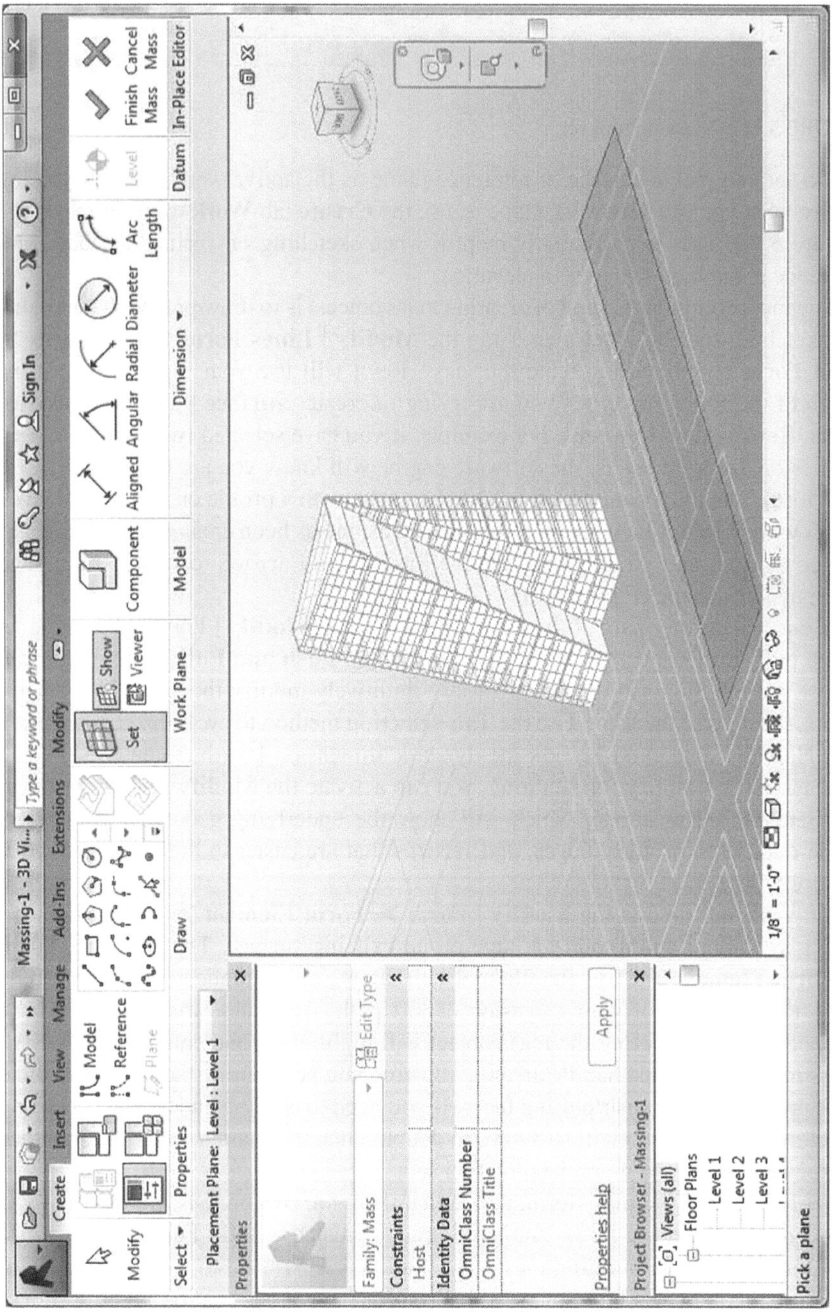

FIGURE 5.12 Setting **Work Plane.**

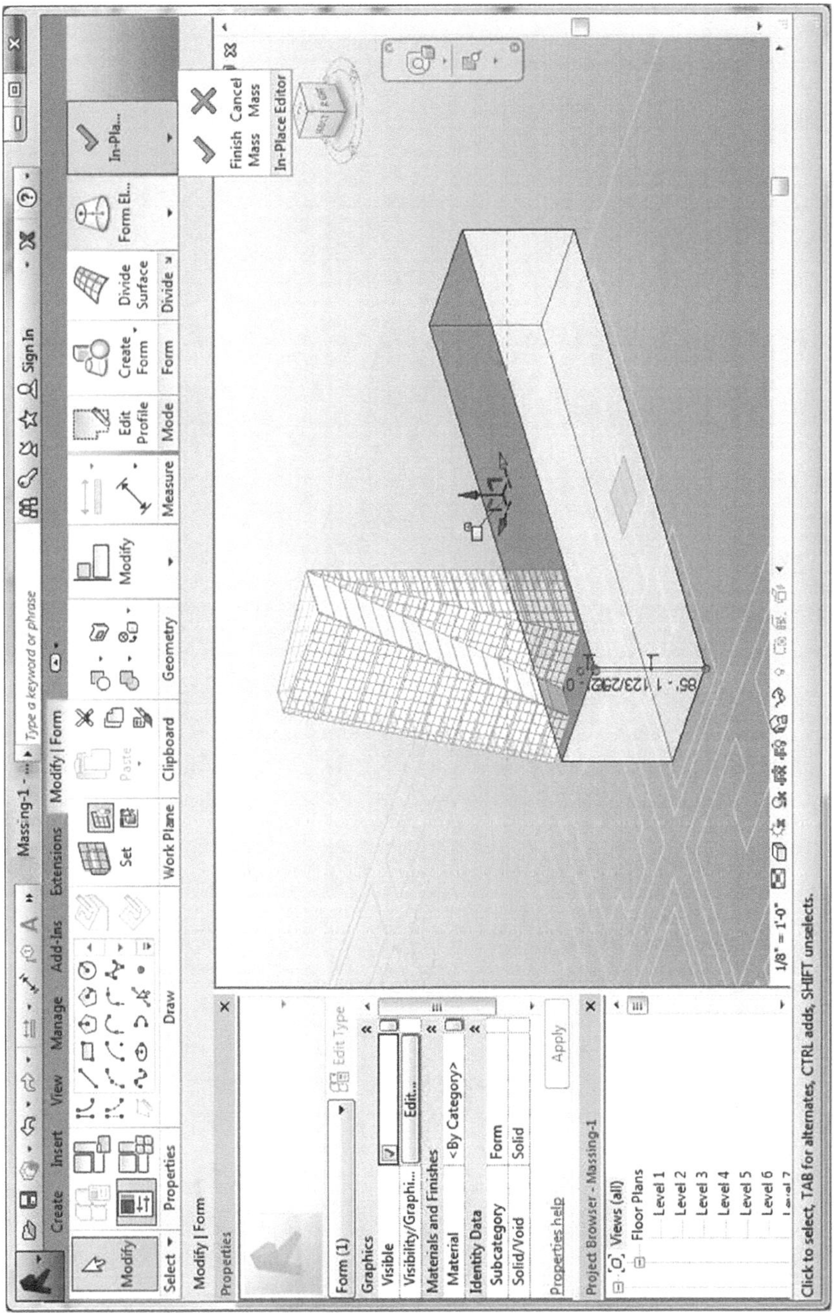

FIGURE 5.13 Creating an extrusion form/3D control arrows.

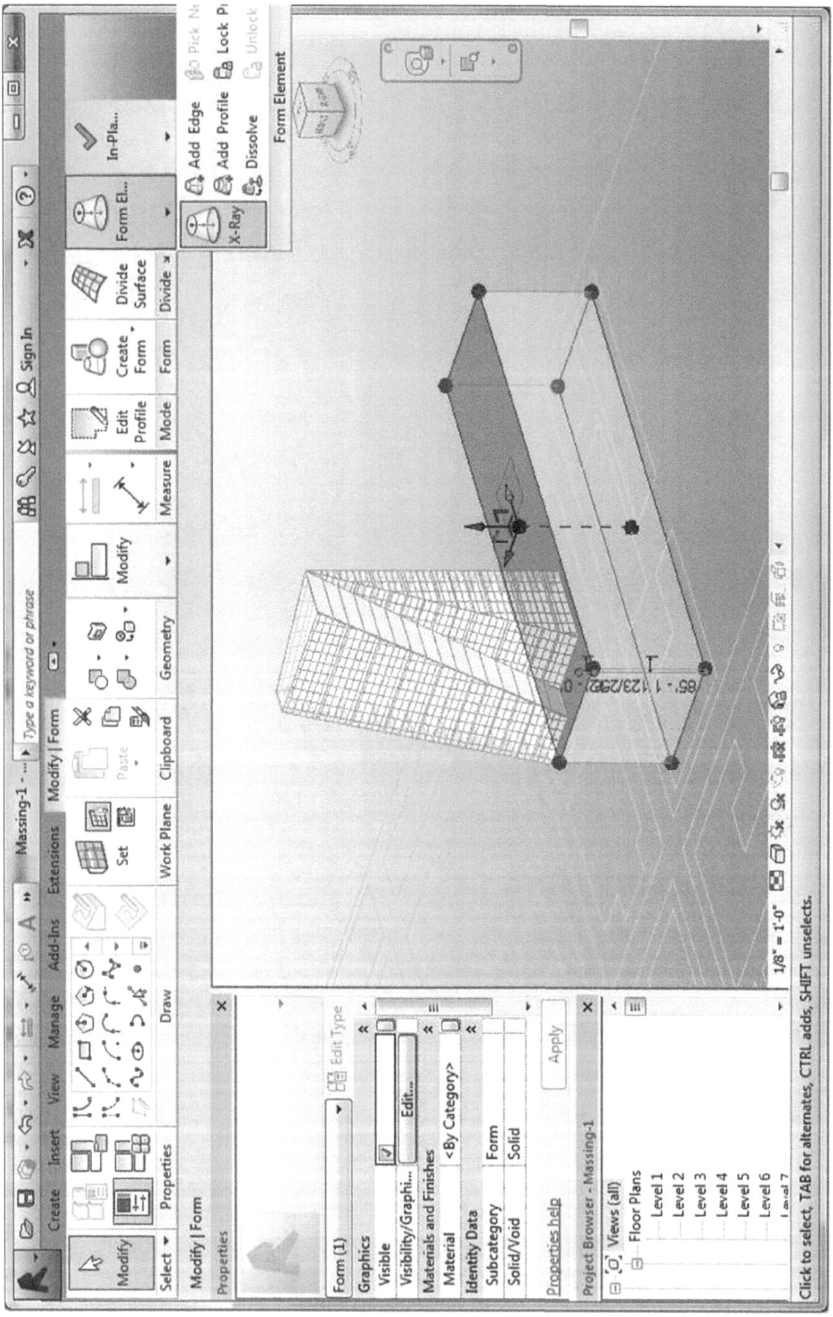

FIGURE 5.14 X-Ray mode.

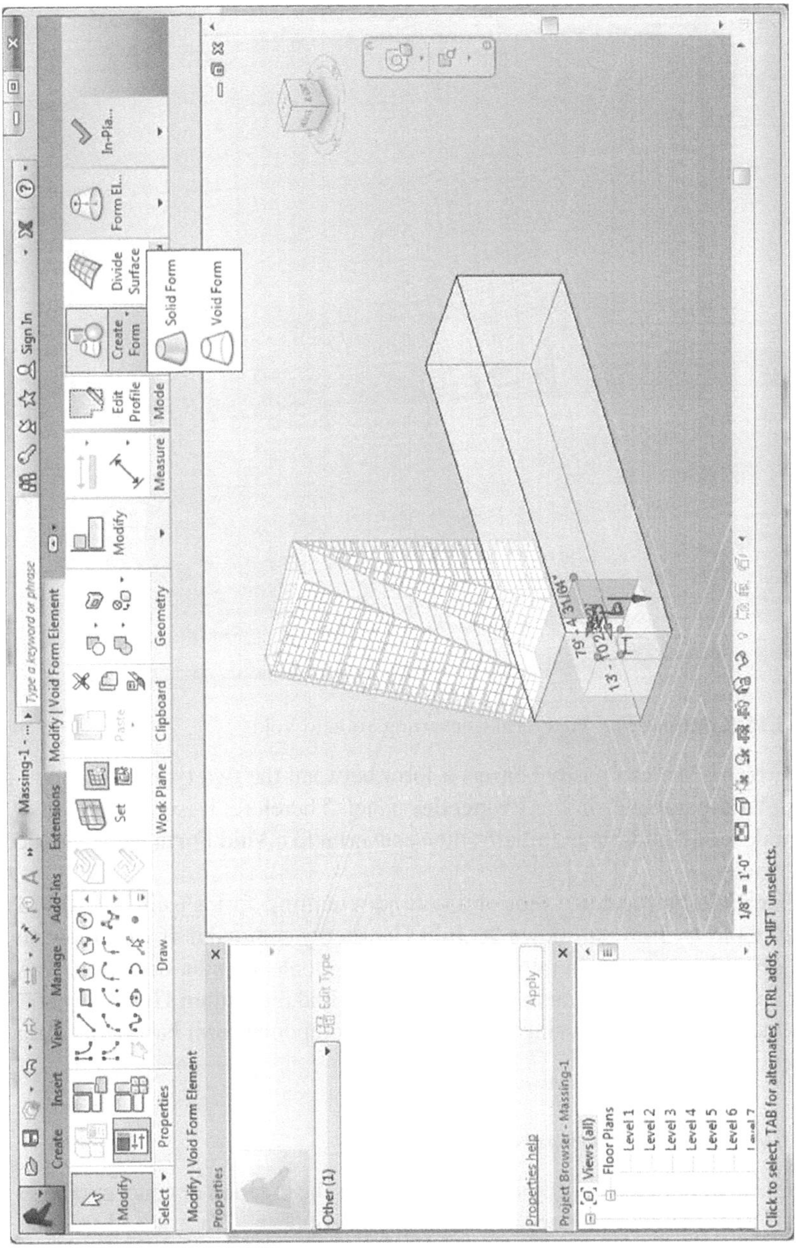

FIGURE 5.15 Void form/converting solid to void.

(Continued)

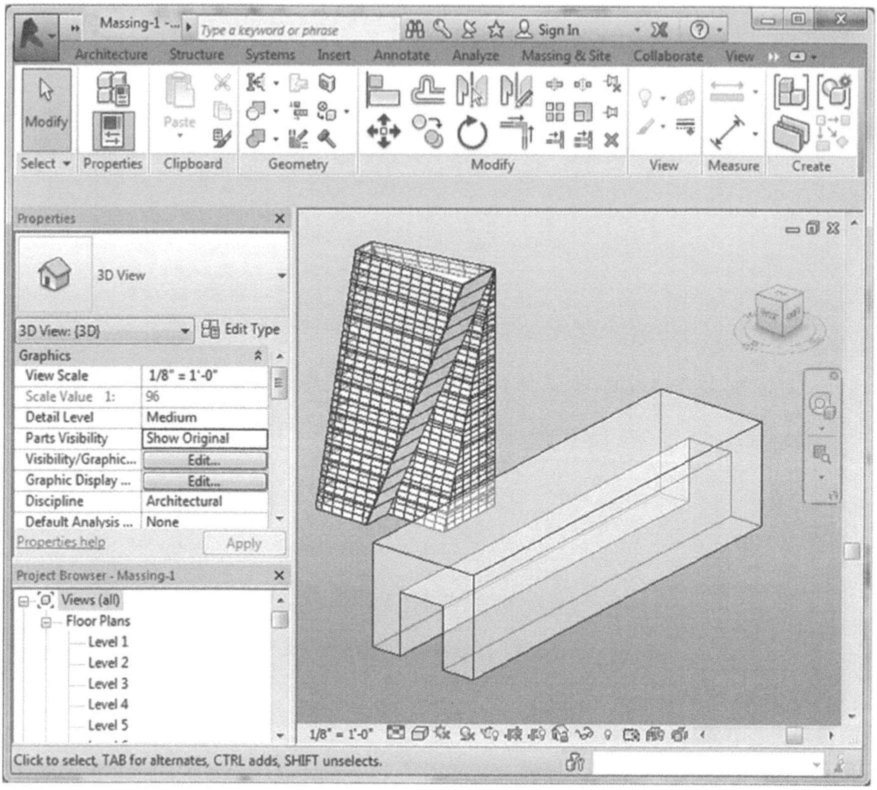

FIGURE 5.15 (*Continued*) Void form/Converting Solid to Void.

Furthermore, one can easily convert a form between the two types by changing the **Solid/Void** parameter in the **Properties** panel. Therefore, it is generally advisable to model as a **Solid Form** initially, then convert it to a **Void Form** when satisfied with the results.

Another important feature is the ability to join multiple forms using the **Modify | Geometry** tab ➤ **Join** pull-down ➤ **Join Geometry** command. It will allow you to select several forms to join into one, similar to a Boolean union operation, while cleaning up the intersections and overlaps. You will find the **Unjoin Geometry** command in the same place, allowing you to revert a compound form back to its initial component forms.

Conceptual Design Environment

The CDE is a family editor specifically designed for the creation of conceptual mass families. The same environment is also used in Autodesk's Vasari, and the mass families created can be imported into Revit for further development and analysis. Its interface is not that different from what you see in the project environment; the major difference you will see is a gradient background to visually remind you that you are in the CDE, as well as a default level and two vertical and two horizontal reference planes (Figure 5.16).

To access this editor, you can go to the **Application** menu at the upper lefthand corner, and select **New ➤ Conceptual Mass**. You will be prompted to select a template file, which should be the default Mass Revit family template (*.rft). After finishing, you can save it as a Revit family (*.rfa) or insert it into a project using the **Family Editor | Load into Project**, which is at the very end of the ribbon in all the tabs.

Adding Mass Floors

Mass floors can be created automatically from your **Mass** objects at every specified level. Note that you will need to be in the project environment to do this, so if you created a **Mass** family, insert it into a project to start adding mass floors. Mass floors need to be hosted on levels. Therefore, you need to add various levels before adding mass floors. To create new levels, activate the **Create** tab ➤ **Datum** panel ➤ **Level** command. You can click and drag or specify an elevation height for the level in the **Properties**, as described previously in Chapter 4. Alternatively, go to an elevation view and **Ctrl-drag** on an existing level to duplicate and move it. Select the mass you wish to create mass floors for and then invoke the **Modify | Mass ➤ Model** panel ➤ **Mass Floors** command. A selection window will allow you to choose the levels where you wish to create mass floors; they are surfaces without thickness but eventually can be turned into floor slabs with the **By Face** method.

Scheduling Masses and Mass Floors

The Revit software engine provides analysis tools to assist in quickly calculating the **Exterior** surface areas, building volume, and floor areas. They can assist in making decisions regarding the validity and proportions of a design option. The process for any of these analytical schedules is the same; thus, the example of scheduling the **Mass Floors** just created is shown. These are all accessed via the **View** tab ➤ **Create** panel ➤ **Schedules** drop-down ➤ **Schedule/Quantities**. You will have the **New Schedule** dialog box; while the majority of these are related to **Architectural** elements, scroll down to find the **Mass** categories and expand by clicking on the plus symbol. Choose the **Mass Floor** category and click **OK** to elicit the **Schedule Properties** dialog box (Figure 5.17).

One may highlight the desired fields and add them to the **Scheduled** fields; usually, this includes **Mass: Family**, **Floor Area**, **Exterior Surface Area**, **Floor Perimeter**, and **Floor Volume**. This **Mass Floor Schedule** (Figure 5.18) will now show up in the **Project Browser** under the **Schedules/Quantities**. The important issue to realize is that these values are parametrically linked to your massing model; therefore, if you modify your **Mass** object, these values will be updated automatically to reflect these changes. This makes it a powerful and flexible analytical tool in the early conceptual design phase.

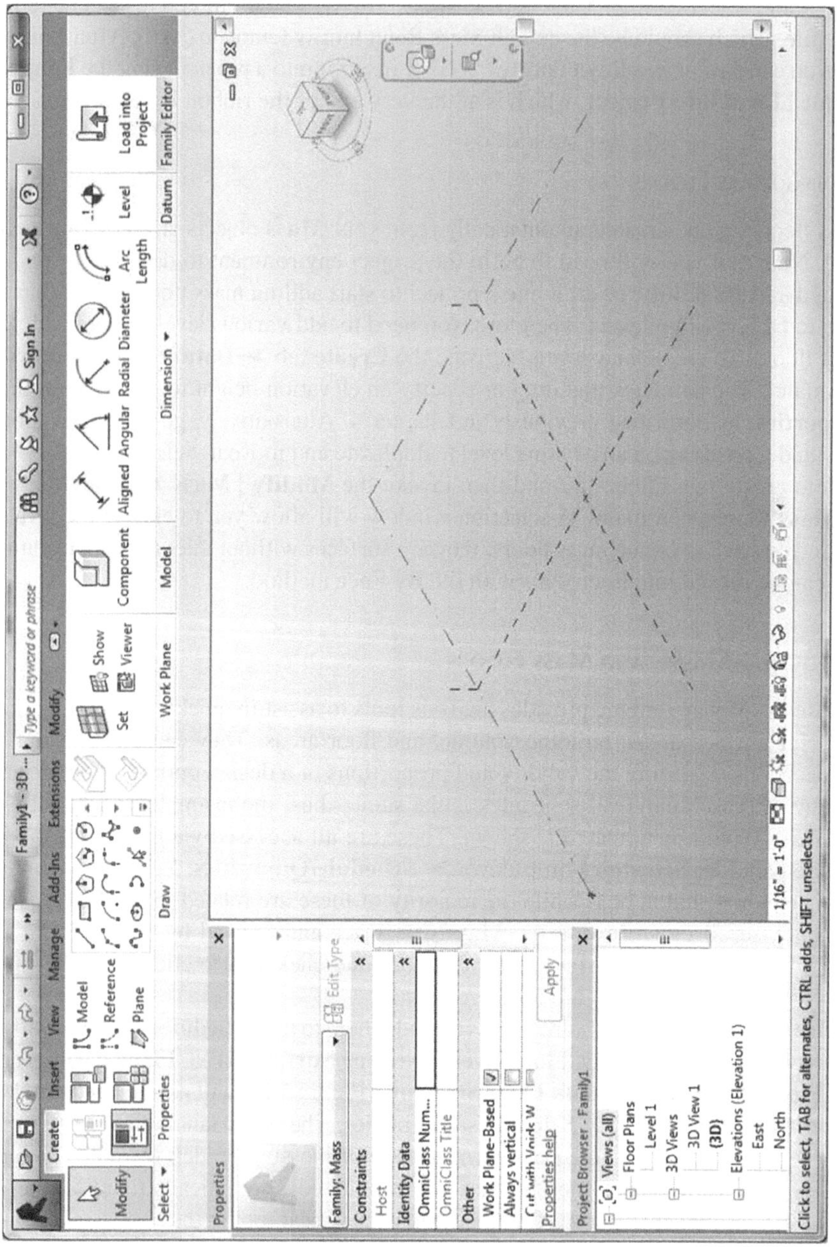

FIGURE 5.16 CDE environment.

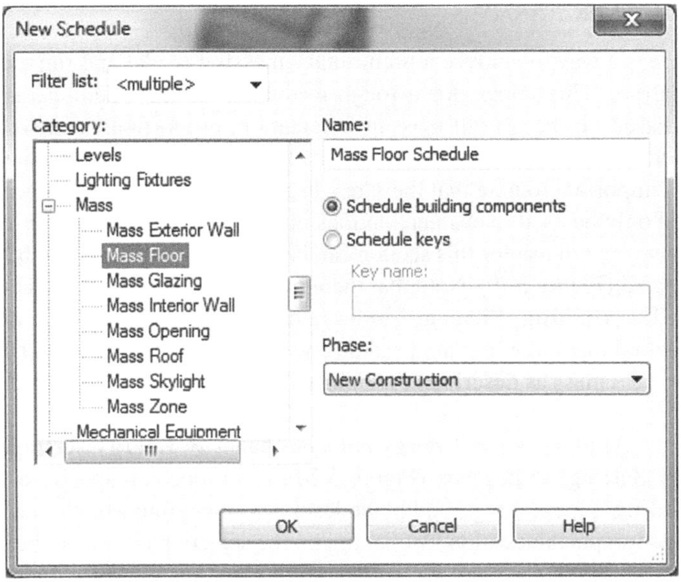

FIGURE 5.17 **New schedule** dialog box/**schedule properties** dialog box.

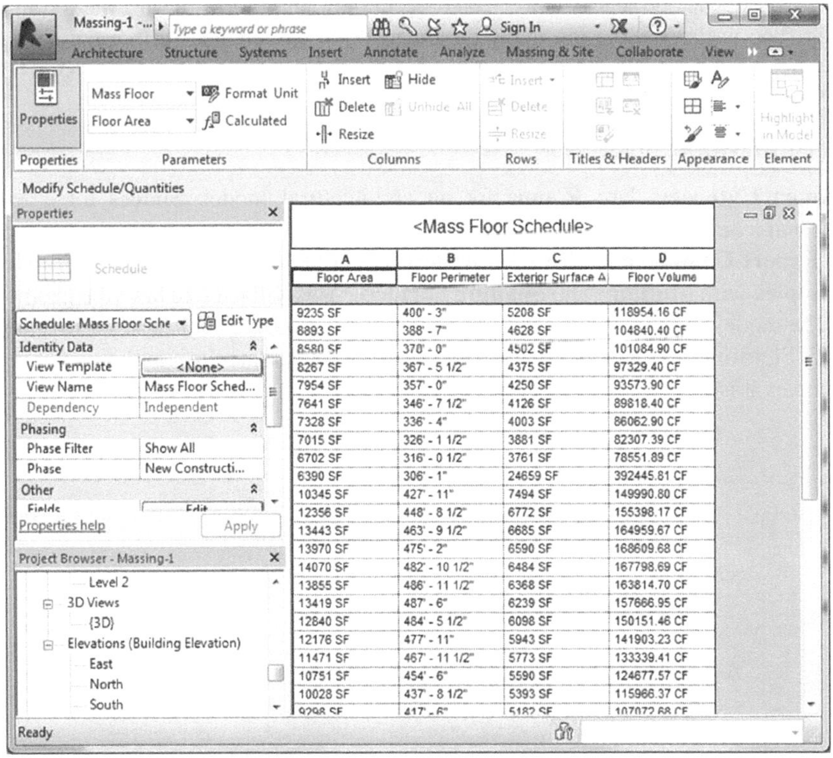

FIGURE 5.18 Mass floor schedule.

Conceptual Energy Analysis

Revit provides a way to analyze a preliminary massing model and run a conceptual energy analysis. This allows for a rough evaluation of the building's energy performance based on the overall massing orientation, proportions, and some general construction material assumptions before developing details of any architectural elements. It is important to note that these results should not be treated as accurate, and they should only be used to evaluate options on a comparative basis.

The tools you will use for this section can be found in the **Analyze** tab ➤ **Energy Analysis** panel (Figure 5.19). Note that there are two modes, **Use Conceptual Mass Mode** and **Use Building Element Mode**. Because we will be running the analysis on a conceptual mass, the first mode is the one to utilize here. Use the CDE environment to create a mass as described previously in the chapter and then load it in a test project.

Enable the **Analyze** tab ➤ **Energy Analysis** panel ➤ **Energy Settings** to obtain the **Energy Settings** dialog box (Figure 5.20). Here, one can specify a wide variety of **Building Types**, the ground plane level, and most importantly the **Location.** Specify a geographical location by clicking on the selection button on the right-hand side of the parameter box; here, in the **Location Weather and Site** dialog box, you can type and search a location under the **Location** tab. Under the **Site** tab, you can specify an **Angle** from **Project North** to **True North**, which is important to enter correctly if your project is not oriented accurately with north toward the top in the plan. You will see there are many parameters you can change further down in the dialog box. The most commonly used parameters of these options are discussed next.

Detailed Model Subsection

Export Category: Use **Rooms** for an architectural model, **Spaces** for a MEP (mechanical, electrical, and plumbing) model.

Export Complexity: There are five levels of complexity, ranging from **Simple** to **Complex** with **Mullions** and **Shading Surfaces**. You will need to have the geometry in the model to utilize the more complex settings, so by default it is set to **Simple with Shading Surfaces**. It will also take longer to calculate the simulation the more complex it is.

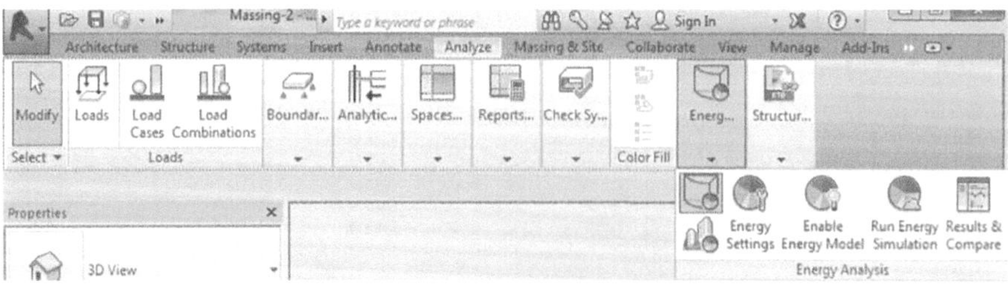

FIGURE 5.19 **Energy analysis** panel/**energy settings** dialog box.

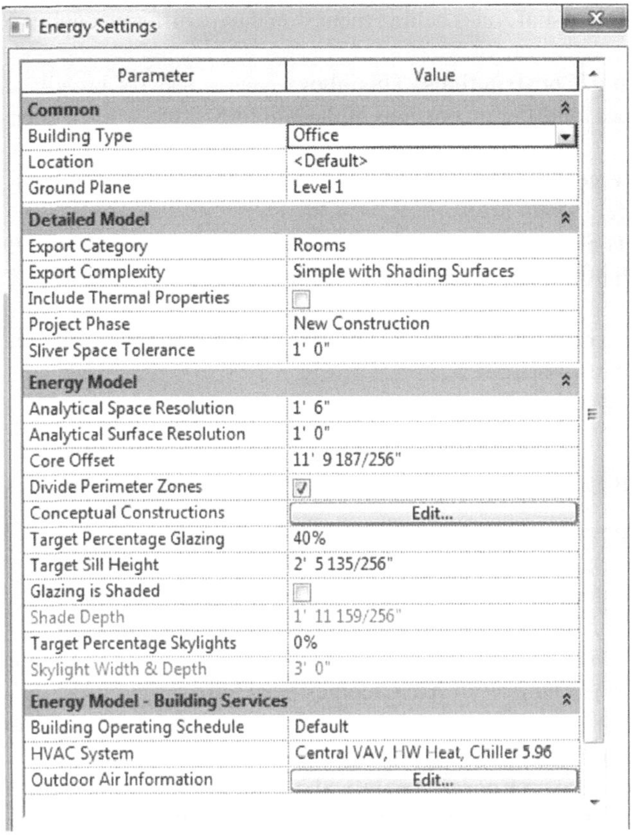

FIGURE 5.20 Energy settings detailed model section.

Include Thermal Properties: Check this box if you have assigned thermal values (R values) to your materials and façade glazing, and Revit will take them into account. If not, just leave it unchecked.

Energy Model Subsection

Analytical Space Resolution: This defines the minimum gap distance between Revit elements that will be ignored when identifying and calculating spaces for the **Energy Analytical Model**.

Analytical Space Resolution: This defines the minimum dimension of a surface that will be included by Revit for calculation in the **Energy Analytical Model**; surfaces smaller than this dimension will be ignored.

Core Offset: This is the distance offset inward from the perimeter of a building; it is used to create a provisional building core volume.

Divide Perimeter Zones: This allows you to divide each mass floor into four equal zones based on their facing orientation. For example, south- and east-facing facades and the spatial zones behind them would have different cooling loads because of solar gain.

Conceptual Constructions: This allows you to specify broadly different construction types for different portions of the building (exterior wall, interior wall, roof, slab, etc.).

Target Percentage Glazing: You can specify the amount of glazing as a percentage of the overall wall area.

Glazing is Shaded: This specifies if the glazing has any form of sun shading.

Shade Depth: This indicates the depth of the sun-shading elements specified previously.

Target Percentage Skylights: The amount of skylight as a percentage of the overall roof area is specified.

Skylight Width & Depth: The size of skylight units is specified.

Energy Model - Building Services Subsection

Building Operating Schedule: Customize the occupation schedule of your building with this option.

HVAC System: Like **Conceptual Constructions**, different HVAC (heating, ventilating, and air-conditioning) system types are specified broadly.

Outdoor Air Information: Air exchange values are specified in cubic feet per minute (**CFM**) or number of **Air Changes per Hour**.

Once all these settings are set to your satisfaction, invoke the **Analyze** tab ➤ **Energy Analysis** panel ➤ **Run Energy Simulation**, and you will reach a dialog box informing you that an **Energy Analytical Model** will be generated to run the analysis. Alternatively, you can choose to invoke the **Analyze** tab ➤ **Energy Analysis** panel ➤ **Enable Energy Model** command first (see Figure 5.19) and then preview the building assumptions based on the chosen **Energy Settings** (core, glazing and skylight openings, perimeter zones, etc.). If you did not create mass floors beforehand, Revit will inform you that it will need to create mass floors automatically to facilitate the analysis.

You need an Autodesk ID, and you must be signed in to Autodesk 360 to run the analysis. After Revit verifies account credentials and connects to the Autodesk cloud servers, one receives a **Run Energy Simulation** dialog box in which one can give a name analysis run. It is good practice to name these descriptively for comparison later. Revit will send the project information to the Green Building Studio cloud service for processing. One will also see a progress bar in the lower left corner of the interface. You may click on the **Analyze** tab ➤ **Energy Analysis** panel ➤ **Results & Compare** command to see the progress or view the results. Part of the energy analysis results shown in Figure 5.21 are for a 200-ft (66 m) hexagon tower in Miami, Florida. A complete report can be downloaded from the companion website (http://www.crcpress.com/product/isbn/9781482240436).

Once complete, you will see the analysis report listed under your project name. You can select it and scroll down to view the results. If you go back to change some of the options in the **Energy Settings**, the **Energy Analysis Model** will update to reflect the changes, and you may run the analysis again, saving it under another name.

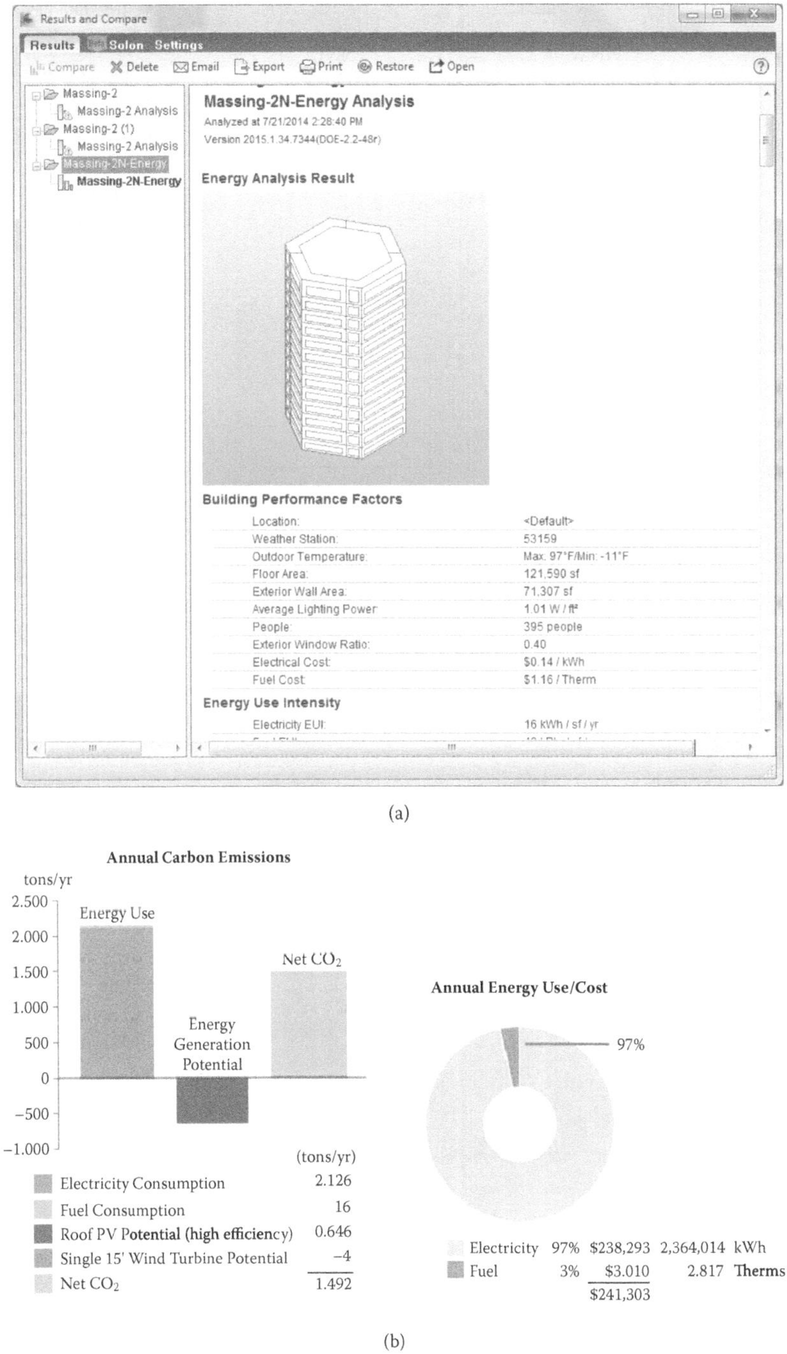

FIGURE 5.21 Part of the analysis results report. (a) Energy analysis results. (b) Annual carbon emissions and energy use/costs. (c) Annual wind speed distribution. (d) Monthly cooling load. (e) Monthly heating load. (f) Annual peak demand.

(Continued)

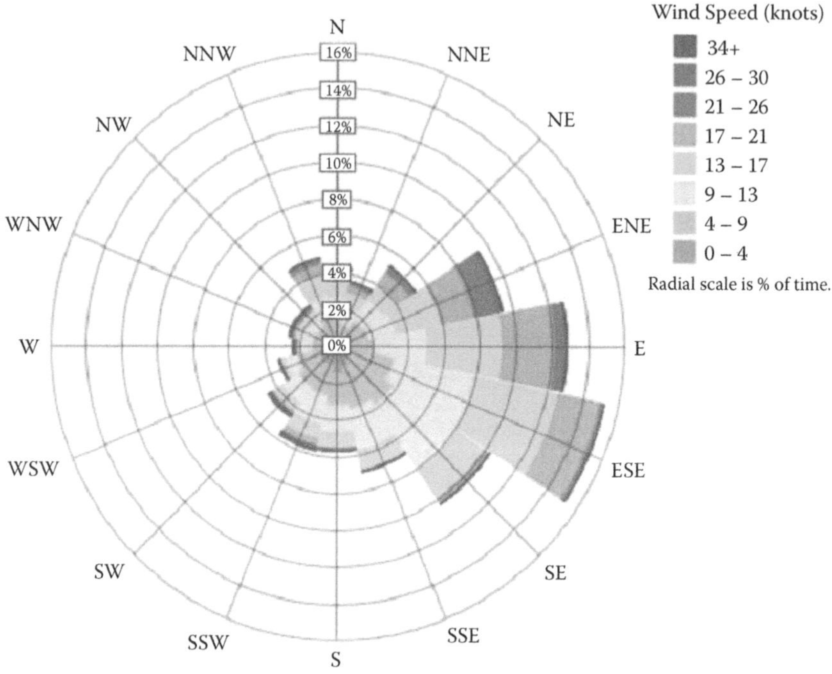

(c)

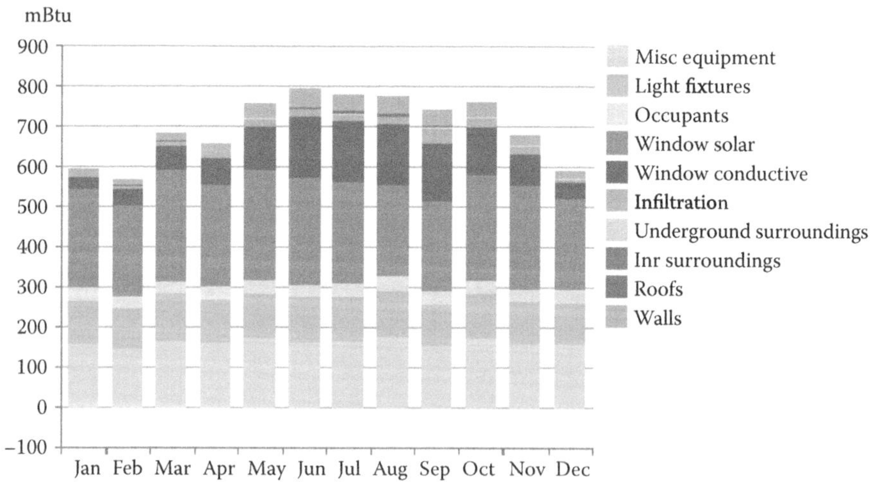

(d)

FIGURE 5.21 (*Continued*) Part of the analysis results report. (a) Energy analysis results. (b) Annual carbon emissions and energy use/costs. (c) Annual wind speed distribution. (d) Monthly cooling load. (e) Monthly heating load. (f) Annual peak demand.

(Continued)

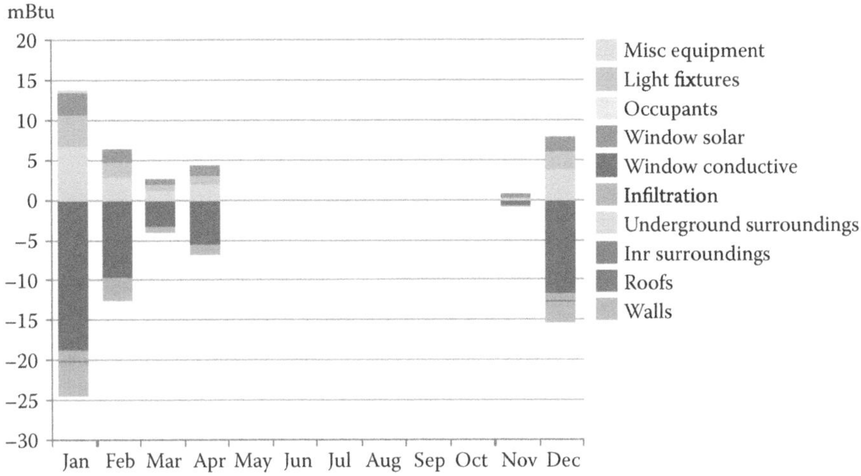

(e)

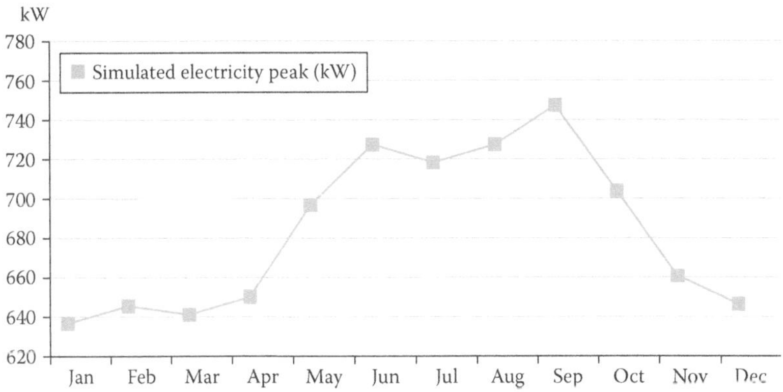

(f)

FIGURE 5.21 (*Continued*) Part of the analysis results report. (a) Energy analysis results. (b) Annual carbon emissions and energy use/costs. (c) Annual wind speed distribution. (d) Monthly cooling load. (e) Monthly heating load. (f) Annual peak demand.

Both reports will show up under your project name. Hold down the **Shift** key and select to highlight both reports; you can then use the **Compare** button on the upper left to place the reports side by side for easier comparison. In fact, you can **Shift-select** multiple reports and compare them all simultaneously.

Solar and Shadow Studies

Revit has a built-in visual display function that can help you visualize natural lighting and shadows based on your project location and orientation, as well as time/date.

But, before you start the analysis, you would probably want to make sure your project location is correct. To achieve this, launch the **Manage** tab ➤ **Project**

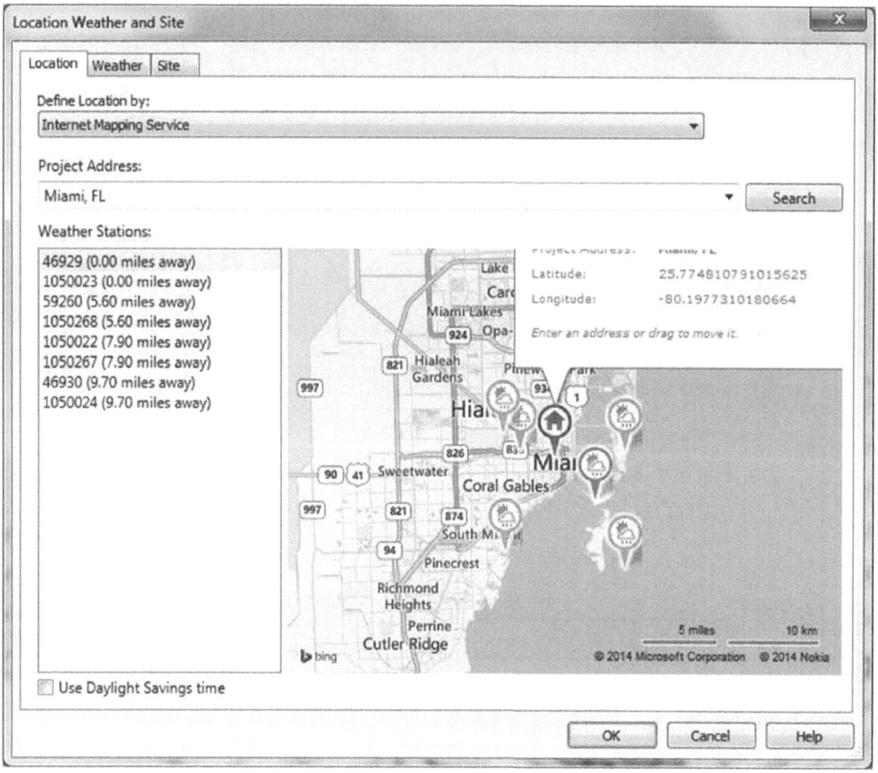

FIGURE 5.22 Project location.

Location panel ➤ **Location** command; this will bring up the **Location Weather and Site** dialog box (Figure 5.22). You can type and search a location in the **Project Address** under the **Location** tab; once a location has been selected, Revit will search and apply the nearest weather station data automatically.

You would normally seek to make sure that your project is oriented correctly in relationship to the cardinal directions because this will affect the sun/shadow study greatly. By default (and drawing conventions), north is usually the top of the view. To examine the project location, launch the **Manage** tab ➤ **Project Location** panel ➤ **Position** drop-down ➤ **Rotate Project North** command (see Figure 5.23).

You obtain the options to rotate the project 90° clockwise or counterclockwise, 180°, or orient a selected line or plane to north/south or east/west, whichever is closer. The last option is used for orienting at arbitrary angles, although it is better practice to sketch a reference line first.

Once that is done, you may switch to a 3D view or a plan/elevation view. The sun path can be activated from the view control bar at the lower left of the view screen. When you try to activate it in an orthographic view, Revit will give you a warning that the sun display is off, and you will need to apply the specified location, date, and time settings to enable the sun path display (Figure 5.24).

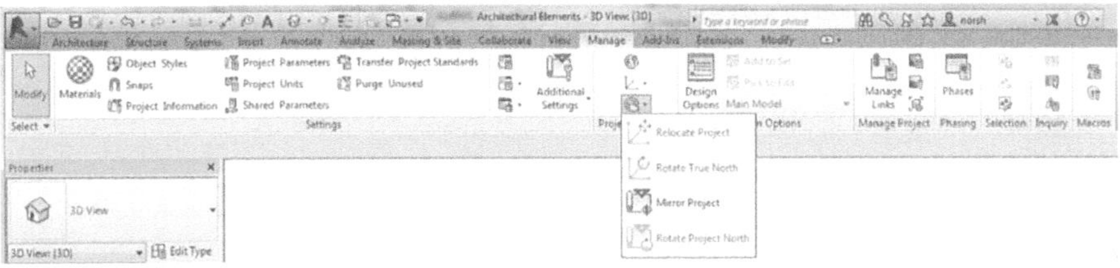

FIGURE 5.23 **Rotate project north** dialog box.

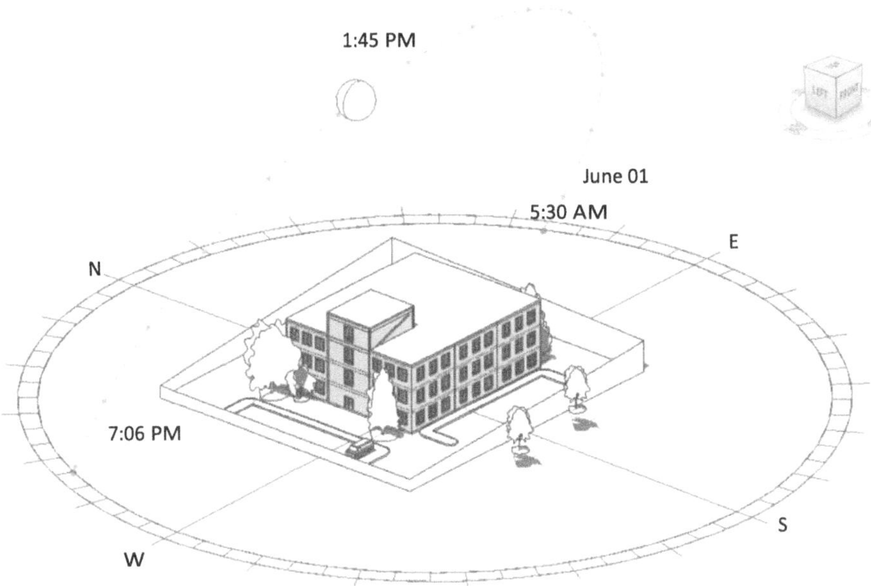

FIGURE 5.24 Sun display/Sun path diagram.

Once that is done, a compass rose and sun path will appear based on your settings. You should also make certain that the **Shadows** are turned on for the current view in the view control bar. Please note that you can click and drag the sun around intuitively to change the time of day as well as the date instead of going into the settings. The sun path size can also be changed as a percentage parameter in the **Properties**. If you click on the **Sun Settings** option (Figure 5.25), besides the date/time/location parameters, you can also change the type of solar study you want to create:

Still: Creates a study at a particular time, date, and location.
Single Day: Creates a study that spans an entire day at 15-, 30-, 45-, or 60-minute intervals.
Multi-Day: Same as **Single Day** except the study can span a range of dates, with time intervals of an hour, a day, a week, or a month.

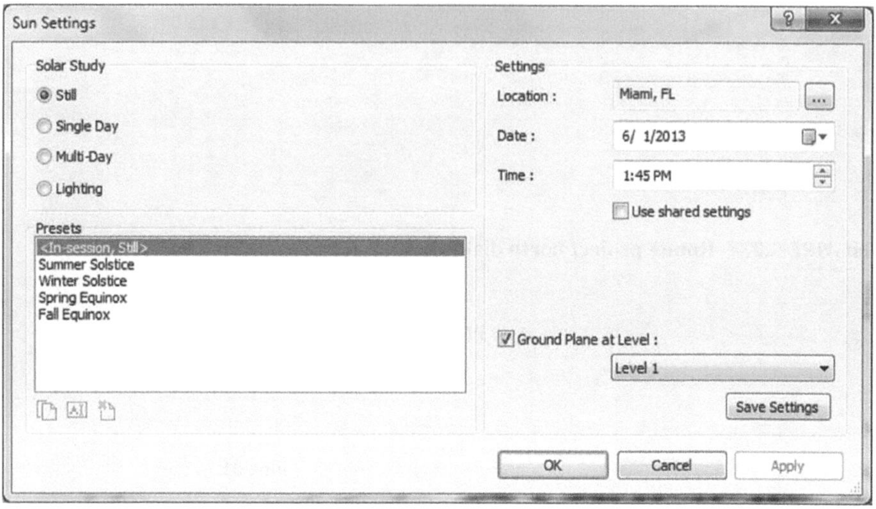

FIGURE 5.25 Sun settings.

Lighting: Creates a study that is not linked to the sun path but is more analytical in nature with a parallel light source coming from the **Top Right** or **Top Left**.

Also, note that for the single and multiday options, once you change the settings, you will obtain an additional **Preview Solar Study** option in the view control bar. Activating that will engage an animation control option bar that will appear under the ribbon, allowing you to play animation or step through it frame by frame. To save a particular frame, right-click on the view for which you set the solar study up in the **Project Browser** and choose **Save to Project as Image**. You can rename the view and modify the output pixel dimensions.

WALLS AND CURTAIN WALLS

Basic Walls

Basic architectural walls can be created using one of two methods:

- By launching **Architecture** tab **Build** panel **Wall** drop-down ➤ **Wall: Architectural**
- By launching **Structure** tab **Structure** panel **Wall** drop-down ➤ **Wall: Architectural**

Walls are modeled in the plan views, with a wide variety of sketching options available. It is good practice to define the following parameters on the **Option** bar before you start sketching walls:

- **Depth**: Select a level for the wall's bottom constraint or enter a value for the default setting of **Unconnected**. Or, if you want the wall to extend upward from the base constraint, select **Height**.
- **Location Line**: Select which vertical plane of the wall assembly you want to align with the cursor as you draw or with the line or face you select in the drawing area.
- **Chain**: Select this option to draw a series of wall segments connected at end points. In the plan representation, wall joins will autoclean to create a continuous wall by default.
- **Offset**: Optionally, enter a distance to specify how far the wall's location line will be offset from the cursor position or from a selected line or face (described in the next section).

Generally, it is advisable to separate and predefine the several different wall types in a project (exterior, interior, partition, curtain wall). To represent the different wall types, one may also wish to define in the **Floor Plan Graphics** settings the **Wall Join Display** behavior to **Clean all wall joins** or **Clean same type wall joins**. If you wish to isolate specific wall joins, you may right-click on the end grip of the wall you wish to change and select **Disallow Join** from the menu. To obtain further customization options for editing wall joins behavior, invoke the **Modify** tab ➤ **Geometry** panel ➤ **Wall Joins** command to see all the options.

Wall By face

In terms of wall modeling, it seems that it is difficult to create complex walls in the standard project environment. However, Revit offers a method to create walls and curtain systems with complex geometry via the **Mass** tool and the **Wall by Face** command (Figure 5.26). This can be accomplished by one of the following two methods:

- By invoking **Massing & Site** tab ➤ **Model by Face** panel ➤ **Wall by Face**
- By invoking **Architecture** tab **Build** panel ➤ **Wall** drop-down **Wall by Face**

Simply highlight and select the massing face to which you will apply the wall type; you can press **Tab** to cycle through the highlight options where faces join.

These wall properties can be easily modified after placement by editing the constraints under properties. To edit an architectural wall's elevation profile, you must navigate to a parallel section or elevation view. You may also double-click on the wall in a plan view, which will bring up a dialog box to choose a parallel view for editing the profile sketch of the wall elevation selected. This invokes the profile sketching mode, in which you can edit the shape or add openings.

It is important to note that doors and windows are wall-hosted component families that automatically create wall openings to fit the component size; it is not necessary to create a wall opening for such elements beforehand. Modeling door and window elements is covered further in the chapter.

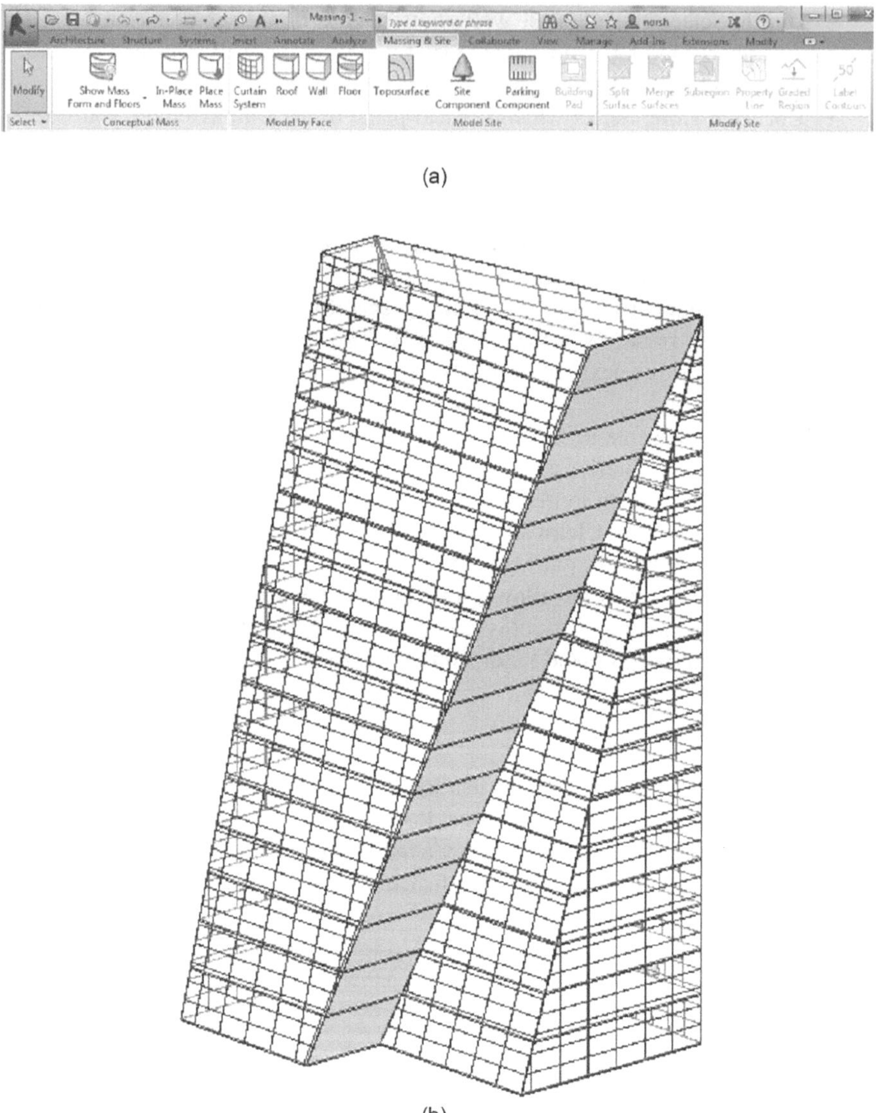

FIGURE 5.26 Wall by face. (a) Selecting the wall tool from the mass panel. (b) Selecting the massing face that is going to be modeled as a wall.

CURTAIN WALLS AND CURTAIN SYSTEMS

In Revit, *curtain walls* and *curtain systems* are special wall types that integrate the mullion and panel subdivision logics into the wall type. Simple curtain walls are modeled and specified as a wall type with the same process used to model basic walls. Curtain systems are used when you want to apply a curtain wall type to a face, usually in situations where the faces are irregular, like a vault or have double curvature. Because these complex geometries are usually modeled as a generic mass, the process is similar to creating a **Wall by Face**.

Curtain Wall

To create a simple curtain wall, one invokes the **Architecture** tab ➤ **Build** panel ➤ **Wall drop-down** ➤ **Wall: Architectural**. With the **Modify | Place Wall** tool active, you may then choose from the **Properties** panel and specify a curtain wall type. Alternatively, any basic wall you have already modeled can be modified to a curtain wall type as well. The basic generic curtain wall type has no preset subdivisions; you will have to specify these using the **Architecture** tab ➤ **Build** panel ➤ **Curtain Grid** tool. These are simply reference lines to help define the vertical and horizontal subdivision centerlines for placing mullions later. With the tool active, hover over an edge to see a preview of where the grid line will be placed. There will be some snapping behavior at certain dimensions to help you subdivide equally or in thirds, but it is always possible to edit the location by dragging the gird line or directly editing the interval dimensions. After specifying the **Curtain Grid**, select the **Architecture** tab ➤ **Build** panel ➤ **Mullion** tool (Figure 5.27).

You may then preselect the type of mullion you wish to place in the **Properties** panel and click on a grid line to insert it into the curtain wall.

Alternatively, one can also edit the mullion type afterward by selecting and changing the mullion type in the **Properties**. Please note that after the mullions have been inserted, they will be subdivided into short segments according to the **Curtain Grid** previously defined. This provides the ability to change the mullion type of one segment individually or delete the mullion outright to allow a glass door to be inserted, for example. In addition, by using the **Tab** select functionality, you can now select a single subpanel of the **Curtain Wall** and change the properties. These are called the **System Panel** type, and the default **Glazed** panel type is applied; it can be changed to a **Solid** panel type in the **Properties**.

In case of two curtain wall segments joining at a corner, Revit supplies special mullion types that can be applied to these corner situations. Choose the corner edge of one of the wall segments and apply one of the predefined corner mullion types. Note that the corner mullion type will adapt its profile shape depending on your corner angle; although the profile cannot be edited, the thickness and offset dimensions can be modified.

Curtain Systems

As described previously, **Curtain Systems** are used when it is necessary to create a curtain wall on a complex surface, usually a mass face. By invoking the **Architecture** tab ➤ **Build** panel ➤ **Curtain System by Face** command, you can select any mass face for curtain system application (Figure 5.28). After clicking **Create System** to complete the command, a curtain grid will be generated that can be customized by modifying the grid spacing in the **Edit Type Properties** dialog box. This can be further customized by using the **Curtain Grid** tool and editing the subdivided segments in similar fashion to the normal curtain wall.

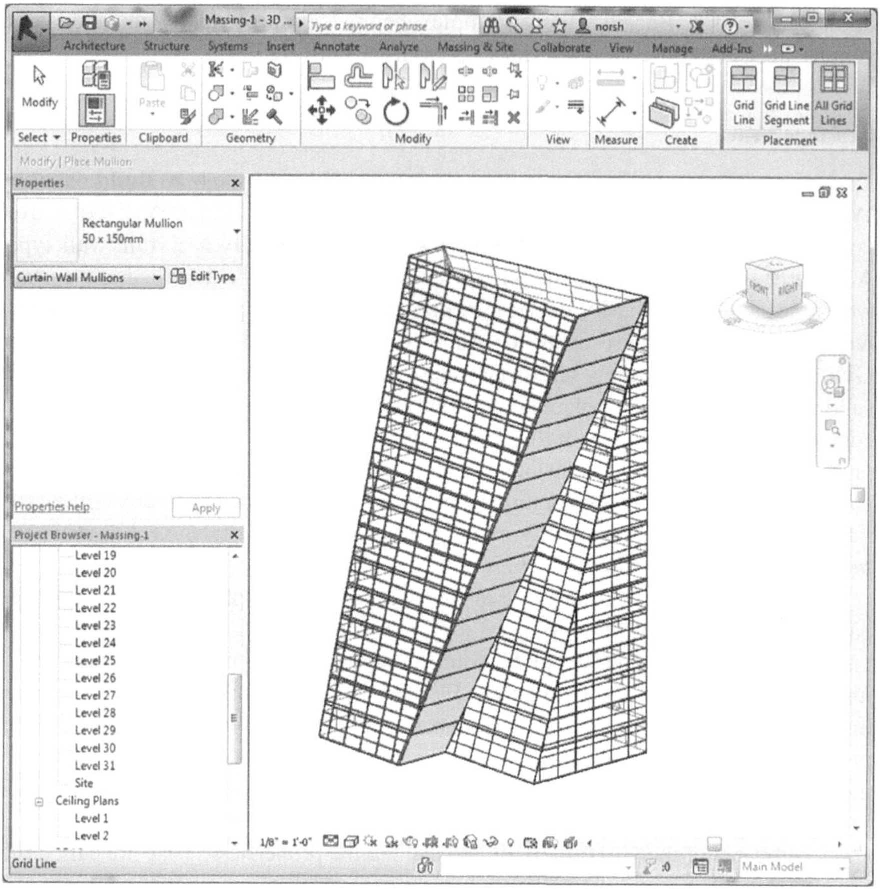

FIGURE 5.27 Curtain grid/inserting mullions.

Application of mullions follows the same process as simple curtain walls. Please note that, in cases of extreme double curvature, the **System Panels** will not automatically triangulate to follow the surface curvature; they will stay as planar quads; therefore, seams will appear between panels.

COLUMNS

As covered previously in Chapter 4, there are two kinds of columns that can be modeled in Revit: architectural columns and structural columns. Structural columns are load-bearing elements and are considered part of the structural analytical model; architectural columns are non-load-bearing elements and are used primarily to complete the spatial element organization of a building or represent the exterior finish surface of a structural column.

To model an architectural column, you would start in a plan view and then invoke the **Architecture** tab ➤ **Build** panel ➤ **Column** drop-down ➤ **Column:**

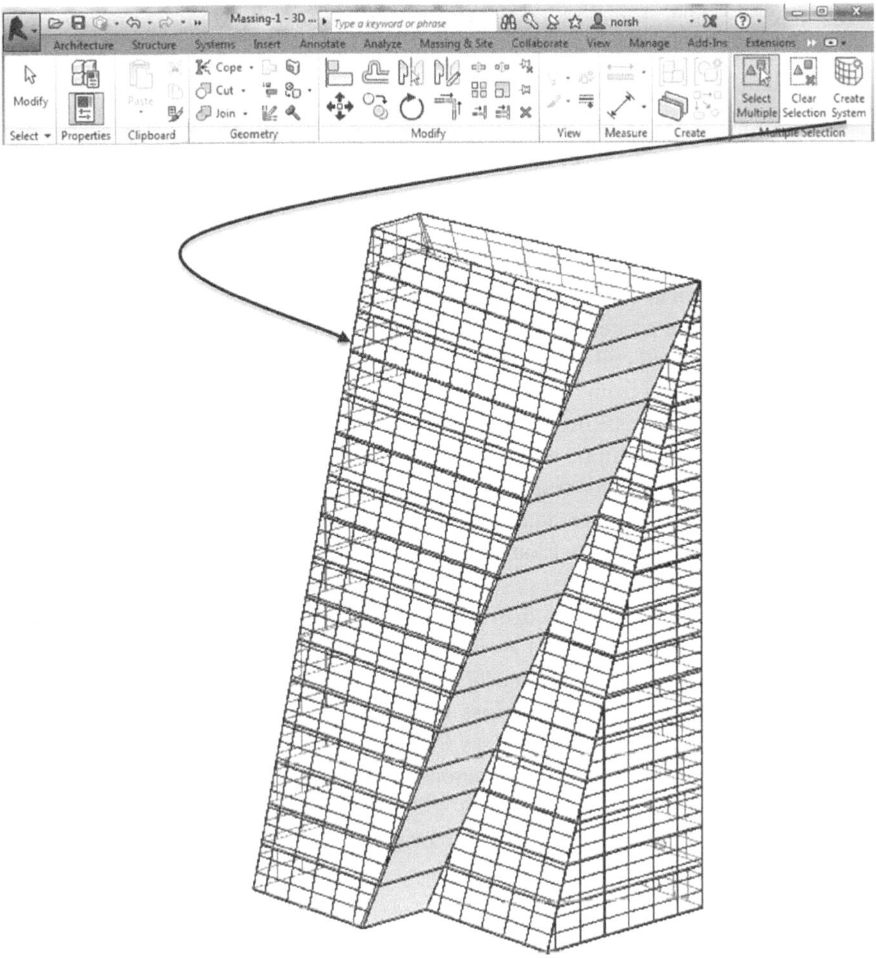

FIGURE 5.28 Curtain system.

Architectural command to place architectural columns constrained to the active work plane or grid lines and grid intersections. As a generic type, you may use the **Edit Type** properties dialog box to modify the **Depth** and **Width** parameters and change the dimensions to the expected outer finish surface of the column.

As a primarily structural building element, any further detail will require the structural column type to be specified. However, there is the option to use the architectural column type as a placeholder that can be populated with structural column type later. To achieve this, invoke the **Architecture** tab ➤ **Build** panel ➤ **Column** drop-down ➤ **Structural Column** command. In the context-sensitive ribbon, pick the **Modify | Place Structural Column** tab **Multiple: At Columns** tool to select the architectural columns into which you wish to insert a structural column.

The modeling and editing of structural columns as well as the wider variety of options are covered in detail in Chapter 4. Please note that there may be specific situations for which it may be more flexible to use a wall type instead of the architectural column type to represent nonstructural columns as spatial elements because you will have more options regarding the wall construction layers and finishes.

FLOORS, CEILINGS, AND ROOF OBJECTS

Floors

As discussed previously in Chapter 4, two options for modeling floors are available under the **Structural** and **Architectural** tabs. The architectural floor can be used earlier in the design process with generic floor types to represent the overall volume of the floor, beam, and ceiling assembly. Keep in mind that the depths of these layers are created downward from the host level; this usually means the top of the structural floor slab. Floor finishes (tiles, wooden floors, linoleum) are added upward. Layers that are more detailed can be added later, or specific assemblies can be specified by converting the floor type to structural floor in the **Properties** panel.

Architectural floors are created by launching the tool: **Architecture** tab ➤ **Build** ➤ panel **Floor** drop-down ➤ **Floor: Architectural**. Otherwise, the primary method of defining the floor boundaries, slab edges, and drop panels are the same as for structural floors and are covered in detail in Chapter 4. The one exception is the **Architecture** tab ➤ **Build** panel ➤ **Floor** drop-down ➤ **Floor by Face** command. This is used when you have previously created a **Massing** as part of the conceptual design phase and have generated **Mass Floors**. In this case, you can use the **Floor by Face** command and select the **Mass Floor** faces that you wish to apply the floor type to and **Create Floor**; Revit will automatically define the floor boundaries using the **Mass Floor** profile. You can always select the created **Floor** object and **Modify | Floors** ➤ **Edit Boundaries** to enter the profile sketching edit mode if you wish to further refine the shape.

Ceilings

Ceilings define the upper boundary of an interior space, usually a dropped ceiling assembly if the structural beams and slabs are not exposed. Ceilings also serve to host interior component families such as ceiling light fixtures. There are two basic kinds of ceiling types, the generic **Basic Ceiling** and the **Compound Ceiling** (Figure 5.29). The **Basic Ceiling** is a generic delineation of the ceiling height and is represented as such in section, but you can specify a basic material parameter for it to display in reflected ceiling plans and 3D section views. The **Compound Ceiling** is a layered material assembly that is visible with thickness in a 3D section view, as well as detailed specification of panel dimensions, thickness, and material layers via the **Edit Type** dialog box.

To start modeling a ceiling, you would invoke the **Architecture** tab ➤ **Build** panel ➤ **Ceiling** command to activate the tool. Ceilings can be defined in one of two ways: **Automatic Ceiling** and **Sketch Ceiling** (Figure 5.30). Because dropped ceiling assemblies are usually part of an interior space, the **Automatic Ceiling** method automatically detects spatial boundaries based on your modeled wall geometry and

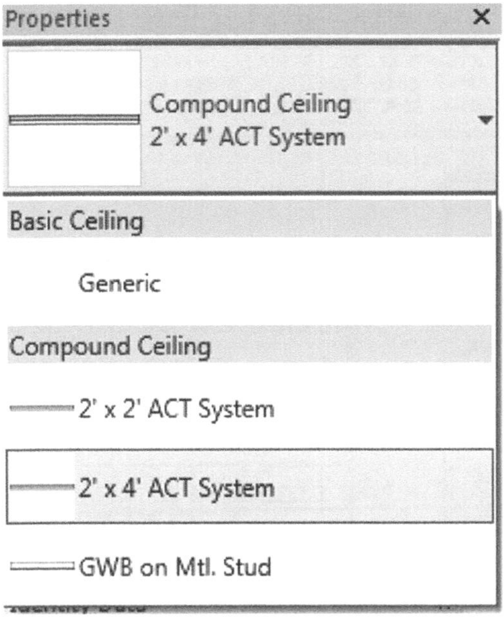

FIGURE 5.29 Basic ceiling versus compound ceiling.

highlights the area with a red boundary line. The **Sketch Ceiling** method works similarly to sketching floor boundaries, with the same sketching tool box. Note that if you use the **Automatic Ceiling** method, the ceiling boundaries will be associated with the walls. This means that if you modify the wall location or dimensions, the ceiling will update automatically to reflect any changes in the boundary. Similar to floors, one can modify the ceiling type in the **Properties** panel.

As you create ceilings in a plan view as you typically should, Revit will notify you that the elements created are not visible in the current view; this is normal because the ceiling should be located above the plan cut in elevation (depending on the **Height Offset** parameter specified). It will be visible in the corresponding reflected ceiling plan and applicable section views. It is also important to note that the ceiling elements are used as the boundaries for the volume calculation of interior spaces, which is particularly critical for environmental analysis programs such as Autodesk Ecotect or Green Building Studio.

ROOFS

Roof by Footprint

Architecture tab ➤ **Build** panel ➤ **Roof** drop-down ➤ **Roof by Footprint**

The **Roof by Footprint** approach is used to create standard roofs that adhere to the wall boundary footprint of your building. After invoking the command, the **Modify | Create Roof Footprint** mode will be active, and the sketching tool box can be used to define the boundary of the roof (Figure 5.31). You may either manually

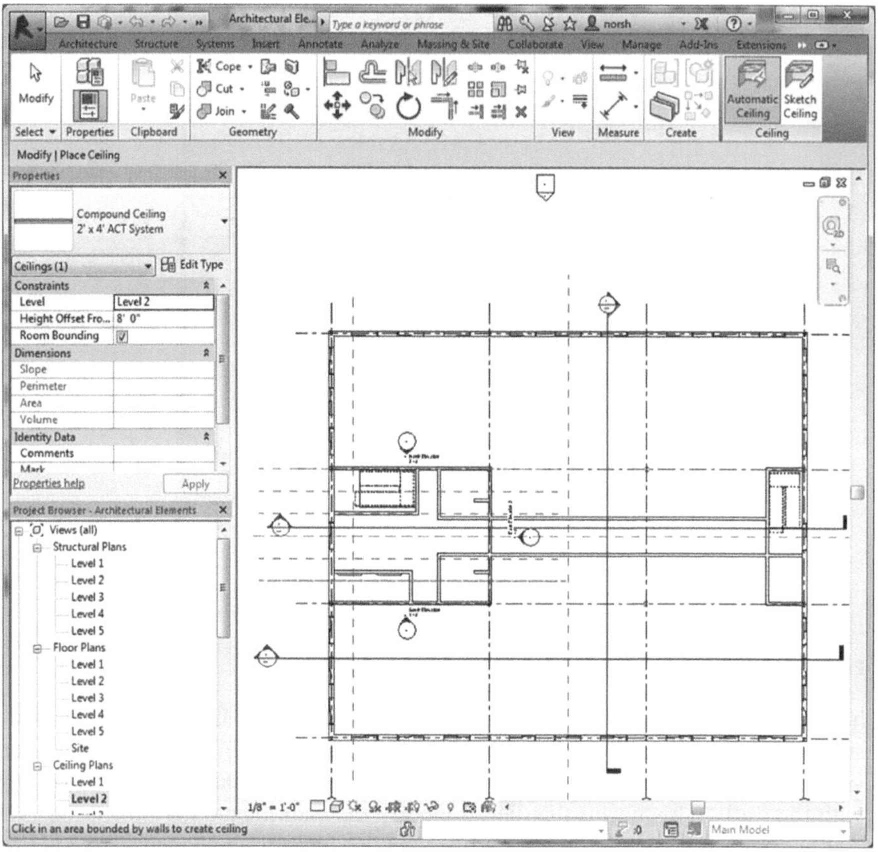

FIGURE 5.30 Automatic ceiling versus sketch ceiling.

sketch a boundary or hover the cursor over a wall section and use the **Tab** key to select all the connected walls. The **Option** bar will allow you to choose if the roof geometry is flat or sloped, as well as the overhang amount measured from the exterior face of the wall. After completing the command by finishing the edit mode, Revit will detect the walls that fall within the roof boundary and ask if you want to attach these highlighted walls to the roof. This will automatically extend or trim the corresponding wall profiles to match the roof geometry, which is useful especially if you selected the sloped roof option.

By default, all the boundary sides will define the beginning a roof slope, shown with a small magenta triangle icon. If you edit the roof footprint using **Modify | Roofs** tab ➤ **Edit Footprint**, you can select individual segments and uncheck the **Defines Roof Slope** check box in the **Option** bar, allowing the specification of single, double, or compound sloping roof geometries. Here, you can also specify the roof slope amount, shown as a percentage rise (rise/length). Alternatively, with the roof selected in a 3D view, you can use the blue shape handles to intuitively modify the roof pitch (Figure 5.32).

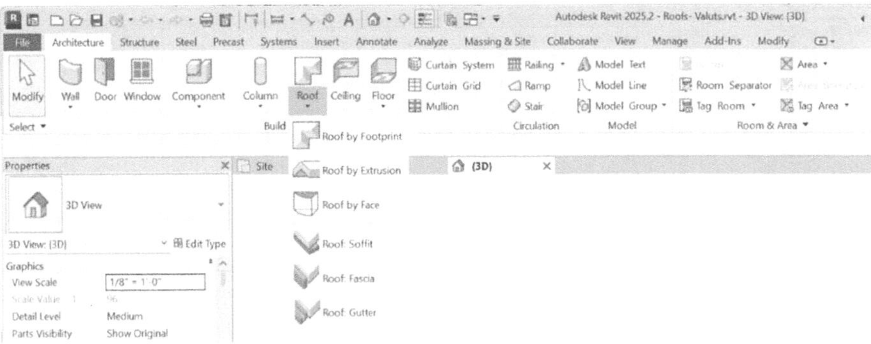

(a)

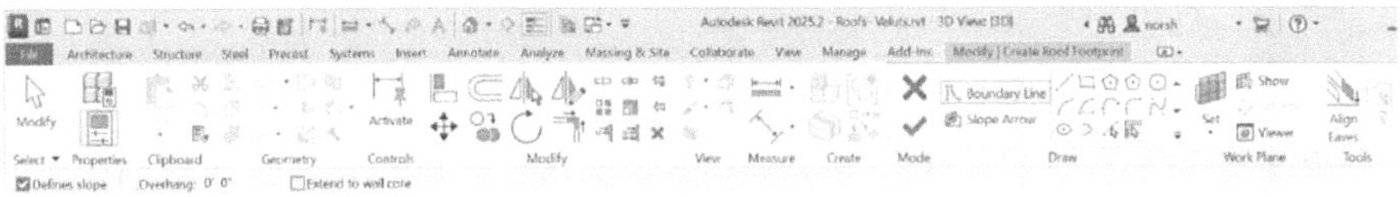

(b)

FIGURE 5.31 (a) Roof modeling tool. (b) Create roof footprint.

The **Roof by Extrusion** method is best used with roof geometries that can be modeled by extruding a roof profile, such as a barrel vault (Figure 5.33). After invoking the command, Revit will prompt you to select a **Work Plane** on which to host the sketch; pick an elevation wall face that is perpendicular to the extrusion direction of your roof profile. With the **Work Plane** chosen, the **Modify | Create Roof Extrusion Profile** mode will be active, and the sketching tool box can be used to sketch the profile of the roof on your chosen **Work Plane**. The profile curve needs to be an open curve, as the offset thickness of the roof is specified by the **Roof Type**. As the command is completed, the blue shape handles will be available to control the location as well as the length of the extrusion. It is important to note that, in this method, one has to attach the walls below to the extruded roof profile. To achieve this, select the walls that need to be attached to the roof; invoke the **Modify | Walls ➤ Attach Top/Base** tool and select the roof. This will automatically extend or trim the corresponding wall profiles to match the extruded roof geometry.

Roof by Face

Architecture tab ➤ **Build** panel ➤ **Roof** drop-down ➤ **Roof by Face**

Roof by Face is typically used when one has created a **Mass** family in the **Conceptual Design** phase and is using it to generate exterior walls, floors, and roof through the **By Face** method (see Figure 5.26). After invoking the command the **Modify | Place Roof by Face ➤ Create Roof**, the editing mode will be active and prompt you to select the mass faces on which you wish to place the roof type. Complete the command by clicking on **Create Roof**, and the blue shape handles will allow you to edit the roof if you want to create overhangs. Because using this method usually means your exterior walls, floors, and roofs are created using the **By Face** method, the individual elements generally are able to integrate well.

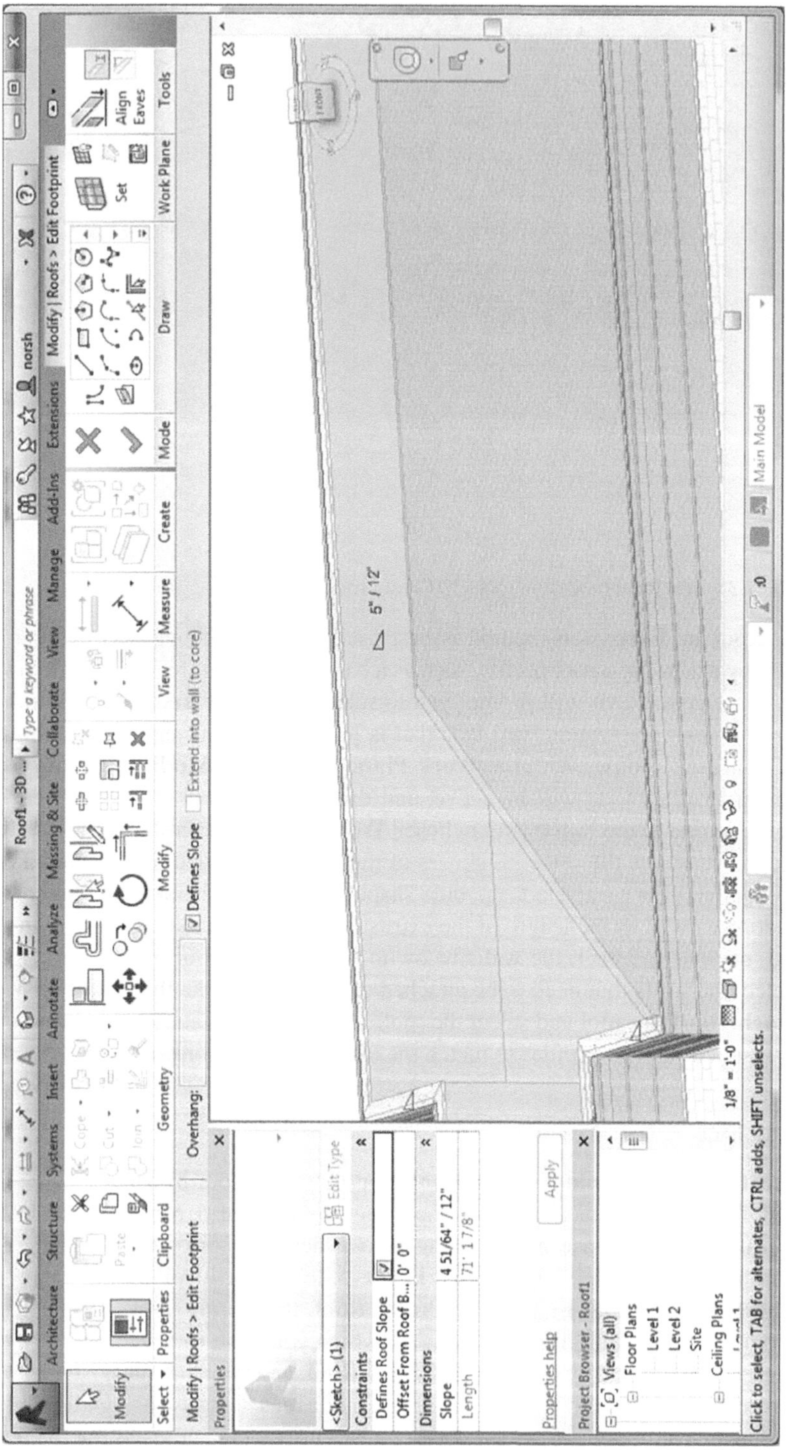

FIGURE 5.32 Defines roof slopes.

FIGURE 5.33 Roof by extrusion.

STAIRS AND ELEVATORS

Stairs

Stairs are an extremely complicated topic, with a wide variety of modeling methods, ranging from the basic to the complex. For the purposes of this text, the basic method, **Stair by Sketch**, is covered. However, if used correctly and by editing the produced outline profiles, it is possible to create a wide variety of stairs.

Sketching the Run

In the **Sketching the Run** method, you specify the start point of the stairs and Revit automatically determines the number of treads based on the level-to-level height defined in the **Constraints** (**Base Level to Top Level**), as well as the **Maximum Riser Height** defined in the **Type Properties** for the stair. Switch to a plan or 3D view and invoke the **Architecture** tab ➤ Circulation panel ➤ Stair drop-down ➤ Stair by Sketch command to enter the **Modify | Create Stairs Sketch** mode. Using the **Draw | Run** tool, define the run trajectory for the stair by clicking the starting point of the stairs (Figure 5.34). As you move the cursor, Revit will show a rectangular preview outline for the stair as well as a text display noting how many risers have been created and how many are left (based on your stair parameter settings). Note that the stair length automatically snaps to **Tread Depth** increments. For a straight-run stair, complete the run by clicking at the end of the stair preview or beyond; Revit will automatically end the stair at the correct length as calculated and defined by your parameters. For a stair with landings, click in the middle of the run preview, and Revit will expect you to input a second segment to complete the second flight. At this point, you may turn the stairs 90°

to obtain an L stair or sketch back toward the starting point of the stairs to create a switchback stair.

Notice that Revit will attempt to automatically connect and fill in the landing between the two runs, as well as add separate **Railing** components to both sides of the stairs. As for these railings, you can **Tab-select** to remove or replace with custom railing families or tweak the parameters in the **Type Properties**. Use the blue shape handles to edit or move elements around if necessary. There are also many customization options present in both the **Dimensions** and the **Type Properties** for the stair itself. You can also always edit the stair outline by double-clicking on the stair object to enter the **Modify | Edit Sketch** mode (Figure 5.35). Also, note the last **Constraint** in the **Properties** panel, the **Multistory Top Level** parameter that will allow you to duplicate the stairs you just created while applying identical parameters and positioning all the way to the top level of the project.

Sketching Boundary and Riser

If you desire more control over the footprint of the stair, you may use the **Sketching Boundary and Riser** method. One can explicitly define the **Boundary** of the stair by sketching a closed curve loop, then using the **Riser** tool to sketch in each riser. The same preview text will appear telling you how many more risers are needed to complete the stair. This gives you more control over individual tread dimensions and the overall stair outline.

Elevator Shaft Openings

In the case of elevators, one would create shaft openings in floor slabs for the elevator to move through. Instead of sketching every single opening on every floor, shaft openings are defined by a void volume that penetrates through all the floors where needed. Any additional floors you add will reflect the shaft location as long as they intersect with the void. To create a shaft opening, go to a plan view and invoke the **Architecture** tab ➤ **Opening** panel ➤ **Shaft** command to enter the **Modify | Create Shaft Opening Sketch** mode (see Figure 5.36). Use the sketching tool box to draw the shaft outline (which must be a closed loop) and complete the command. A blue volume representing the void will appear, and you will be able to use the blue shape handle arrows to intuitively manipulate the void or use the **Properties** panel to specify **Top/Base Offset** constraints or **Height** constraints. Note that you may sketch multiple shaft openings simultaneously, and the void volumes can be selected afterward for further editing. One important fact to keep in mind is that the shaft opening will only work vertically on floors and roofs; geometry such as walls and structural frames will not be affected.

Elevator components can be inserted just like any other family by loading and inserting them into the project; the shaft walls can be added just like normal walls. The wall openings for the elevator doors can be created afterward by using the **Architecture** tab ➤ **Opening** panel ➤ **Wall** command. Keep in mind that you can create the shaft walls first and then use the walls to define the shaft opening outline. Because most of the relevant dimensions for creating all the architectural elements associated with the elevator begin with the elevator component, the ideal creation

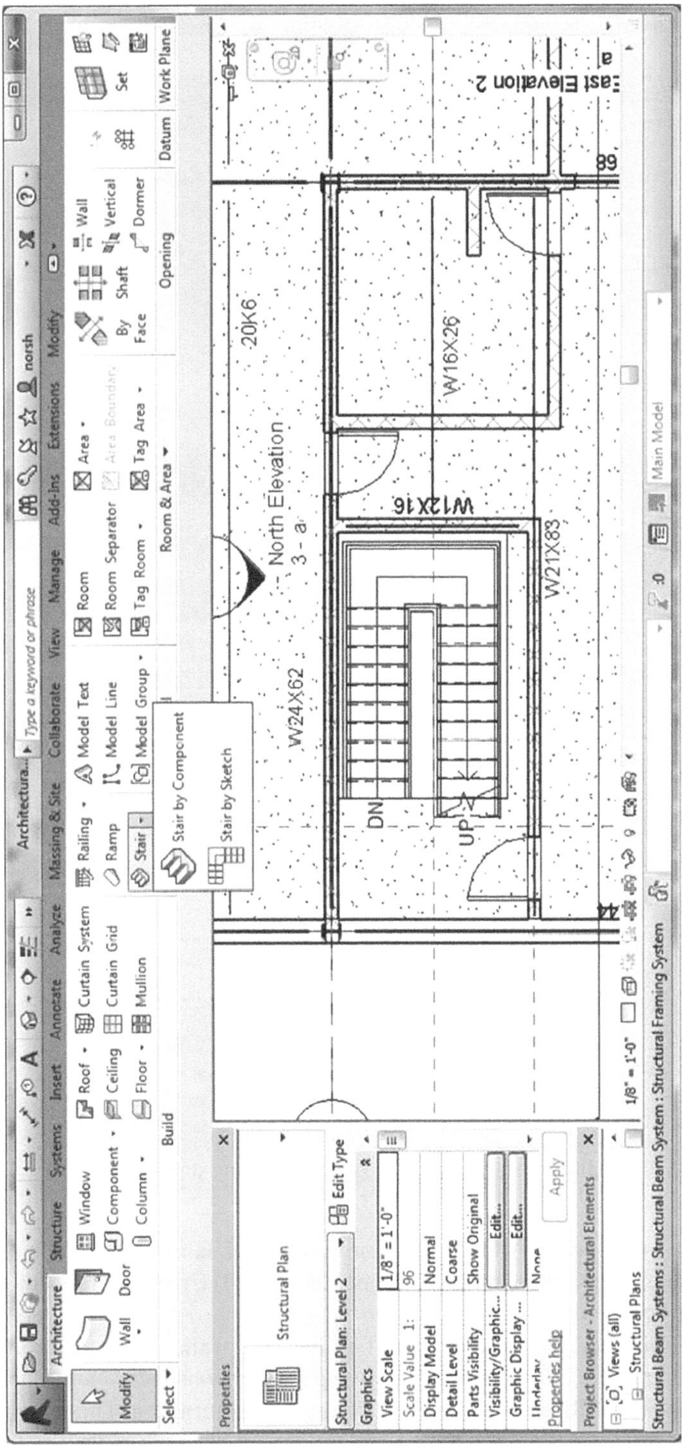

FIGURE 5.34 Stair by sketch.

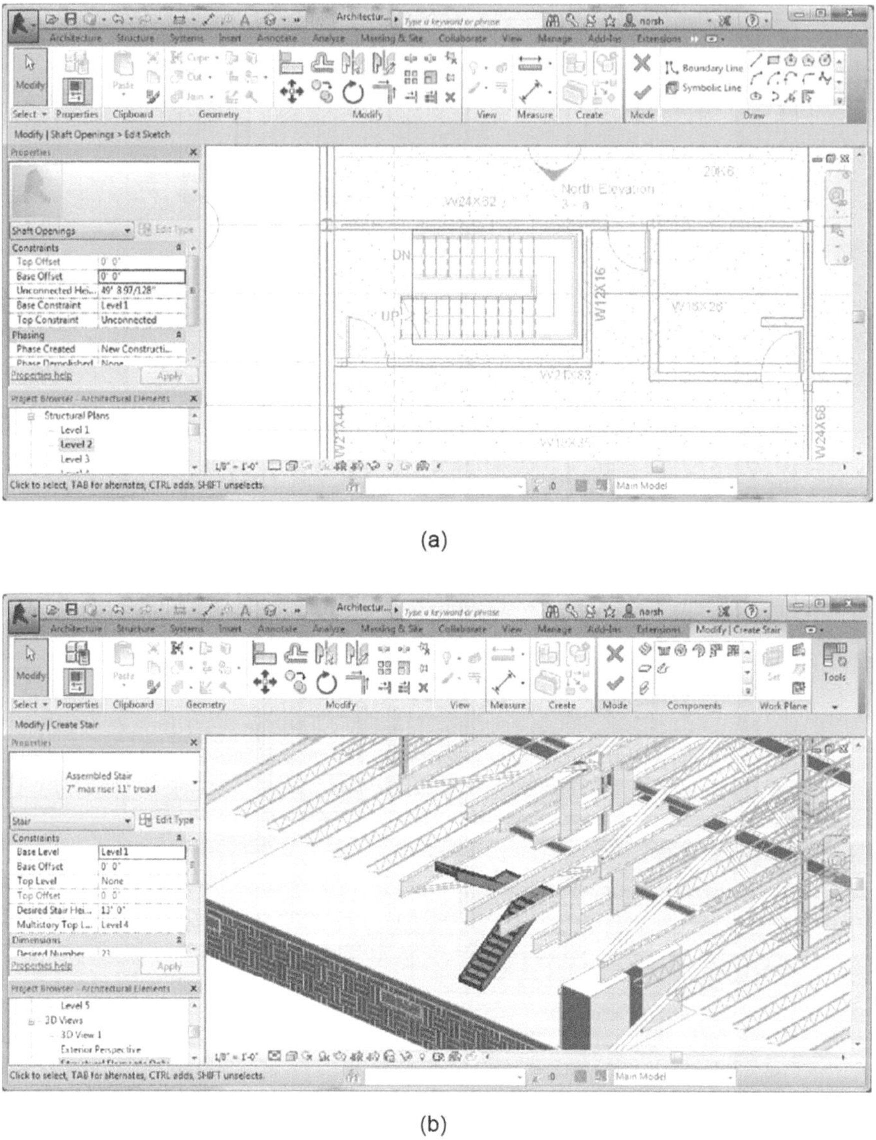

FIGURE 5.35 (a) Editing a stair using plan view. (b) Editing stair using 3D view.

sequence is as follows: (1) Place elevator component family onto floor slabs. ➤ (2) Create shaft walls and elevator door openings. ➤ (3) Create shaft opening based on the shaft walls.

Previously in this chapter the system families (i.e., walls, floors, and roofs) were covered that are project-specific; the following two sections focus on loadable families (component families) such as doors, windows, and furniture. These are families

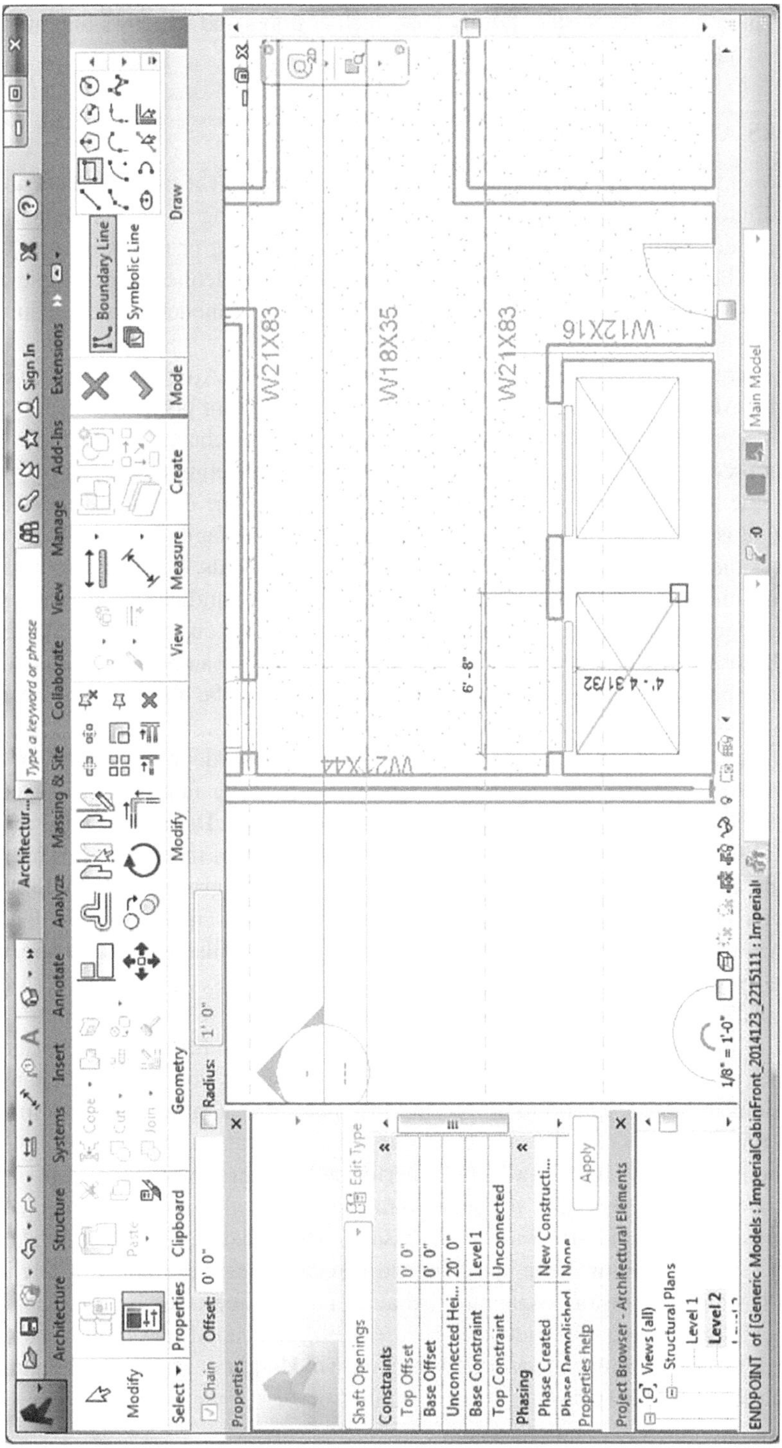

FIGURE 5.36 Creating a shaft opening.

that are more self-contained in nature and can either stand-alone like furniture or be dependent on a specific system family host, such as doors and windows that need to be hosted on walls.

DOORS AND WINDOWS

Windows and doors in Revit are wall-based hosted component families that can be inserted into walls while creating the corresponding opening in the wall automatically. Hosted families are meant to cut or create an opening or recess in their host geometry. Because these family types are parametrically defined, it is simple and easy to insert, modify, or move these elements around. For modeling a door, simply invoke Architecture tab ➤ Build panel ➤ Door.

Doors are usually inserted in the plan view (Figure 5.37). After invoking the command, the **Modify | Doors** mode is active and will prompt for placement location in the wall segment you select. Smart dimensional snapping behavior will be active to assist in precise placement, but as is always true with modeling elements, you may modify the location by dragging or clicking and editing the dynamic dimensions. Sill height is set to 0′0″ as defined by the active level by default as it usually should be unless there is a level change between the room thresholds. You can also reverse the hinge and swing directions by clicking on the small parallel arrows when you have the door selected and the **Modify** mode is active. If you toggle the **Modify | Place Door ➤ Tag on Placement** option, door tags will be automatically added as you place them; there also are additional options regarding the horizontal or vertical orientation and leaders for the tags.

By default, Revit will only load the basic **Single-Flush** door type with varying dimensions; to access other door types, the corresponding families will need to be loaded into the project. To achieve that, **Modify | Place Door ➤ Load Family** to obtain the **Load Family** dialog box (Figure 5.38a). From there, you can browse to the default Revit component families (found in the **Program Files/RVT 2014/ Libraries/Doors** folder) or any RFA family files you have created or downloaded from manufacturers. These loaded families will then be available in the **Properties Family Type** pull-down menu (Figure 5.38b).

The **Edit Type** button in the **Properties** panel will also allow access to a wide variety of modifiable door dimensions. However, if any modifications are made to the type, it is advisable to rename the family type unless you want the changed dimensions to be propagated to all present instances of that family type within the project file.

Modeling windows into a wall can be performed in plan, elevation/section, or 3D views; however, it is generally advisable to model them in the plan view and change the vertical dimensions afterward if necessary. After invoking the command, the **Modify | Place Window** mode is active, and you can place the window family into the wall; smart dimensional snapping behavior is exhibited to assist in placement alignment depending on the view (see Figure 5.39). You can always drag to modify the window placement in plan, as well as edit the **Sill Height** or **Head Height** parameters in the **Properties** panel.

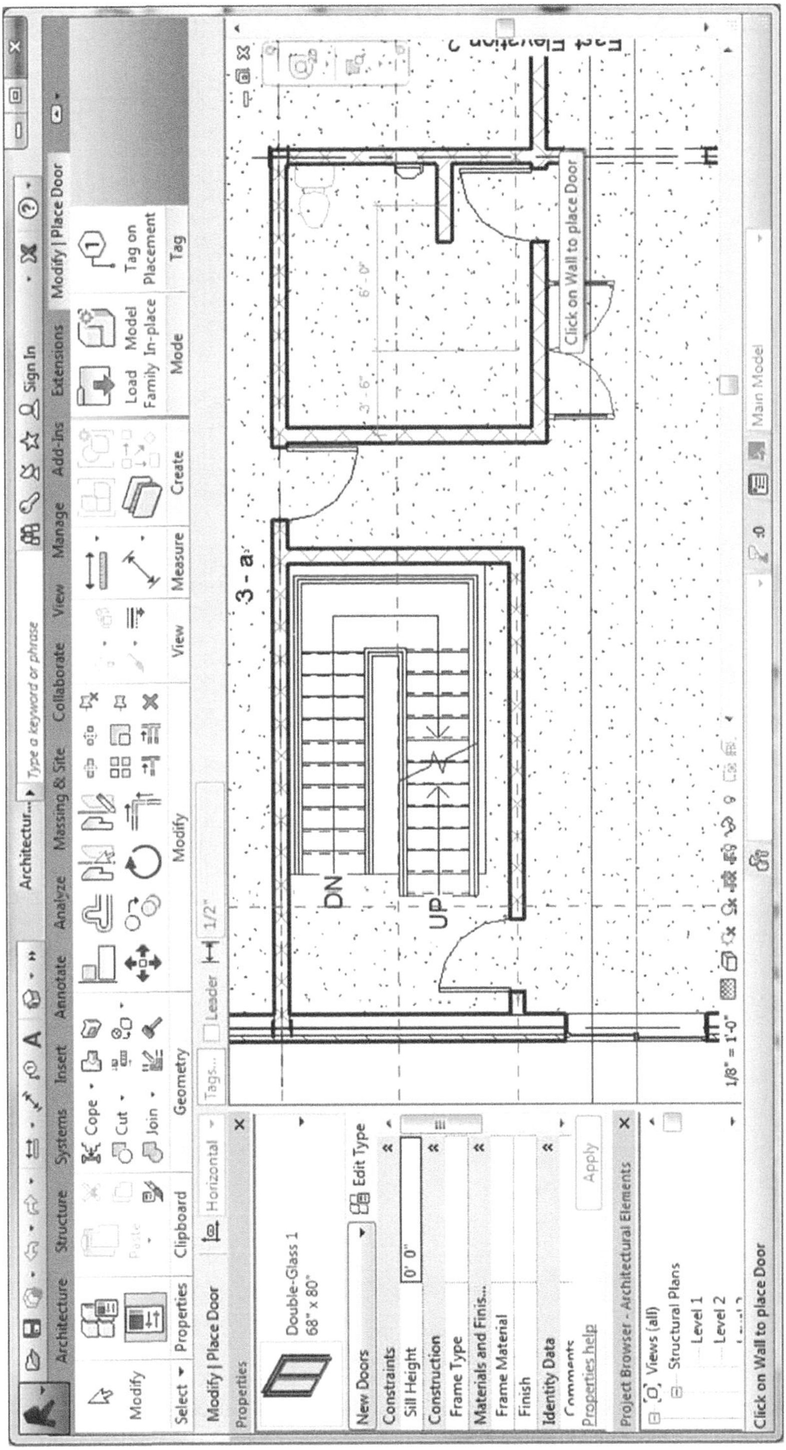

FIGURE 5.37 Modeling doors.

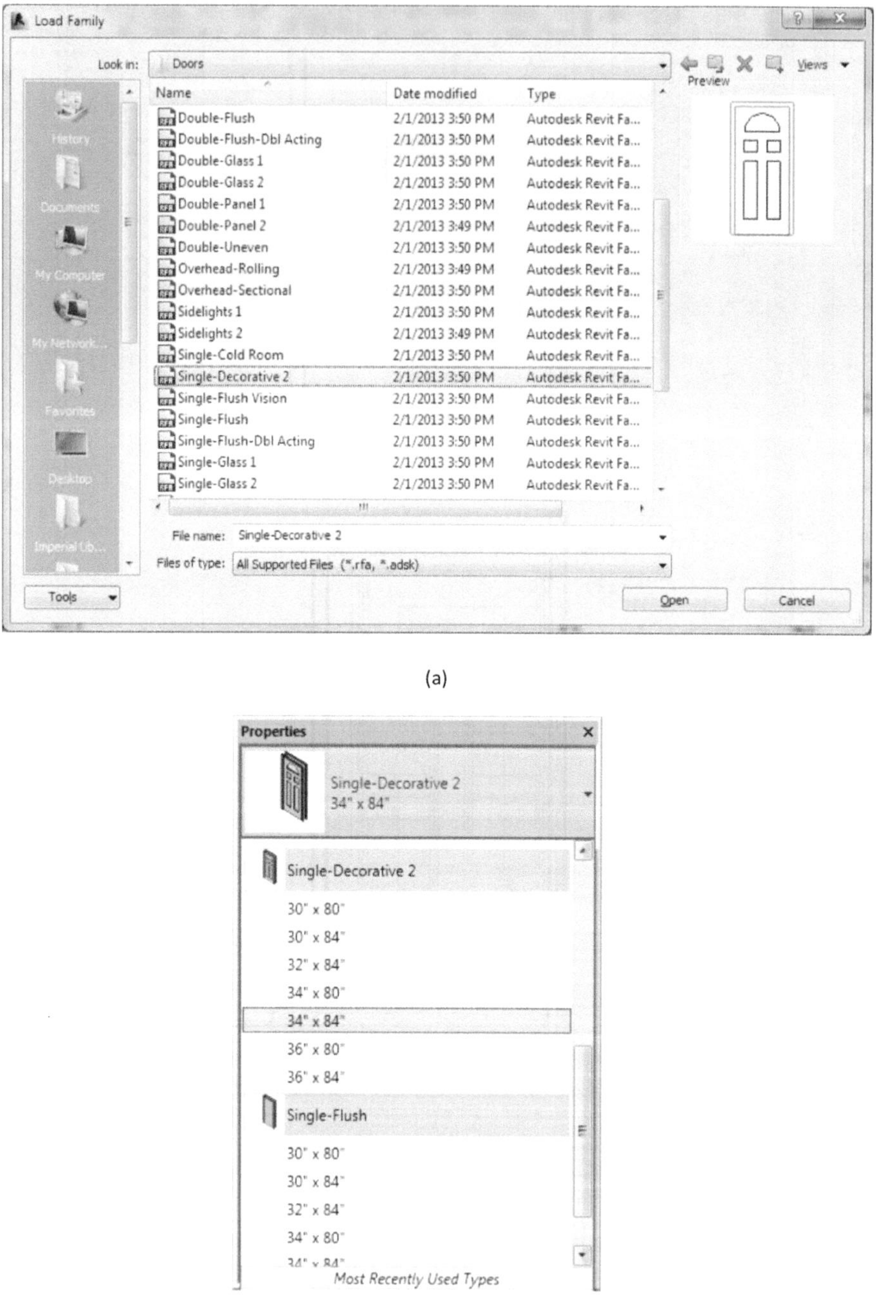

FIGURE 5.38 Loading door families/door parameters. (a) Selecting the door family to download. (b) Types of the downloaded door family.

As is the case with doors, by default Revit will only load the basic **Fixed** window type with varying dimensions; to access other window types, the corresponding families will need to be loaded into the project. To do this, **Modify | Place Window ➤ Load Family** to obtain the **Load Family** dialog box. From there, you can browse to the default Revit component families (found in the **Program Files/RVT 2014/ Libraries/Windows** folder) or any RFA family files you have created or downloaded from manufacturers. These loaded families will then be available in the **Properties Family Type** pull-down menu.

For some specific window types (i.e., slider windows with interior/exterior orientations), you can reverse the window inset direction by clicking on the small parallel arrows when you have the window selected and the **Modify** mode is active. Similar to doors, if one toggles the **Modify | Place Window ➤ Tag on Placement** option, window tags will be automatically added as you place them; the additional options controlling the horizontal or vertical orientation and leaders for the window tags are present as well.

If you need specific dimensions that are not available, you can use the **Edit Type** dialog box to modify the **Height**, **Width**, **Window Inset**, and **Material** parameters. However, it is important to rename this modified version with a new type name; otherwise, you will overwrite the current family type with the updated dimensions, as well as propagate all the changes to all instances of the component family that are present within the project.

FURNITURE

A wide variety of furniture families is available in RFA format, both in the default Revit component families and from various manufacturers online. This makes adding furniture to the spaces quick and simple, with a process similar to loading door and window families. Although for the most part furniture is freestanding and placed on a horizontal level or floor, there are some component types that can be placed on vertical walls or ceilings, such as wall-mounted shelves. Therefore, it is important to recognize the type of reference geometry that your component needs for correct placement.

To place furniture (or any component family, for that matter), invoke the **Architecture** tab ➤ **Build** panel ➤ **Component** drop-down **Place a Component** command. Revit will automatically prompt the placement of the last active component family. Choose the **Modify | Place Component Load Family** tool to bring up the **Load Family** dialog box, where you can browse to any saved RFA family files. These families can then be accessed through the **Family Type** drop-down in the **Properties** panel.

GROUPS

Groups are used for repetitive elements that cannot be organized into component families or clusters of model elements that may require exceptions in terms of placement, such as furniture. Grouping is useful when you need to create objects that

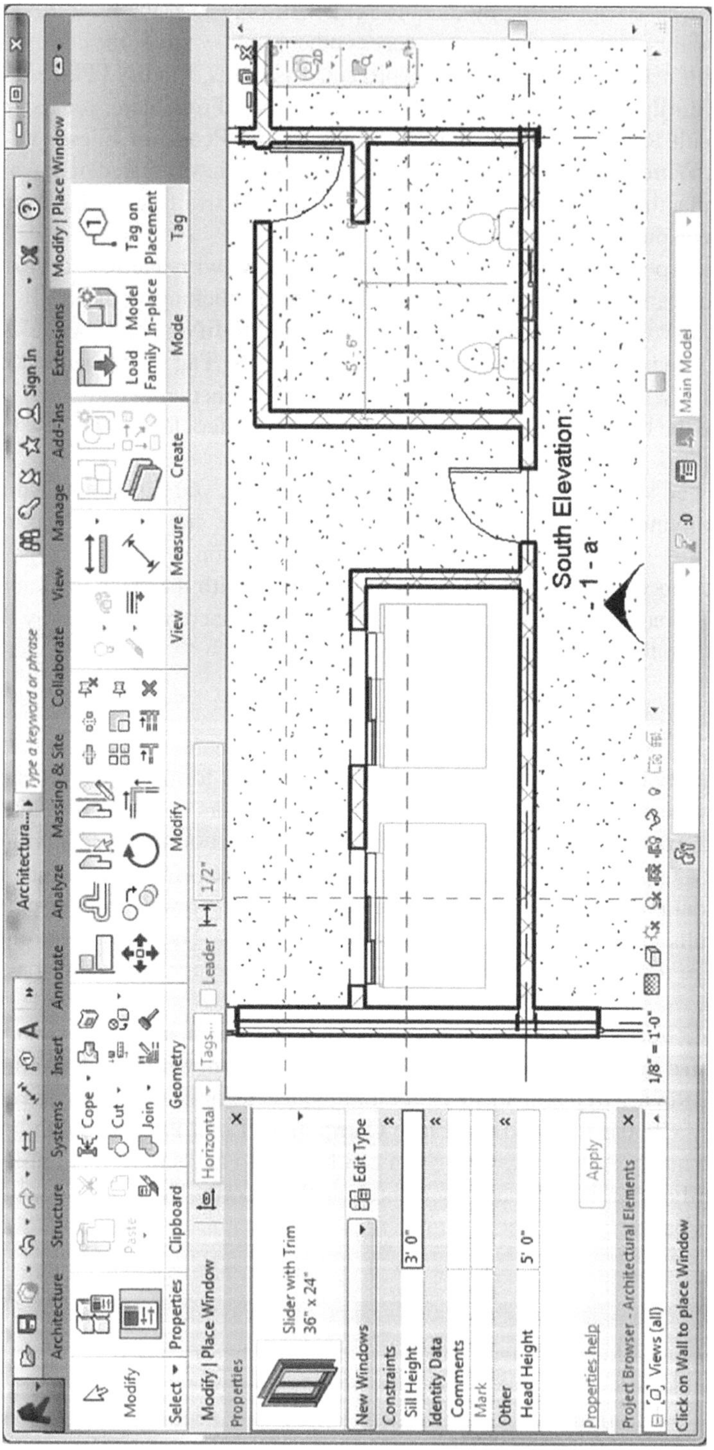

FIGURE 5.39 Modeling windows.

represent repeating layouts or are common to many building projects. There are three main types of groups: **Model**, **Detail**, and **Attached Detail** groups. **Model Groups** contain only model elements (Figure 5.40) that are visible in all views, including wall, doors, windows, and furniture. **Detail Groups** are only annotation elements that are view specific, such as tags, dimensions, symbols, and text; they are useful for maintaining consistency in annotation and documentation. **Attached Detail** groups are a mixture of both, occurring when the **Create Group** tool is used to select model elements along with the annotations that associated to it.

To create a group, invoke the **Architecture** tab ➤ **Model** panel ➤ **Model Group** ➤ **Create Group** command. Revit will give you a dialog box to name the group and to choose the **Group Type** (**Model**, **Detail**) you want to create. It is always a good idea to name your groups descriptively and accurately; it will make group management much easier as you proceed. When you complete the dialog box, the **Edit Group** toolbar will be visible in the upper left corner of the viewport, and the view is lightly grayed out (Figure 5.41). One can also preselect elements before you invoke the command; in this case, Revit will detect which type of group you are creating but will still ask you to name the group.

To add an element to an existing group, you need to launch **Edit Group | Add**. While you are in this **Edit Group** overlay mode, any new elements you create are added to the group as well. Launching the **Edit Group | Remove** command will remove elements from the group. They will not be deleted but will remain in place within the project (Figure 5.42). If you delete the element while in the **Edit Group** overlay mode, then it will be removed from both the group as well as the project. Invoking the **Edit Group | Attach** command will allow you to attach a **Detail** group to a **Model** group. Check **Edit Group | Finish** to complete editing the group; you will see the group you just created highlighted in blue, with a **Group Origin** icon in the center that can be moved or rotated. To place a group you have created, invoke the **Architecture** tab ➤ **Model** panel ➤ **Model Group** ➤ **Place Model Group** command. You will enter the **Modify | Place Group** mode, which will allow you to choose the insertion point as well as which group to insert via the pull down menu in the **Properties** panel. Alternatively, you can simply drag and drop a group from the **Project Browser** (Figure 5.43).

SUSTAINABILITY ANALYSIS

The long-term goal of green building design is to create sustainable, self-sufficient structures by optimizing resource usage and minimizing environmental impact. Enhancing building designs with this objective in mind can significantly improve resource efficiency and environmental sustainability.

Throughout the entire lifecycle of a structure, green design aims to reduce resource consumption and minimize emissions and waste. Many strategies can be employed to improve the sustainability of our building designs. There is no single best approach, but the steps typically include:

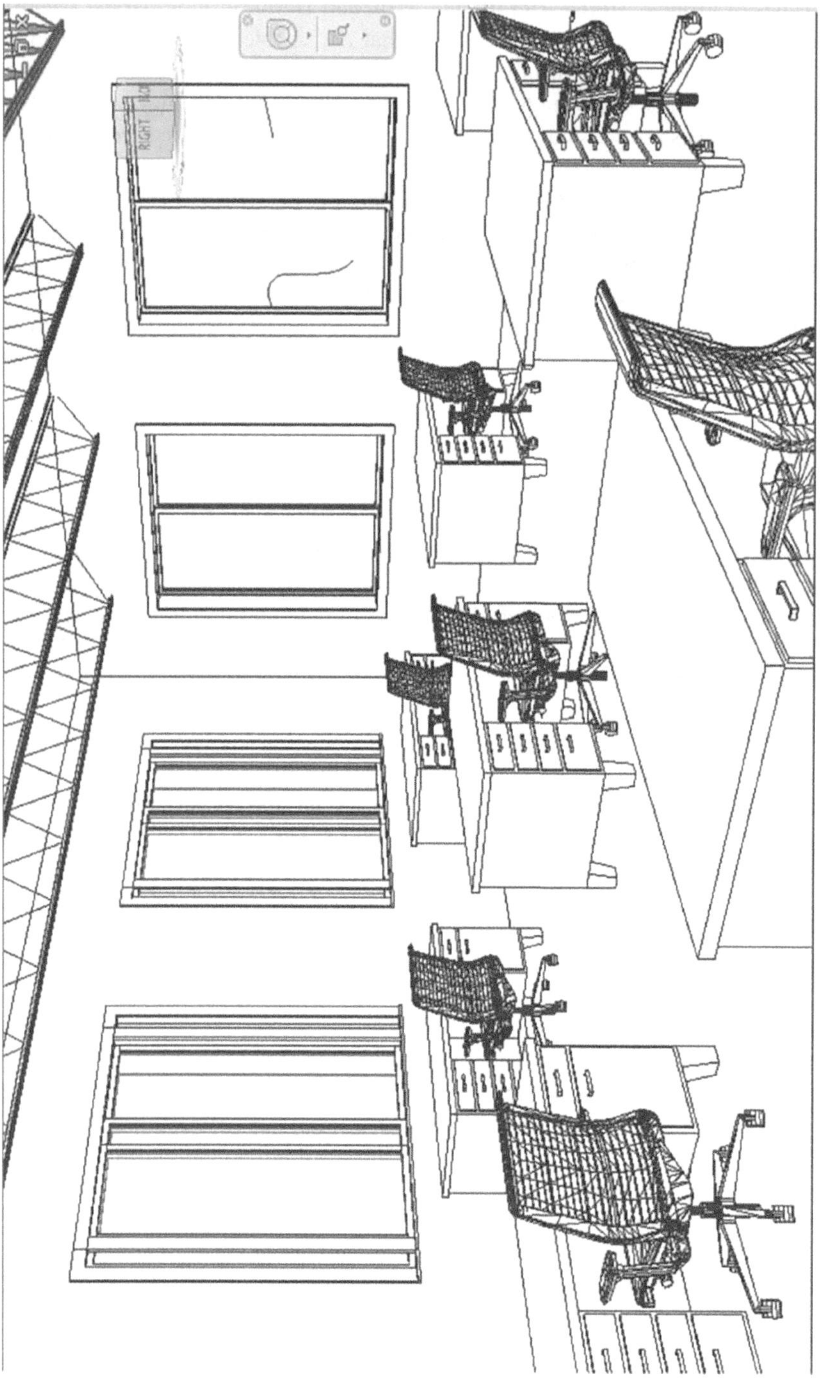

FIGURE 5.40 Example of a model group containing model elements (furniture).

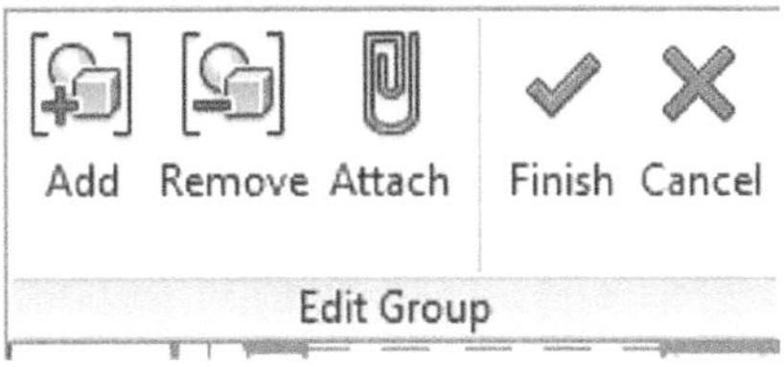

FIGURE 5.41 **Edit group** tool bar.

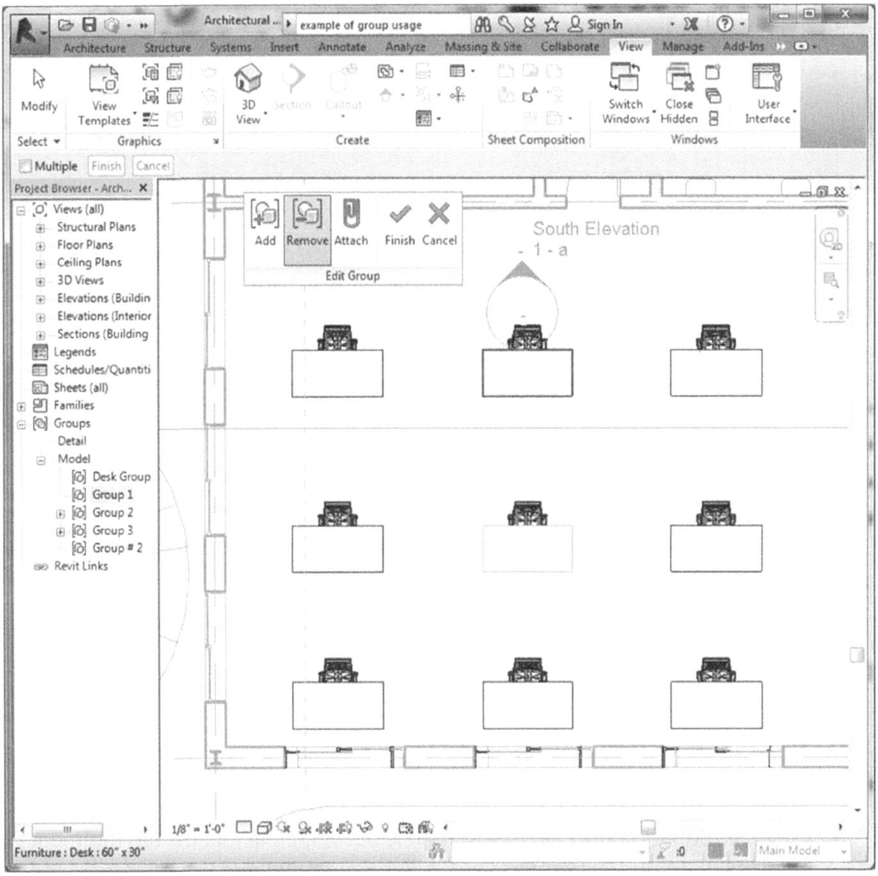

FIGURE 5.42 Removing an element from a group.

- Consider the passive architectural design features, including building placement and orientation, massing and form, and the placement of windows and openings to promote daylighting and natural heating and cooling.
- Design the building envelope using materials that minimize energy waste and help to lower the carbon footprint.

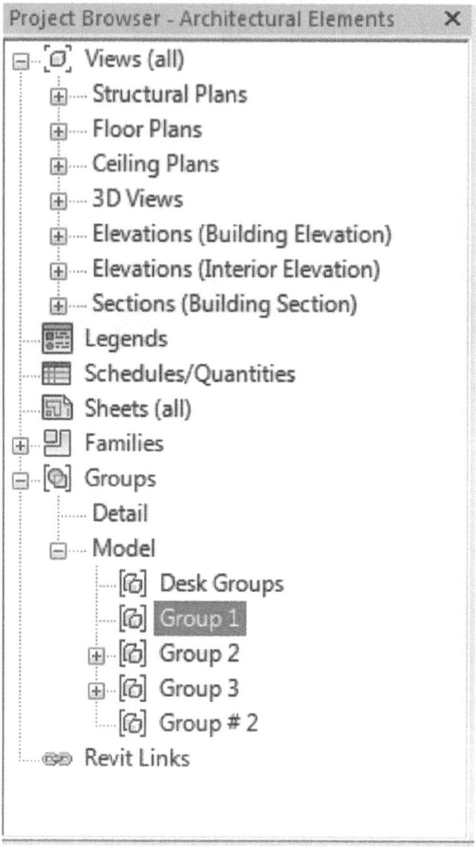

FIGURE 5.43 Placing a group mode from the **project browser**.

- Improve the efficiency of the active building systems, including lighting and power, heating and cooling, and water supply and waste, to minimize the demand for external resources.
- Use innovative strategies to incorporate renewable and sustainable energy and water resources wherever possible and to reduce reliance on external sources.
- Green buildings contribute to sustainable development and the environment and provide building owners and consumers with many advantages and benefits.
- Lower construction costs, lower running costs, greater comfort, better quality of the indoor climate, improved stability, and lower maintenance costs are characteristics of a traditional green building.

The following sections will list the most often used design strategies for sustainable buildings.

Water Use and Collection

- Estimating the water demand baseline
- Improving efficiency through plumbing fixtures
- Offsetting water use through net-zero measures

Power Use and Generation

- Estimating the electrical demand baseline
- Improving electrical efficiency through fixtures and controls
- Offsetting power use through net-zero measures
- Payback period

Daylighting & Shading

- Analyzing daylighting and shading levels provided through architectural features in the building model
- Implementing daylighting and shading design strategies

LEED Rating System

The United States Green Building Council (USGBC) created a green building rating system to help promote sustainable building practices. This system is known as LEED®, an acronym that stands for Leadership in Energy and Environmental Design. The LEED rating system, which defines standards for measuring how green a building is, encourages building designers, contractors, building owners, and product manufacturers to use sustainable practices by providing recognition for exemplary performance.

The LEED rating system evaluates buildings and assigns points across key categories. In 2009, the system was updated to expand credit opportunities in several areas and recalibrate the points required for each certification level. This revision also heightened consumer and public awareness of sustainable design, encouraging stakeholders to prioritize sustainability in their decision-making processes. Table 5.1 describes the comparison between the old and new rating systems.

LEED predominantly evaluates environmental factors, including Sustainable **Sites**, **Water Efficiency**, **Energy** and **Atmosphere**, **Material** and **Resources**, and **Indoor Environment Quality**. All of the building's lifecycle could be evaluated based on criteria from building design and construction, interior design and construction, building operations and maintenance, and neighborhood development.

The total points accumulated determine the building's overall performance rating, which falls into one of the following categories: **Platinum, Gold, Silver, Classified, or Unclassified**. Table 5.2 summarizes the level of certification and the corresponding points.

TABLE 5.1
LEED Rating Systems

Category	Points Available (Pre-2009)	Points Available (Post-2009)
Sustainable Sites	14	26
Water Efficiency	5	10
Energy and Atmosphere	17	35
Materials and Resources	13	14
Indoor Environmental Quality	15	16
Innovation and Design	5	6
Regional Priority	N/A	4
Total	**69**	**110**

TABLE 5.2
LEED Certification Level and Points

LEED Certification	Points
Platinum	>80
Gold	60–79
Sliver	50–59
Certified	40–49
Uncertified	<40

BIM and LEED

BIM has rich information that is essential for green certification and sustainability aspects that stakeholders can use in different phases of the project. BIM is also used to improve the sustainability of the built environment. For example, the amount of energy consumed in buildings can be measured using tools within BIM platforms. Different design options for sustainability can be implemented and easily traced in the BIM model.

BIM technologies are considered to provide an effective platform for attaining compliance with sustainability rating criteria. In the past few years, the application of BIM-based technologies in sustainability rating certification has attracted increasing research attention. BIM has the ability to integrate and incorporate many of the sustainability facets of the AEC industry, typically regarded as individual green building tasks. Figure 5.44 depicts the sustainability modeling aspects using BIM.

Energy Analysis

Building energy analysis can be performed by the Insight tool within the Revit platform (see Figure 5.41). It is based on the DOE 2.2 dynamic thermal whole building energy simulation engine to estimate building energy use and operating costs, which are founded on the effects and interactions of building form, materials, systems, usage, and climate (Figure 5.45).

Autodesk Revit's whole building energy analysis tools, powered by Green Building Studio, support analysis at all stages of building design and construction, from conceptual to detailed design. Figure 5.46 illustrates the different aspects of the energy analysis that can be performed using a BIM model.

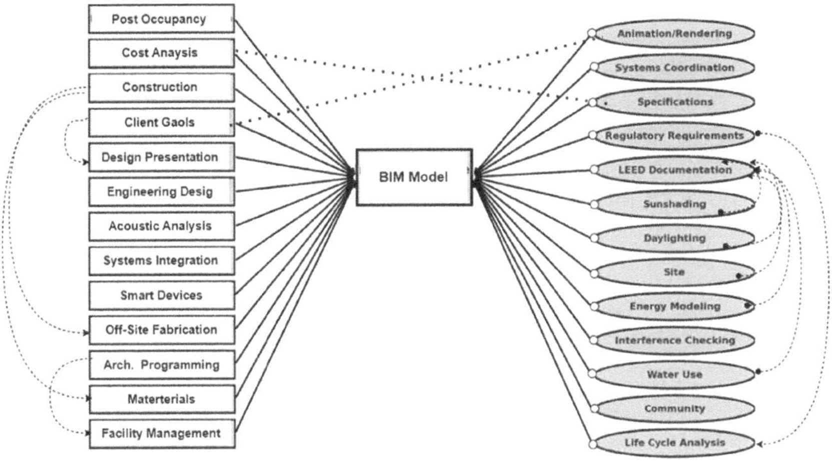

FIGURE 5.44 Sustainability modeling using BIM.

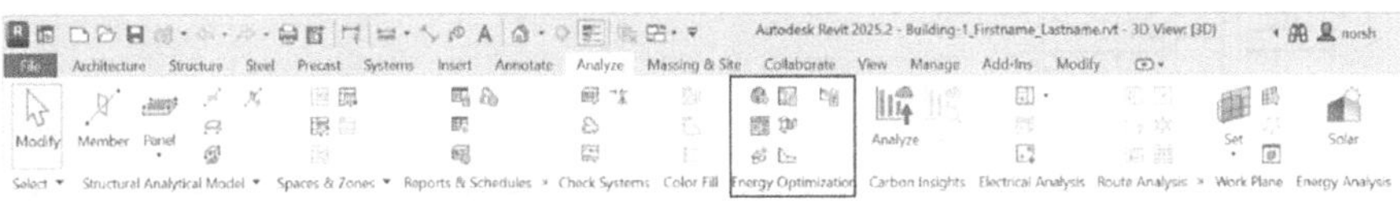

FIGURE 5.45 Energy analysis tool within Revit platform.

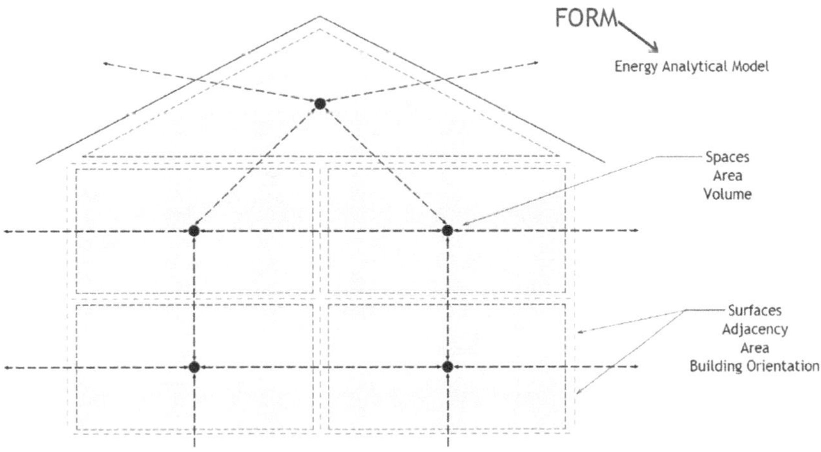

(a)

FIGURE 5.46 (a) Building form in energy analysis. (b) Building materials in energy analysis. (c) Renewable sources in energy analysis. (d) Occupancy and use role in energy analysis. *(Continued)*

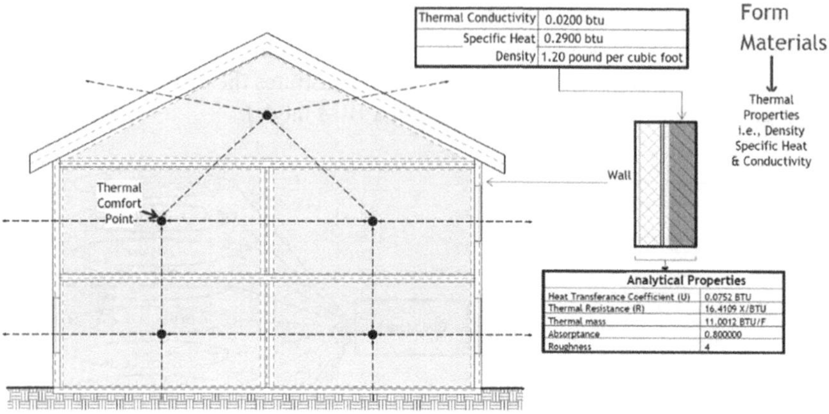

(b)

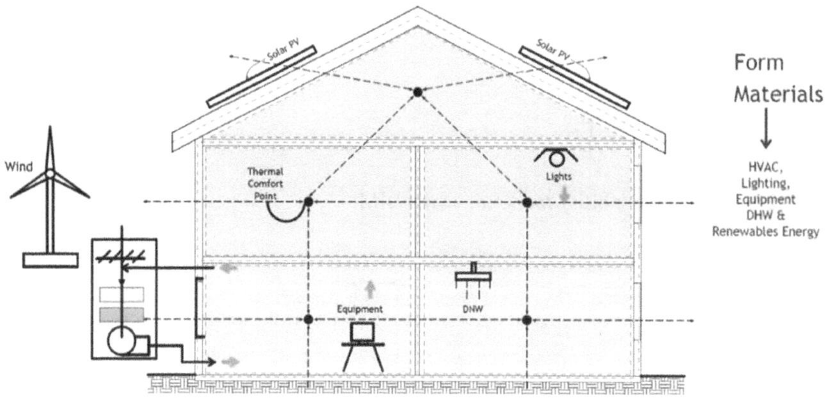

(c)

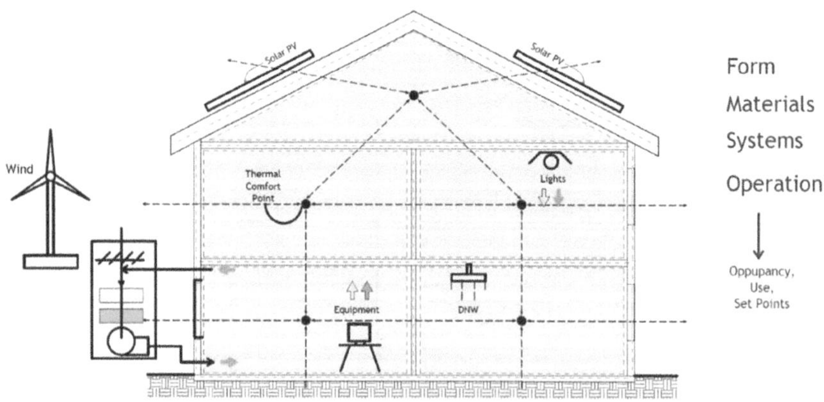

(d)

FIGURE 5.46 (*Continued*) (a) Building form in energy analysis. (b) Building materials in energy analysis. (c) Renewable sources in energy analysis. (d) Occupancy and use role in energy analysis.

There are three key workflows in which one can run energy simulations on Revit:

1. conceptual massing elements,
2. room/space elements, and
3. building elements.

Each has a different level of speed vs. relative accuracy.

Conceptual Massing Analysis: In Revit, creating conceptual masses offers a quick and efficient way to generate a basic **Energy Analytical Model**. The energy analysis is performed using **Insight**, with results readily available within the tool for further evaluation.

Room/Space Elements: In Revit, a full building model can be used to create an Energy Analytical Model. This can then be exported as gbXML and analyzed directly on Green Building Studio.

Building Elements: In Revit, the thermal properties of layered constructions—such as walls, roofs, and floors—can be integrated into the **Energy Analytical Model**. These properties are then analyzed using **Insight** in the cloud, with simulation results readily available within the tool. The following section will demonstrate this process.

To run the energy analysis, activate the **Analyze** tab, then focus on the **Energy Optimization** panel (see Figure 5.41). Then invoke the **Location** command to define the site location of the building (see Figure 5.47).

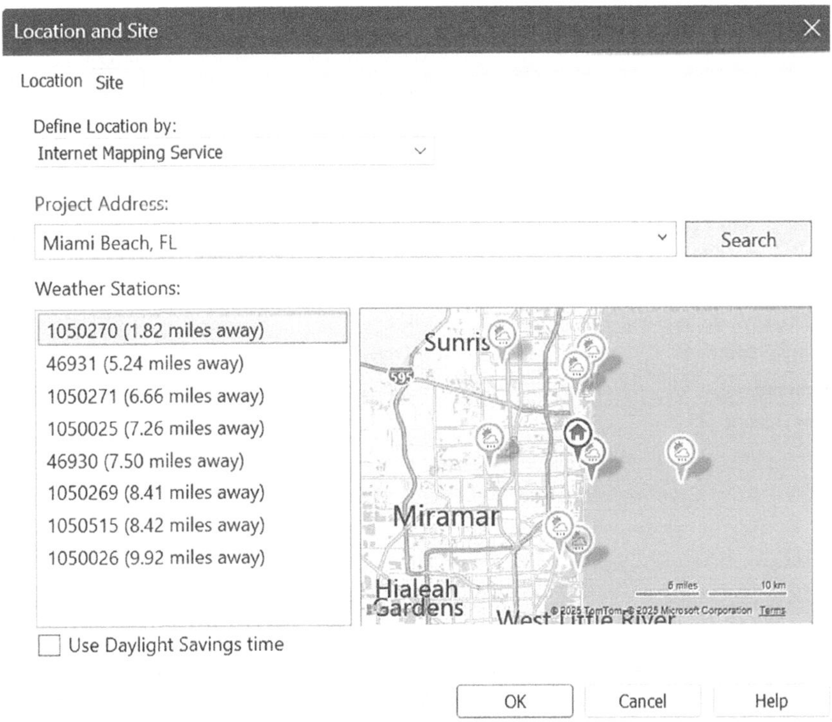

FIGURE 5.47 Site location.

After specifying the site location, you need to invoke the **Energy Settings** command to input the necessary parameters (see Figure 5.48).

The next step is to invoke **Create Energy Model** command (see Figure 5.44). After the energy model is created, one can start running the analysis. To do that, you need to activate the **System Analysis** tool within the **Energy Optimization** panel (Figure 5.49). The system analysis will run on the background and a details report will be generated. The report is comprehensive and includes:

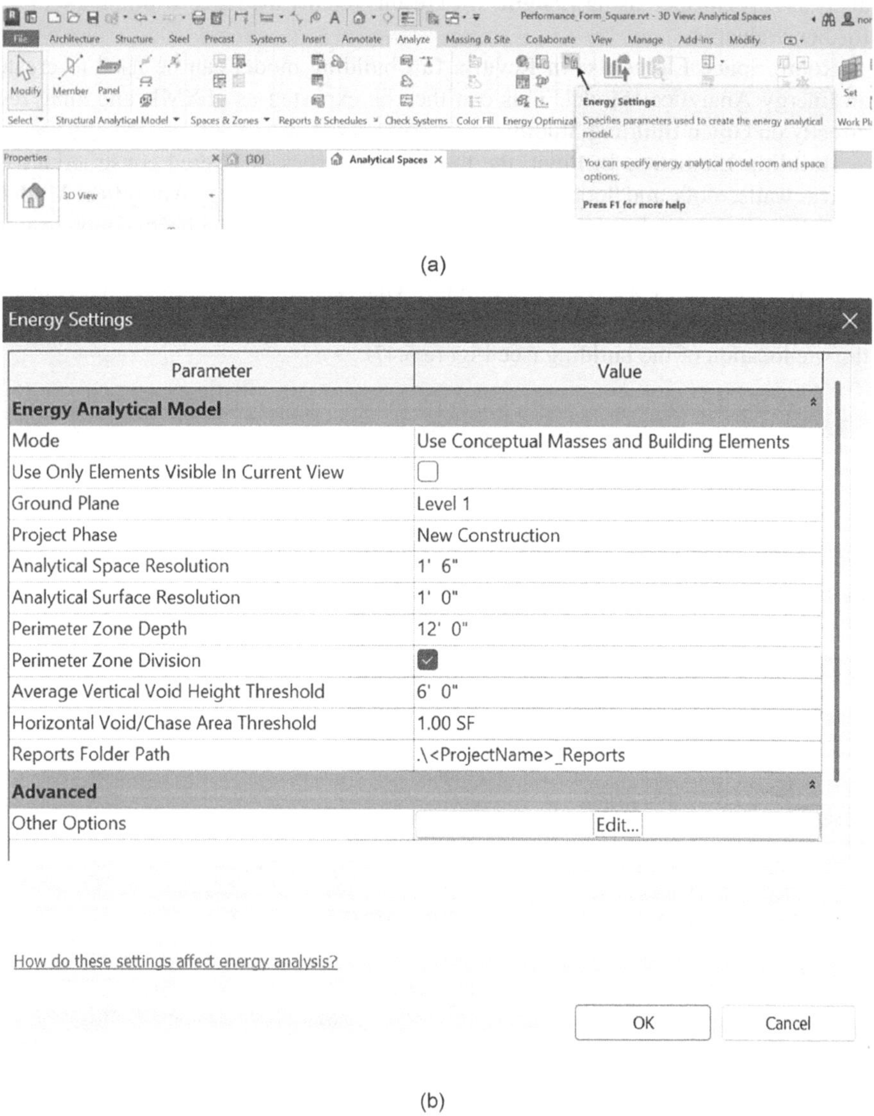

FIGURE 5.48 (a) Energy settings tool. (b) Energy settings parameters.

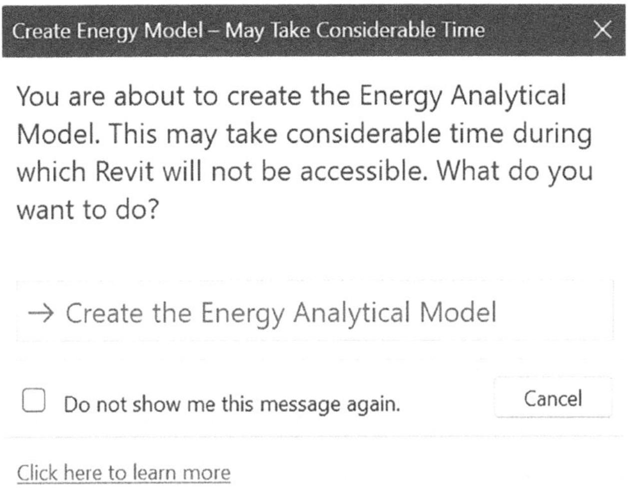

(a)

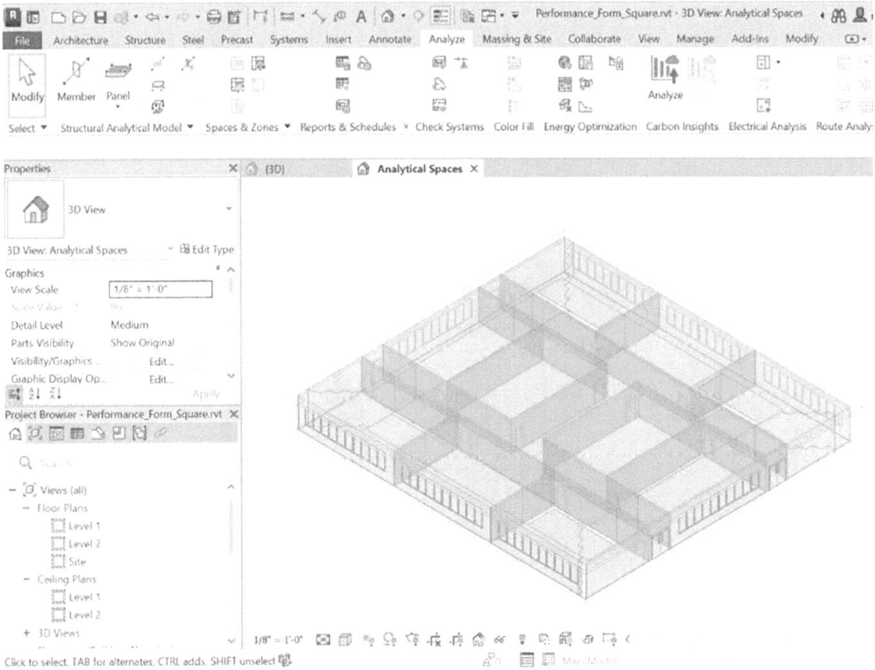

(b)

FIGURE 5.49 (a) Creating a new energy model. (b) Analytical energy model.

- Annual energy consumption overview
- Monthly overview (Figure 5.50)
- Detailed energy consumption analysis (see Figure 5.51)

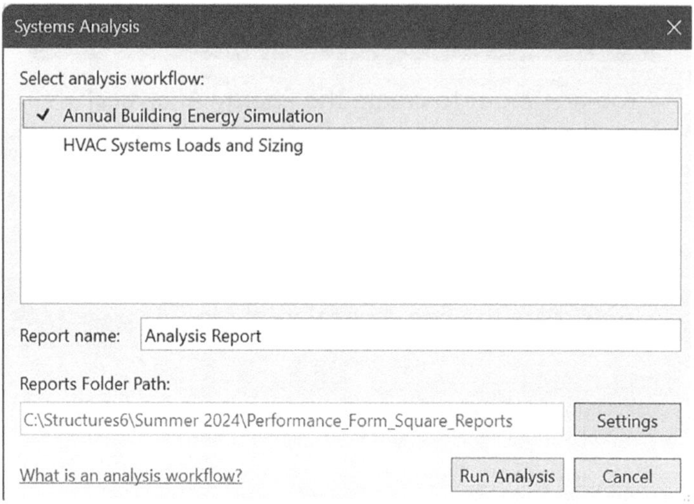

FIGURE 5.50 Energy analysis options.

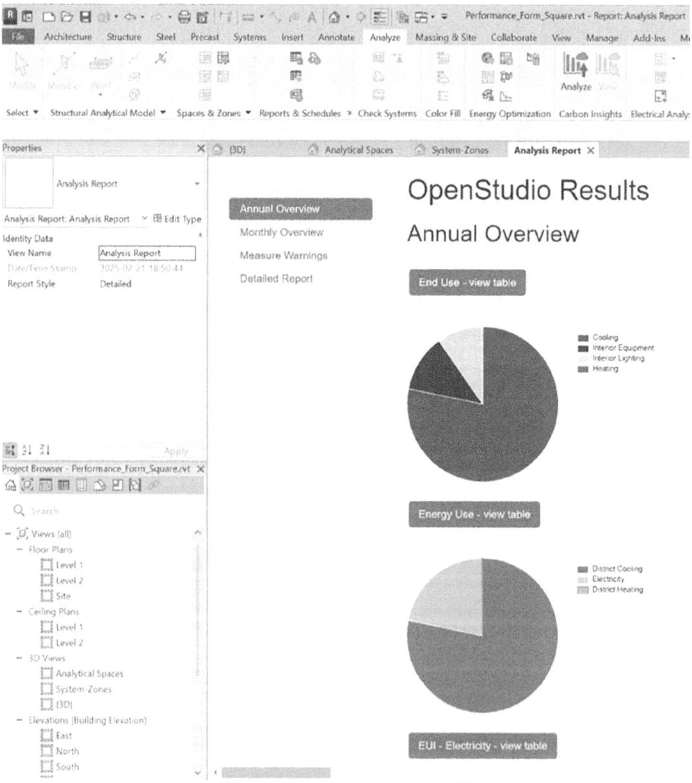

(a)

FIGURE 5.51 (a) The energy analysis report.

(Continued)

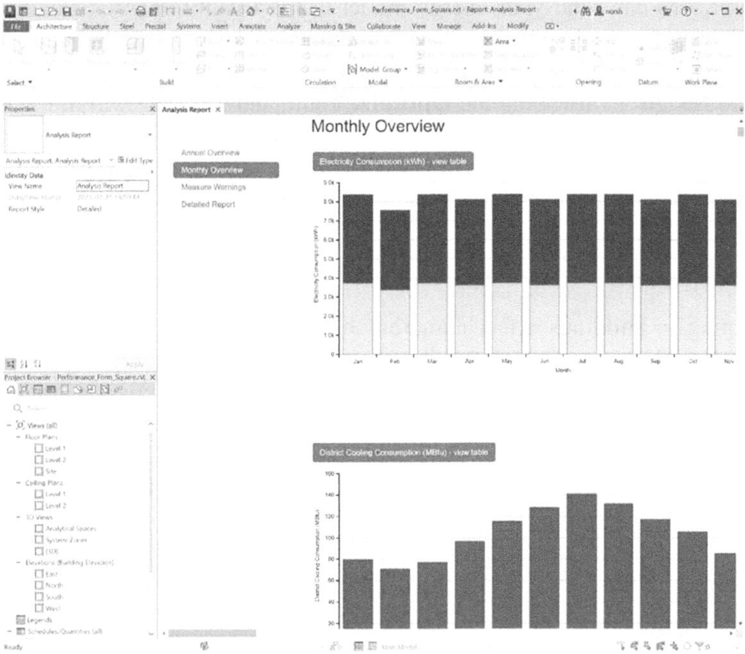

(b)

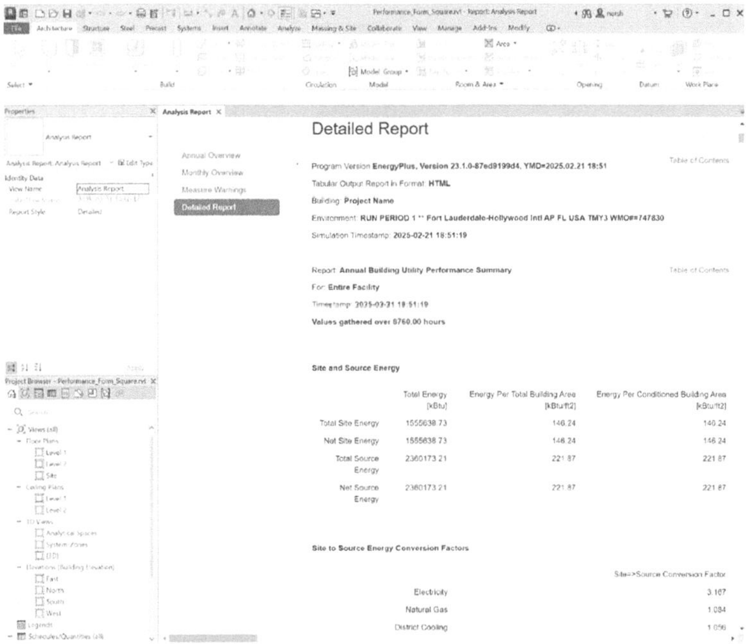

(c)

FIGURE 5.51 (*Continued*) (b) The energy analysis report – monthly overview. (c) The energy analysis report – detailed report.

Moreover, you can send your energy model to the **Insight** analysis cloud tool. At the Insight web portal, optimize the performance of your design by adjusting factors of the design and comparing different scenarios. One can send the energy model to **Insight** by invoking the **Optimize** command available within the **Energy Optimization** panel (Figure 5.52).

By running Insight analysis early and frequently during design, you can ensure your model always meets project efficiency goals (see Figure 5.52).

Lighting Analysis

Daylighting in buildings offers numerous benefits, including enhanced productivity, as natural light creates a more vibrant environment, improving user satisfaction, visual comfort, and overall performance. It also helps lower emissions by reducing reliance on artificial lighting and cooling, thereby decreasing greenhouse gas production and slowing fossil fuel depletion. Additionally, daylighting strategies can lead to significant cost savings by cutting energy consumption for lighting and cooling, potentially lowering a building's operating costs by 10% to 20%.

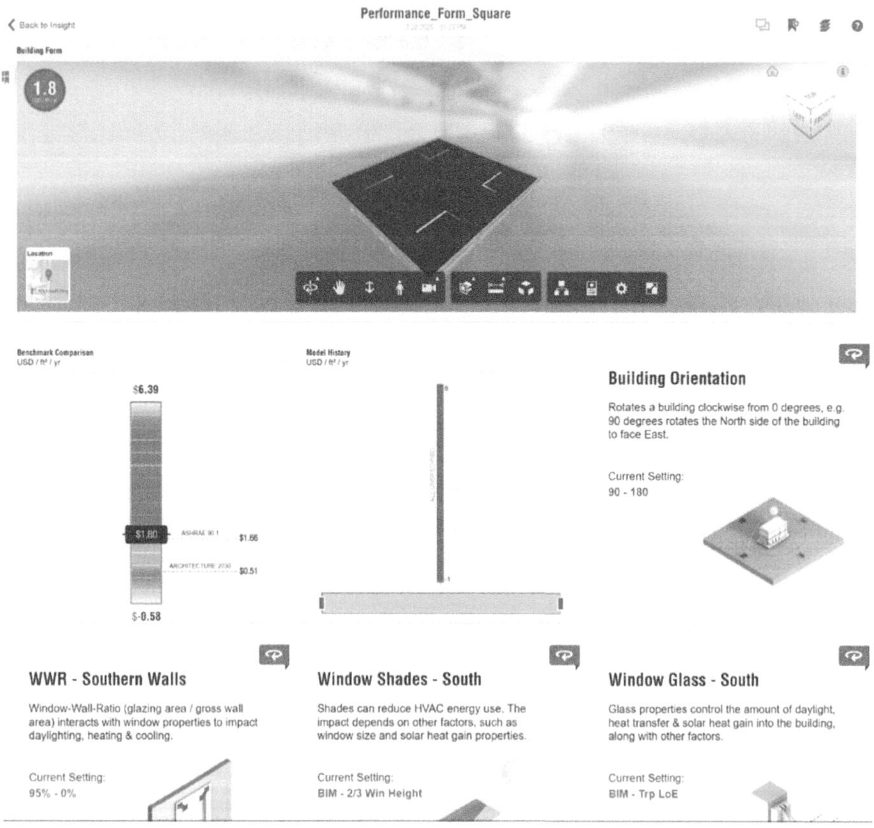

FIGURE 5.52 **Insight** analysis cloud tool.

Proper daylighting design requires balancing multiple factors, including heat gain and loss, glare control, and variations in daylight availability. Successful designs typically account for window sizing, placement, and spacing, along with the selection of glazing materials to optimize light transmission. Shading devices help minimize glare and excessive contrast, while light-reflecting features enhance light capture and distribution. Additionally, the color and reflectance of interior finishes, as well as the location and design of interior partitions, play a crucial role in maximizing daylight penetration and minimizing obstructions.

Lighting Analysis leverages Autodesk® Construction Cloud to provide the results of a lighting analysis in context. The results are applied to 3D and plan views of your model. Schedules are automatically created to track your model against lighting goals.

To perform a lighting analysis your model must be composed of building elements (walls, floors, roofs, windows, curtain walls, etc.). Then, activate the **Analyze** tab to access the **Energy Analysis** panel (see Figure 5.53a). Next invoke the **Lighting** analysis tool (Figure 5.53b) to run the analysis.

The analysis will be performed on the cloud and when completed a message will pop up in Revit platform (Figure 5.53c) indicating the completion of the analysis. Figure 5.53d depicts a summary of the daylighting results.

The definitions of the abbreviations depicted in Figure 5.53c are as follows:

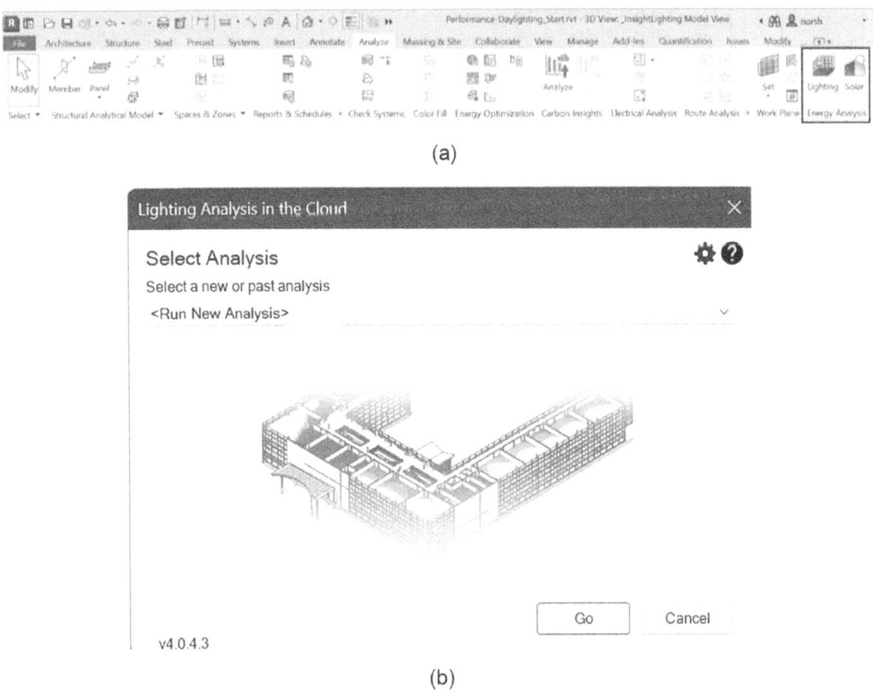

FIGURE 5.53 (a) Energy analysis panel. (b) Energy analysis tool.

(Continued)

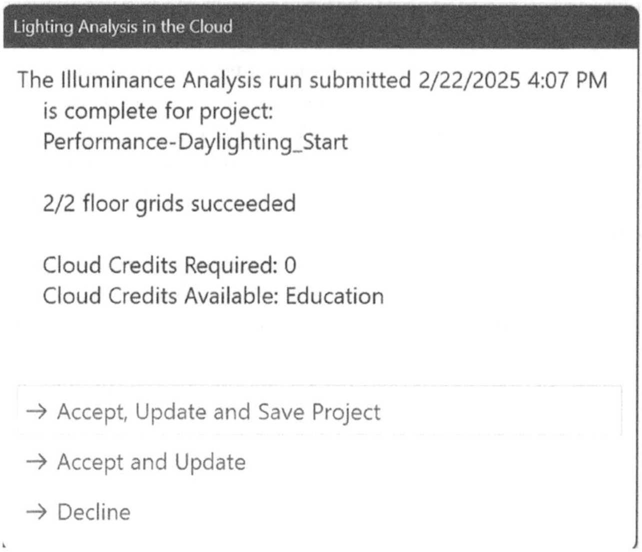

(c)

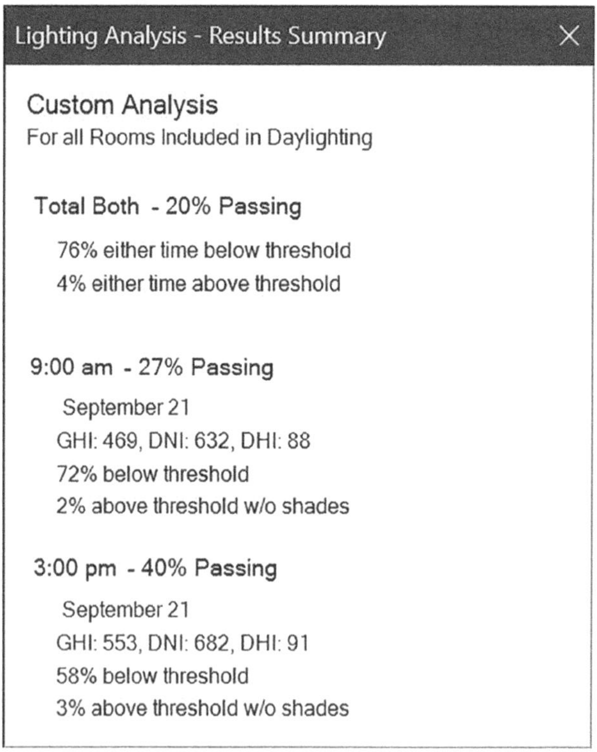

(d)

FIGURE 5.53 (*Continued*) (c) **Lighting** Analysis tool notification about completion of the analysis. (d) **Lighting analysis** tool showing the results summary.

(*Continued*)

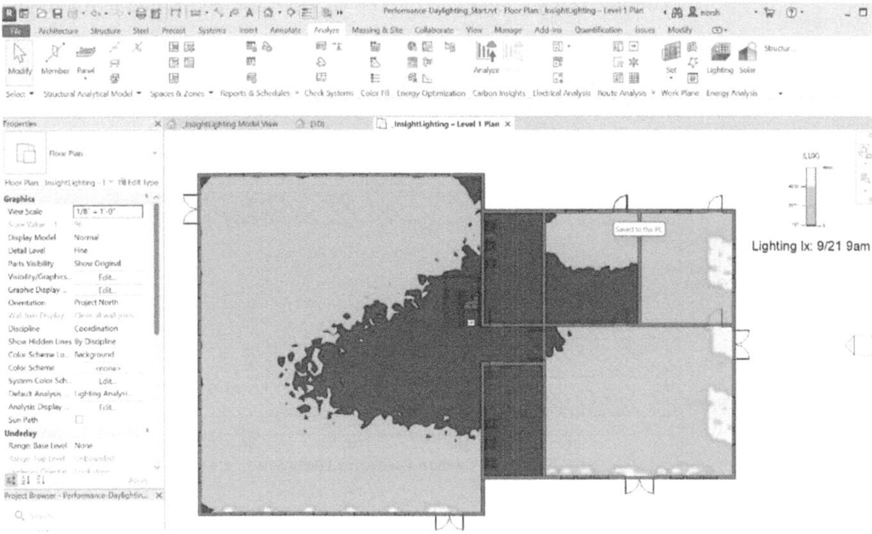

(e)

FIGURE 5.53 (*Continued*) (e) Results of the daylighting analysis on the ground floor plan.

DNI = Direct Normal Irradiance: It is the amount of solar radiation received per unit area by a surface that is always held perpendicular (or normal) to the rays that come in a straight line from the direction of the sun at its current position in the sky.

GHI = Global Horizontal Irradiance: It is the total amount of shortwave radiation received from above by a surface horizontal to the ground.

DHI = Diffuse Horizontal Irradiance: It is the amount of radiation received per unit area by a surface (not subject to any shade or shadow) that does not arrive on a direct path from the sun, but has been scattered by molecules and particles in the atmosphere and comes equally from all directions.

Figure 5.48d illustrates the daylighting illumination on the ground floor of a building. Daylighting illumination is measured in **lux (lx)**, which quantifies the amount of light per unit area. One lux is equal to one lumen per square meter (**lm/m^2**). In the US customary units, it is measured using **foot-candles (fc)**, where **1 foot-candle = 10.76 lux**.

Total Carbon Analysis

Early-phase iteration and design analysis, including energy modeling and embodied carbon assessment, play a crucial role in reducing design decisions' environmental impact and advancing the built environment's decarbonization. **Embodied carbon** refers to the carbon footprint associated with construction materials, while **operational carbon** stems from energy consumption throughout a building's lifespan (see Figures 5.49 and 5.54).

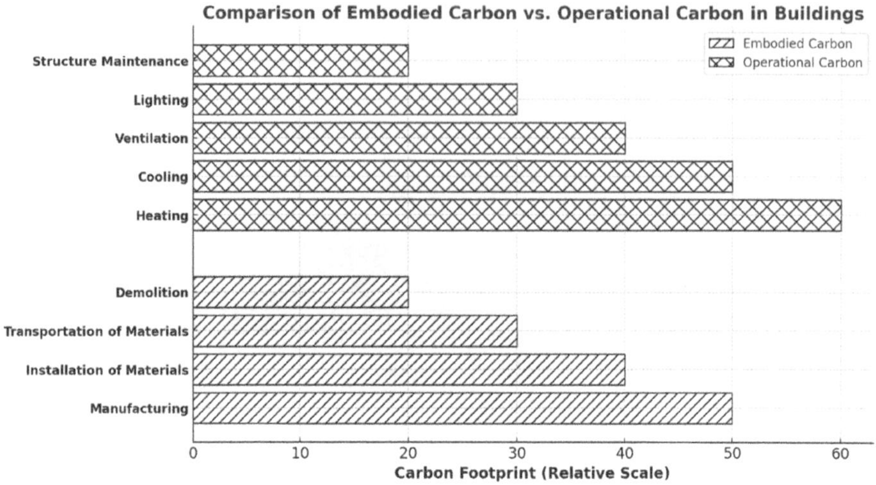

FIGURE 5.54 Embodied and operational carbon emissions.

EXERCISES

5.1. Using the BIM model Assignment5-1.rvt from the companion website (http://www.crcpress.com/product/isbn/9781482240436), perform the following tasks:
- Open the file and rename it to Problem 5.1; then,
  - Add interior and exterior doors to the building model at the locations indicated at the underlay drawing. The door types and sizes needed are shown in the legend that appears in the plan view.
  - Add windows to the east exterior wall at the locations indicated in the underlay drawing.
  - Use the window types and sizes shown in the window type legend that appears in the plan view.
  - Set the head height property for all windows to be 7 ft.

5.2. Create a BIM model for the roof vaults' structures shown below.

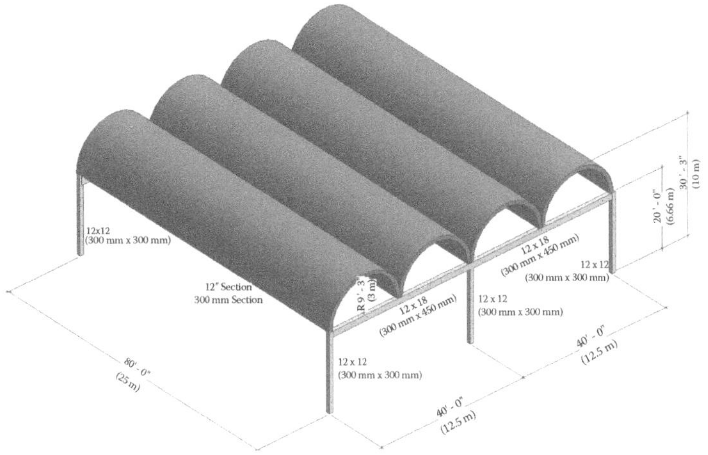

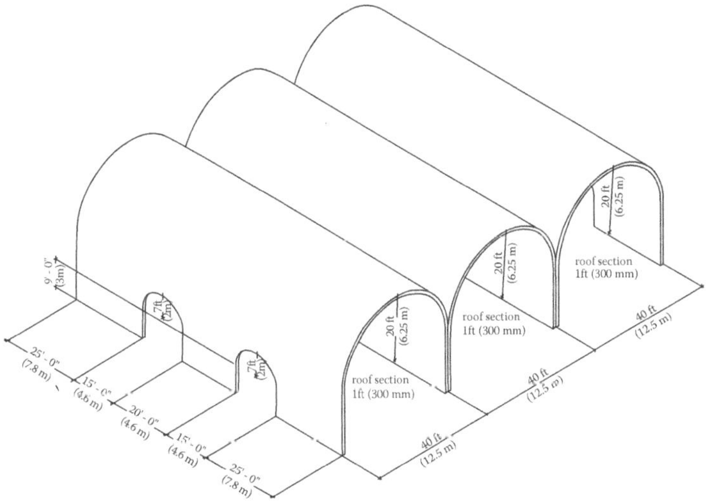

5.3. Using the architectural and structural plans given below, perform the following modeling tasks:

a. Create A grid system along with columns and walls.
b. Add floors and roof.
c. Add interior and exterior doors to the building model at the locations indicated in the given plans.
d. Add windows at the locations indicated in the given elevations.
e. Add staircases at the locations indicated in the given plans.
f. Add beams and beam systems as shown in the structural framing plans.
g. Add a structural floor (concrete on a metal deck) and roof (metal deck).
h. Add a curtain wall at any appropriate location.

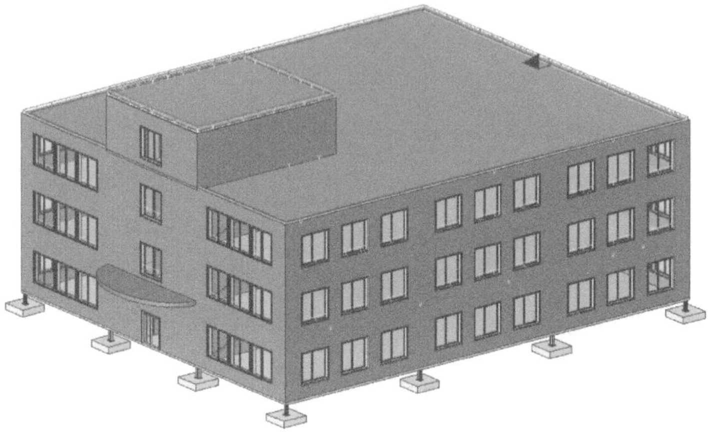

3D view

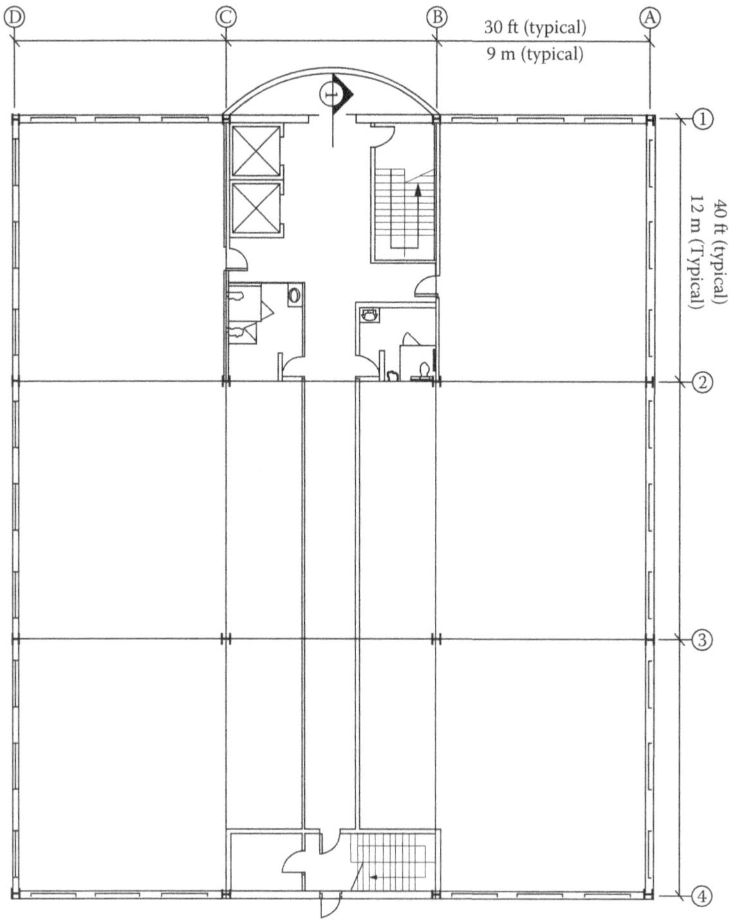

1st Floor plan

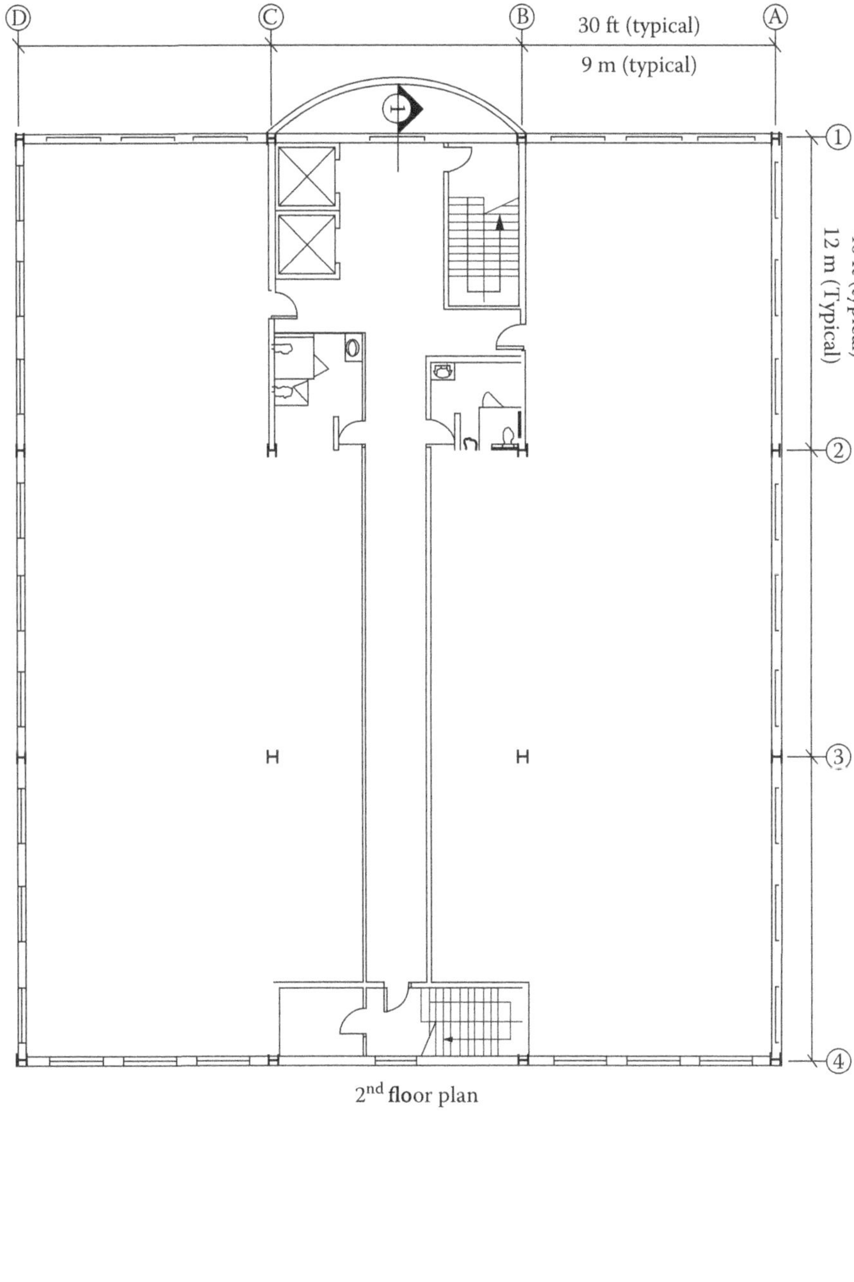

2nd floor plan

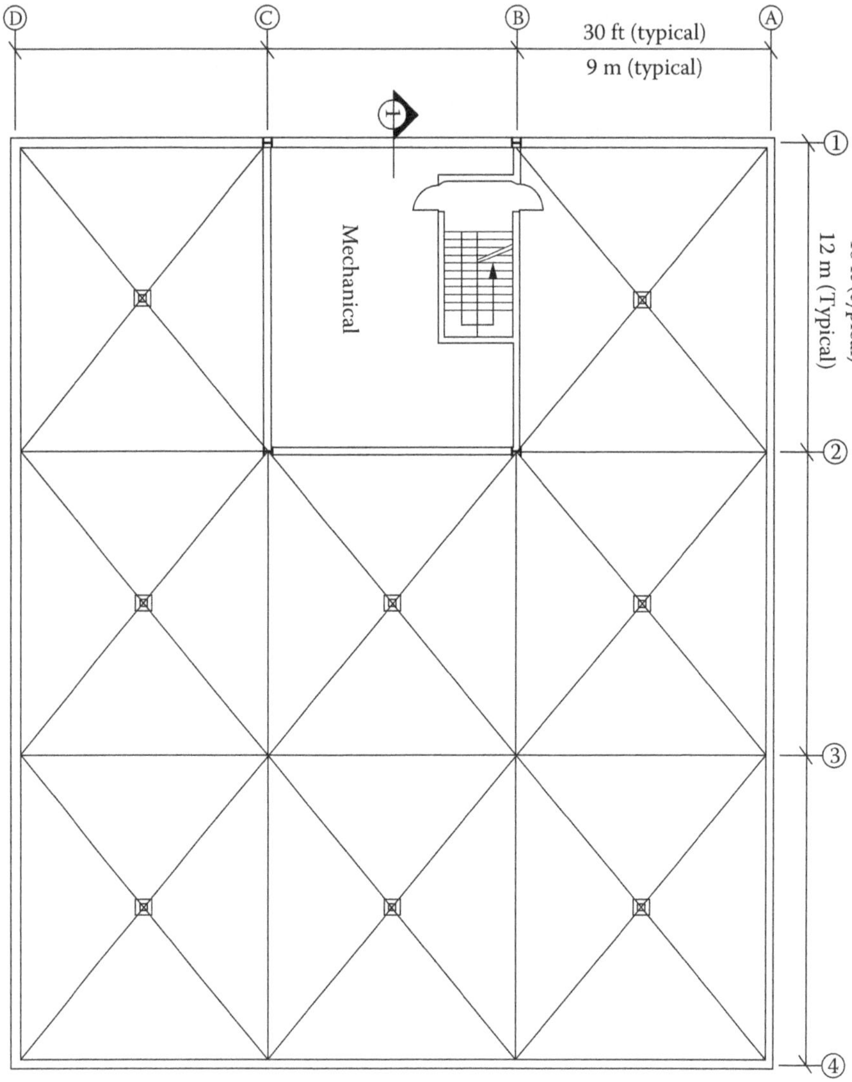
D
C
B
A
30 ft (typical)
9 m (typical)
1
Mechanical
40 ft (typical)
12 m (Typical)
1
2
3
4

Roof plan

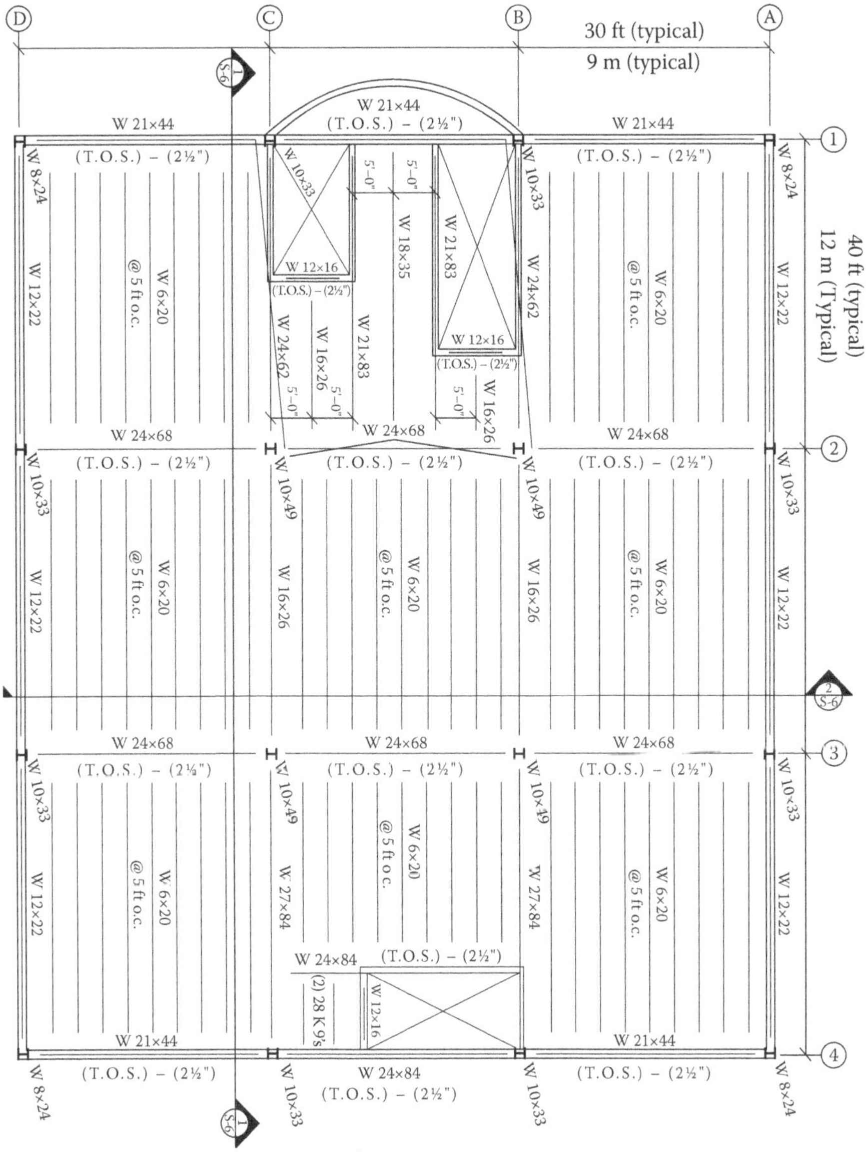

2nd Floor framing plan

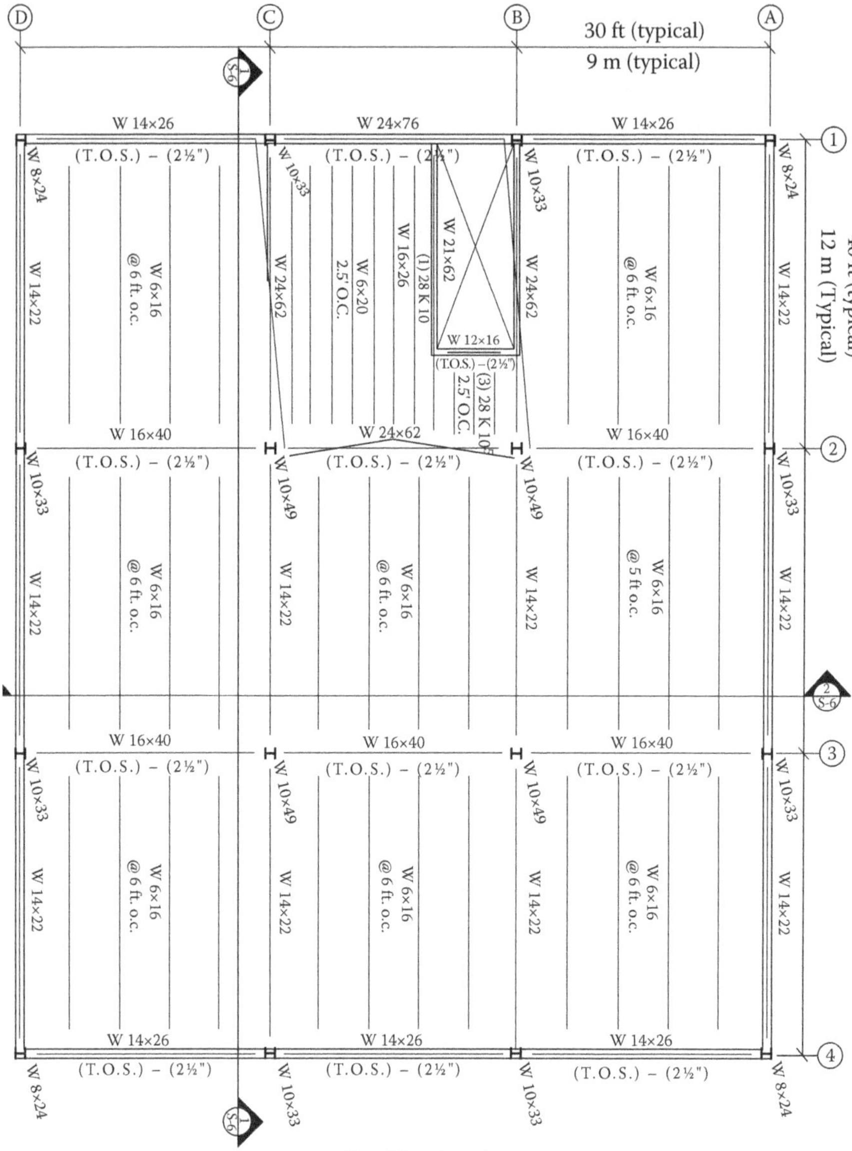

Roof framing plan

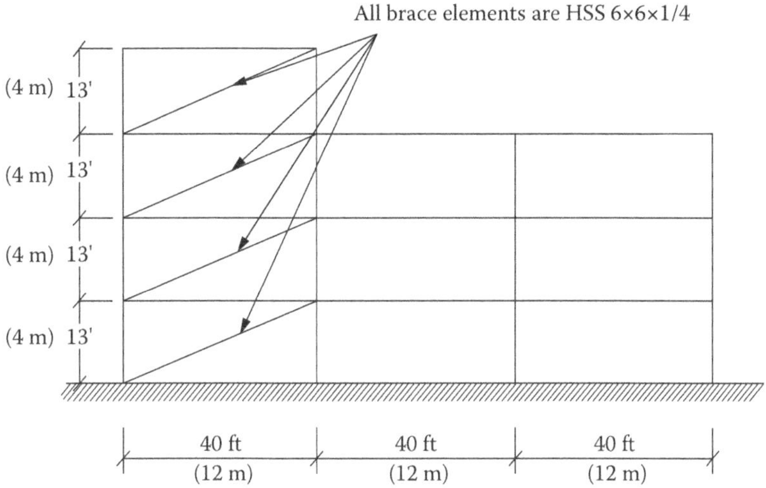

N-S Section framing elevation

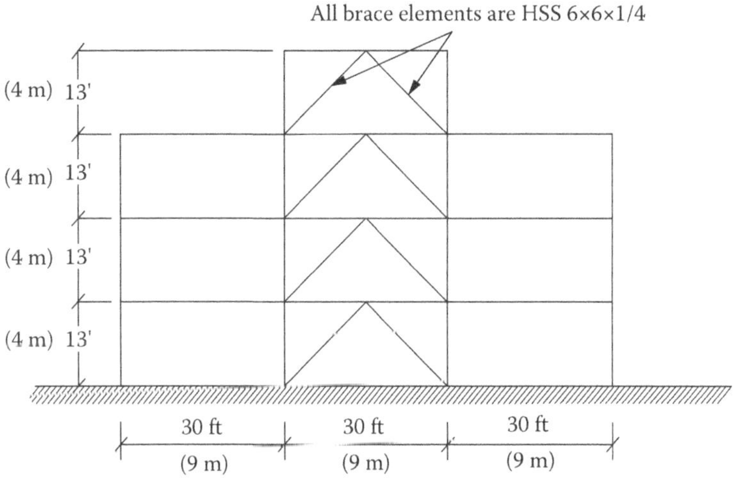

E-W section framing elevation

5.4. Using the BIM model developed in Exercise 5.3, add the following:
   a. Create a spiral stair with 15 risers and run radius of 2 ft 0 in (0.60 m) to connect the first and second floors of the office building at the central bay of the building.
   b. Open the **Stairs** tool and duplicate the **Residential – Open Riser** type; rename the new type to **Residential – Open Riser – Spiral**.
   c. Set the maximum riser height to 10 in (0.25 m) and the minimum tread depth to 11 in (0.28 m).

d. Using the curved run line option, try sketching a spiral stair using this type. The sketch cannot be completed because the rotation required exceeds 360°.
e. Change the instance properties for this stair to set the **Actual Tread Depth** property to 10 in (0.25 m). This value will override the minimum value specified in the type properties.

5.5. Using the BIM model developed in Exercise 5.3, perform the following tasks:
a. Place the **Electric_Lift** elevator component in the studio near the stair.
b. Add 6-in (0.15-m) generic walls to the first and second floors as needed to enclose the elevator with shaft walls.
c. Use the **Shaft Opening** tool to cut an opening from the first floor to the roof within the shaft walls and place wall openings on the appropriate shaft wall to provide access to the elevator.
d. Provide interior rendering of these elements using cameras and walkthroughs.

5.6. For the model of Exercise 5.4, add toposurface by adding points manually and then add various site features that include trees, building pad, cars, street, and sidewalks.

5.7. Download and open the model Massing-Pr5–6.rvt from the companion website (http://www.crcpress.com/product/isbn/9781482240436). Using the site boundaries given in the file, select three site plans; then,
- Create a different mass model for each. Be creative in your conceptual mass and include curvilinear and linear forms.
- Add mass floors and floors (minimum 15 floors).
- Add curtain systems and walls.

5.8. Create a new family **Curtain Panel Pattern Based**. Using the **Family Editor**, create any form *other than a pyramid* for your panel pattern. Test your parameters and make sure the family is parametrically working. Save the family and upload it into your open project (Exercise 5.7 project). Select one of the masses you created in the previous assignment (Exercise 5.7) and edit it. Select one surface of that mass and divide it into panels. Apply the family **Panel Pattern** to that surface.

5.9. What are the possible alternatives if Revit reports that it cannot create a roof by footprint using the boundary sketched?

5.10. What would be the method for creating a barrel vault roof using Revit? Describe it briefly.

5.11. Using the **Conceptual Mass Family Editor**, create a mass family for the mass shown below. The following parameters are required for the mass family:
- Parameters for the width and height of the base plan rectangle
- Parameters for the width and height of the top plan rectangle
- Parameter for the height of the mass
- Parameter for the material of the mass

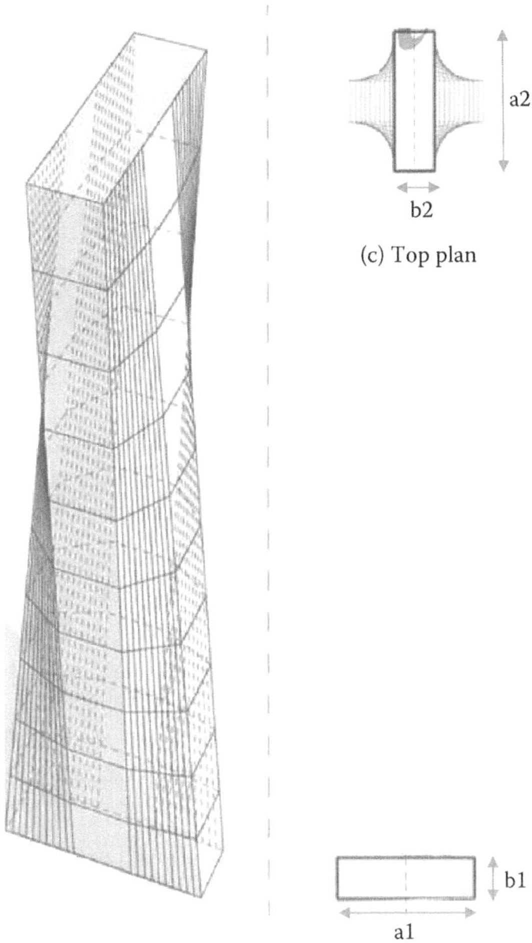

(a) 3D view (b) Bottom plan (c) Top plan

5.12. Repeat Exercise 5.11 for the masses shown below.

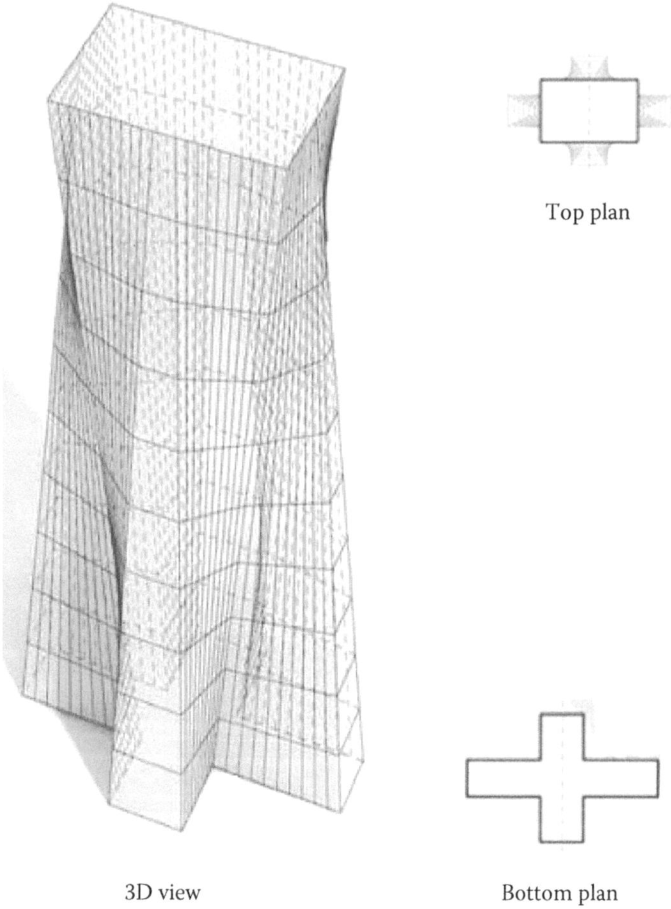

5.13. What strategies can be used to reduce the power density of a building?
5.14. How do lighting controls and sensors work to reduce energy demand?
5.15. How can you encourage a building's users to become active participants in reducing power demand?
5.16. Why is payback used as a driving consideration in determining how many photovoltaic panel areas to provide? Why not cover the entire available area with panels?
5.17. What are the primary factors that determine the payback period for installing photovoltaic panels? How can changes in the values assumed affect the results of our payback analysis?

6 Structural Analysis

INTRODUCTION

Structural analysis within the structure and architecture synergy (SAS) framework establishes the notion that structural analysis computation is a primary tool not only to verify structural decisions but also to provide assistance for design strategies. The structural analysis is thus an integral part of the overall building design.

Structural analysis is conducted to verify that any structure must possess the following:

Strength and Stability: It must be stable and strong enough (i.e., provide necessary strength) to keep the building up under any type of load action so it does not collapse on either a local or a global scale (e.g., because of tension, compression, bending, buckling, instability, yielding, fracture, etc.). The structure makes the building and spaces within the building possible; it also gives support to other building systems, such as mechanical, electrical, and plumbing systems.

Serviceability: It must be durable and stiff enough to control the functional performance, such as excessive deformation, vibrations, and drift.

In building structures, applied loads flow along the members and joints to the external supports (i.e., foundations). Because the applied loads cause internal forces in the members and joints, one can visualize the supporting structural elements as substituted by equivalent, multifaceted three-dimensional (3D) force systems. In other words, under various external load actions, such as live load, wind, seismic forces, thermal stresses, vibration, or settlement, the structure responds by deforming and changing form, thereby causing strains in the members, which in turn correspond to internal forces denoted as stresses or stress results.

The structural framework of a building in general is located in a spatial grid and defined by a single global Cartesian coordinate system *XYZ* that follows the right-hand rule, in which the thumb is the *X* axis, the index finger is the *Y* axis, and the middle finger is the *Z* axis. The *Z* axis is the vertical axis, and the *Z*–*X* plane is the vertical plane (see Figure 6.1). *Z* is positive upward and negative downward.

Each element of the structure (beams, columns, trusses, slabs, surface objects, etc.) in turn has its own local coordinate system, referred to using lowercase letters *xyz* or sometimes 123 (see Figure 6.1). Local coordinate systems are defined with respect to the single global *XYZ* coordinate system. These reference coordinate systems are important in understanding the analytical model of the structure.

ANALYTICAL MODELS

The analytical model of the structure is an abstract mathematical construct of an idealized structure that simulates the real building structure. Analytical models utilize simplified assumptions, such as for connecting members (hinged or rigid).

DOI: 10.1201/9781003598435-6

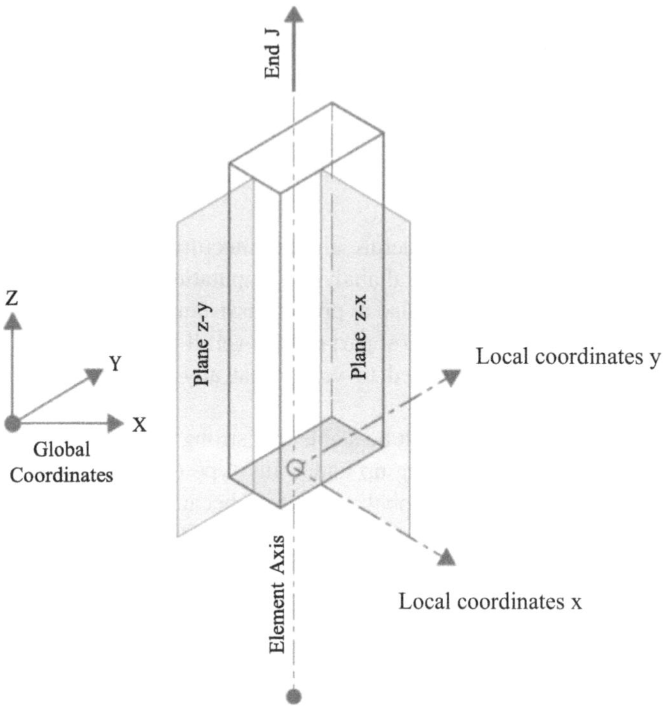

FIGURE 6.1 Spatial coordinate systems.

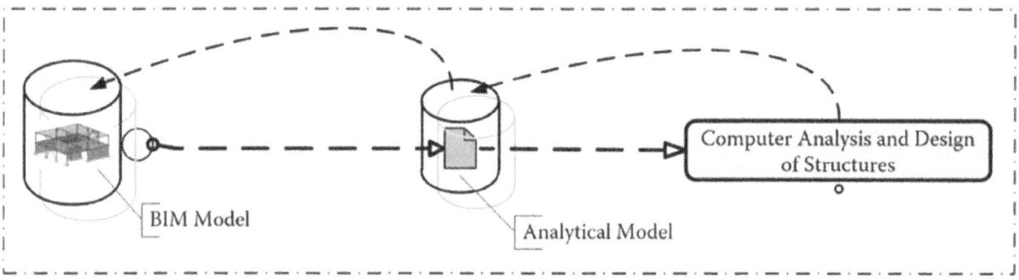

FIGURE 6.2 Relationship between physical and analytical structural models.

It is important to emphasize that only analytical models of the structure can be analyzed and not the real structure (Figure 6.2).

The analytical model treats the structure as an assembly system of its physical objects. It breaks down the building structure into its supporting elements, connections, material, and external loads.

The development of the analytical models is the most important step in the structural analysis phase. This approach practically replaced classical methods of analysis, such as slope deflection, moment distribution, Castigliano, and virtual work. However, the computer and conventional methods of structural analysis are both based on the fundamental principles of structural mechanics (i.e., force equilibrium, force-displacement relationship, compatibility of deformation, and energy balance of the structure).

The computer analysis of the developed analytical model of the structure is based on the finite element method (FEM) as a mathematical tool by which numerical methods approximate the equations describing the behavior of the structure. The FEM analysis procedure was developed during the 1950s in the aerospace industry and further advanced rapidly during the 1970s with the increase of computation processing power, when engineers turned to numerical methods using matrix mathematics rather than differential calculus.

The FEM analysis generally starts by discretizing the continuous structure into a number of one-dimensional (1D), two-dimensional (2D), or 3D elements, with various joint types, and subsequently reassembling them. Thus, when a structural skeleton is modeled as a discrete structure, its behavior is governed by algebraic equations, whereas a continuous structure is controlled by partial differential equations. For example, for a simple truss in which a member carries axial loads, each member can be represented as a 1D element. Similarly, a structural skeleton clearly represents an assemblage of discrete elements and thus is modeled as a 1D element (e.g., columns, beams, arches, cables, etc.) by the FEM. On the other hand, for surface structures and solid structures, such as slabs, walls, and shells, the continuum is discretized into a mesh or finite parts of polygon elements that can have various shapes (Figure 6.3).

In the FEM discretized state, the structural elements are considered connected to nodes located at the ends or corners of elements (Figure 6.3). Nodes are normally positioned where members are connected, at external supports, and at points of discontinuity such as edges, corners, or abrupt changes of materials and sections. Nodes can also be placed along the span of structural elements.

In the FEM analysis, the applied loads on the structural elements are converted to nodal loads, which in turn cause deformation. Deformation equations are then established for each element, typically using stiffness matrices, which is known as the stiffness method of analysis. When the structural elements are reconstructed to form the overall structure, compatibility at the boundaries of the elements must be satisfied. In other words, independent, simultaneous equations are set up that are equal to the number of unknowns and can then be solved by computer software, so the unknown deformations, and hence internal stresses, are determined. Generally, the FEM requires extensive mathematical computations because of the huge number of finite elements, especially in surface structures, and the corresponding number of simultaneous equations to be developed, which can only be solved with the help of computers. Designers must keep in mind, however, that FEM analysis represents an approximation. It does not provide precise results, except for simple structural frame and similar structural problems that denote closed-form solutions.

This section is not intended to introduce a detailed discussion of FEM analysis or other structural analysis methods. Readers may consult other references dealing with FEM and principles of structural design. We do not expect that readers will write their own finite element program but will only use structural software based on finite element analysis. For all readers, a precise understanding of the theory of the FEM is not a prerequisite for working with FEM-based structural software, although the reader should have some background knowledge of the basic concepts of the method.

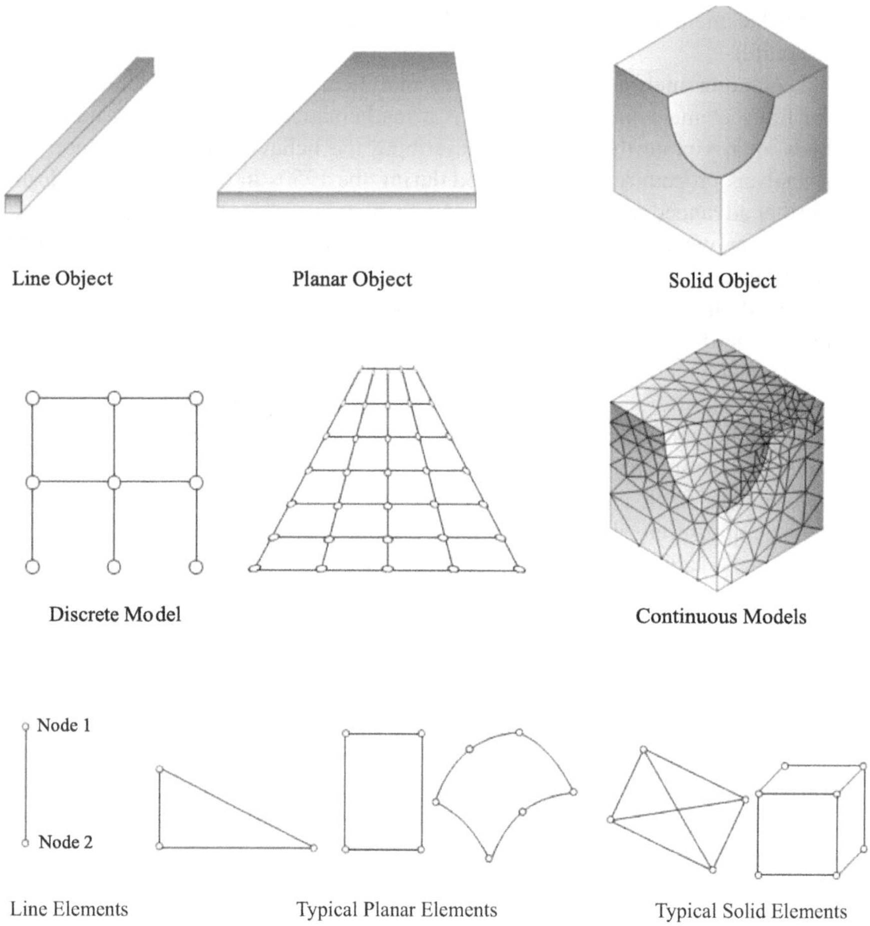

FIGURE 6.3 FEM modeling elements.

One of the clear benefits of the FEM is its ability to make the solution of indeterminate structures accessible to general designers during the preliminary design stage. The study of the effect of stiffness (e.g., size and form of the element) on force flow, as well as the influence of change of member arrangement and support locations in building structure is substantively significant for developing a sense of structural responses and behavior. For example, this involves studying the effect of the type of member assembly with respect to global and local stability by considering rigid joints or bracing. In addition, the FEM aids in becoming aware of structural systems, thus developing a sense of what is admissible and not admissible.

In summary, the FEM is not only a numerical technique involving matrix mathematics to solve large sets of simultaneous algebraic equations but also an abstract model that makes it possible to understand structural behavior in general. It is also a powerful educational tool because it has the unique feature of incorporating all of the structural concepts of loads, deformation, geometry, member behavior, materials, and stability of building. In other words, all the necessary components of structural design are in one system. In the classical teaching approach, all of those topics are handled more or less in separate courses in academia.

ANALYTICAL MODEL AND ELEMENT CONNECTIONS

Connections in building structures consist of internal element connections and external support connections. In the analytical model, the actual support conditions and member connections are idealized as resisting force systems by discounting secondary effects, such as stiffness toward joint rotation, friction, and slip. The true constraints (i.e., restraint to movement) that each element exerts on the adjacent ones, however, can only be approximated. For instance, a connection rarely permits free translational and rotational movement, or it can seldom provide complete resistance to movements.

Structural element connections are generally unrestrained joints whose displacement depends on the resistance of the connected elements; support joints are considered fully restrained against certain movements. In the case of the special loading condition for which rotation and translation (caused by foundation settlement, creep, shrinkage, etc.) are known quantities, the connection is modeled as a spring support. Common structural support joints include the following:

- **Basic Supports:** Roller, pin, fixed. These support types can be modeled in Revit by selecting the **Structural** tab.
- **Elastic Supports:** Spring (e.g., for soil), modulus of subgrade reaction. These types of supports are not yet available in Revit.
- **Full Structure-Soil Model:** 2D plane elements, 3D solid elements. These types of connections currently cannot be modeled in Revit.

Typical member-to-member connections are as follows:

- **Pinned Joints:** All elements attached to a joint rotate independently. This type of connection can be modeled in Revit using the member **Properties** palette as shown in Figure 6.4.
- **Rigid (Fixed) Joints or Moment Connection:** All elements attached to a joint share the same rotation (e.g., monolithic joints in cast-in-place reinforced concrete automatically are continuous, but in steel construction, bolting and welding must be used to make joints monolithic). Similar to a pinned connection, a rigid connection can be modeled in Revit using the member **Properties** palette as shown in Figure 6.4.
- **Semirigid Joints:** These are partial restraint connections; the rotational restraint is only a fraction of the full restraint provided by the connection. The degree of rigidity of member connections is contingent on the stiffness of the attached members. This type of element connection cannot be modeled in the current Autodesk® Revit platform.
- **Hybrid Joints:** In this type of connection, some members are connected rigidly, and others rotate independently. Again, this type of connection is not available in most current building information modeling (BIM) platforms.

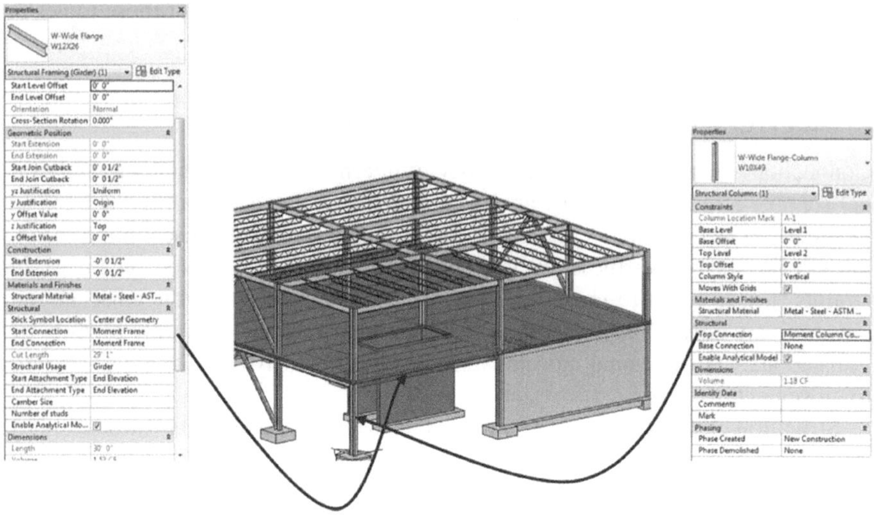

FIGURE 6.4 Modeling element connections in Revit.

SAS APPROACH FOR STRUCTURAL ANALYSIS

Without the traditional learning emphasis on first understanding single elements, such as beams, columns, bearing walls, and so on two dimensionally, using the laws of statics and strength of materials, the SAS framework utilizes BIM tools to help students create whole 3D structures and then investigate structural solutions. Using the principle of structural melodies and poetry, different structural forms exemplifying various structural support concepts, including vertical and lateral resisting systems, can be developed. The simplest types of these forms are referred to as "buildoids," and they are principally developed to stress the 3D nature, order, and organization of structural elements as well as to develop a sense of scale, proportion, and harmony.

Because building structures are in 3D, with the applied loads flowing along the elements and connections to the external supports (foundations), introducing and emphasizing this fact early in structural design education is one of the principal objectives of the SAS framework. BIM tools are then applied to these buildoids to promote the understanding of fundamentals of structural analysis, such as the force equilibrium, support reactions, shear force, and bending moment diagrams; frame and truss analysis; and steel, wood, and concrete design. The Figure below illustrates details about the various BIM structural analysis tools utilized in this book.

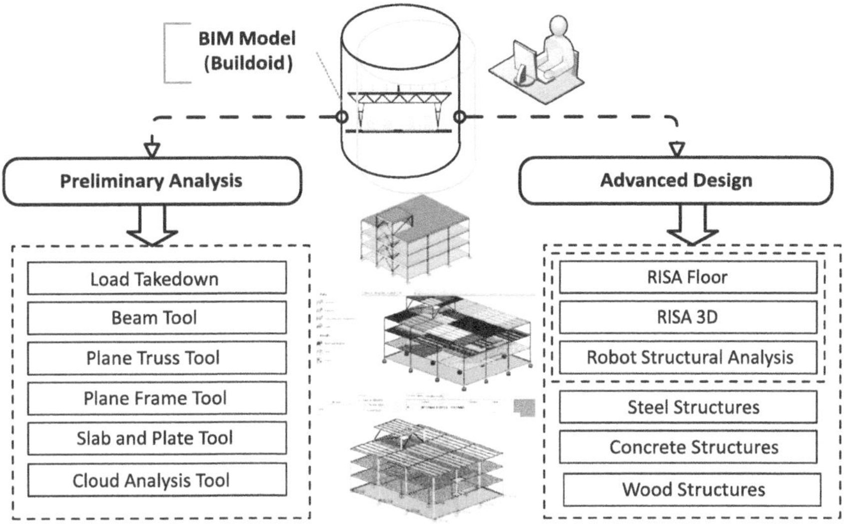

PRELIMINARY ANALYSIS

FEM AND REVIT EXTENSIONS

Traditionally, the workflow for design engineers involved interpreting architectural drawings and making their own analytical model from these drawings, constantly checking and rechecking the correlating models. The FEM from the structural engineer typically runs on its own platforms and does not interface with any BIM software. Autodesk Extensions changed this process by introducing FEM tools that are fully linked to Revit based on Autodesk Robot™ Structural Analysis Professional (Robot).

Autodesk Extensions for Autodesk Revit software provide a full range of tools for preliminary structural analysis (see Figure 6.5). These tools are based on the FEM software package Robot. They are fully integrated with Revit. These extensions provide a vehicle to perform preliminary structural analysis computations in Revit, as well as to produce professional reports for documenting the work.

In addition to the extensions, Revit provides a direct link to Robot and provides a bidirectional flow of data between the two programs. Robot is feature-rich FEM structural analysis and design software capable of modeling and analyzing many types of building structures. Robot enables Revit users to directly analyze their models without oversimplifying or interfering with the 3D building model to satisfy the restrictions of their chosen analysis solution.

These extensions provide an instrument to perform less-complicated structural calculations in Revit, as well as to produce professional reports for documenting the work. Not only does Robot interface with Revit, but it also links with many other software applications using a technique known as interoperability. This enables BIM models to be easily moved on a common platform, ensuring the finite element model is the exact same as the physical model. The following sections illustrate the

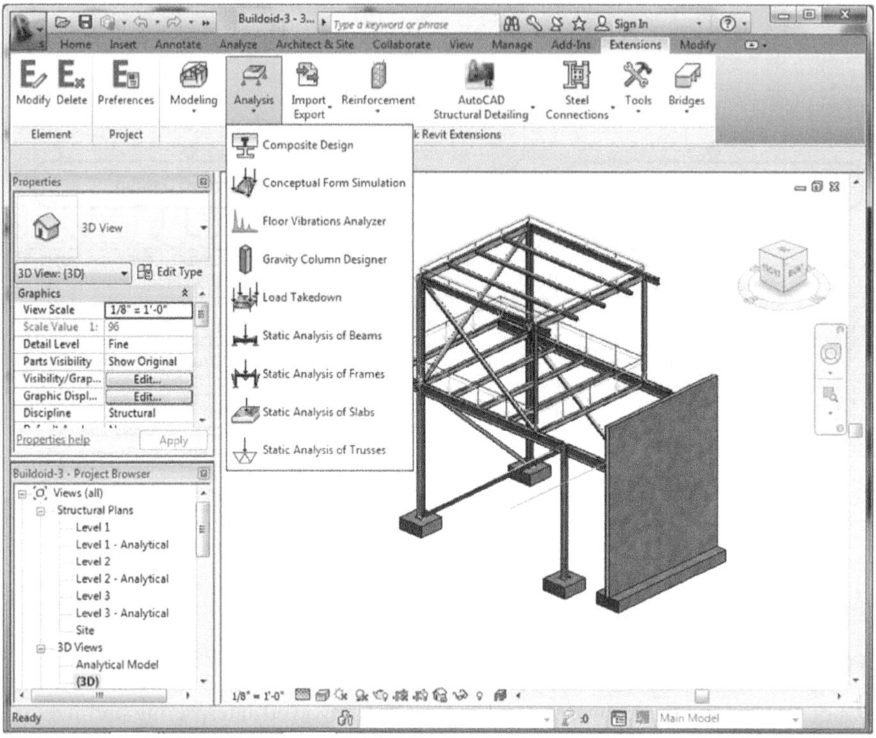

FIGURE 6.5 Revit extensions.

application of these extension tools in performing preliminary structural analysis and design.

Load Takedown

The **Load Takedown** tool allows you to perform a simulation of the flow of forces in different structural elements in a BIM model caused by gravity loads.

Since 2023, The load Takedown tool has been moved from Autodesk Revit to Robot Structural Analysis. In this new edition, we decided to keep the old version discussion for back compatibility and introduce the new tool as well in the next section.

To instantiate the older version of the tool, one needs to launch the **Load Takedown** tool from the **Extensions** tab under **Analyze** (Figure 6.6). This will open the interface shown in Figure 6.6. Users can make further changes to the model if desired by modifying geometry, material parameters, support conditions, loads, and load cases. The user interface for the tool is generally divided into three main parts as illustrated in Figure 6.7.

Part 1 of the dialog displays options for selecting items used to define geometry, supports, load cases and combinations, loads, and parameters; perform analysis

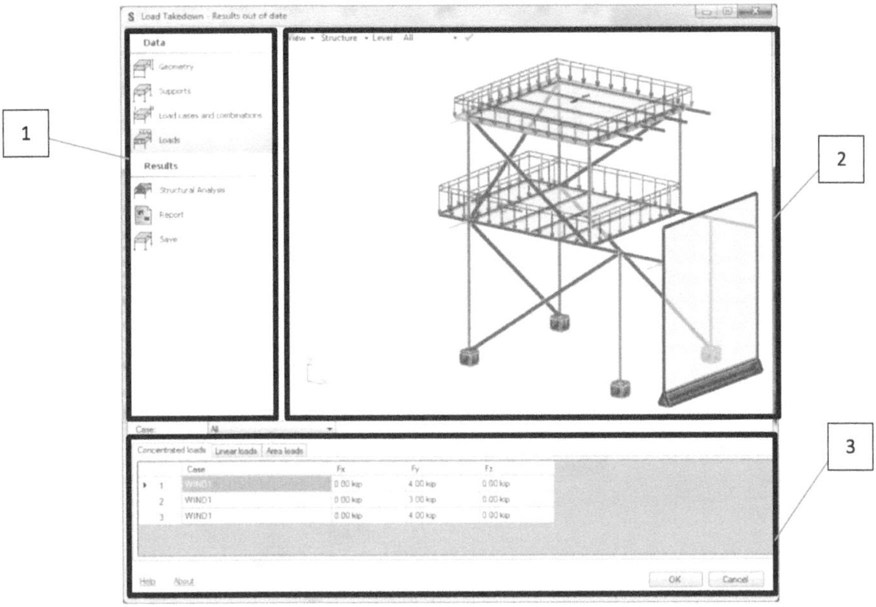

FIGURE 6.6 **Load Takedown** tool for Revit 2014 or earlier.

computations; and receive parameter results and reports. Part 2 in the center is a graphical viewer of the BIM model under consideration. In part 3 at the bottom, there are tabs that display a table of results depending on the tab selection.

Figure 6.7 depicts the results obtained from the **Load Takedown** tool after it has been applied to the structure shown. The tools give the tributary area for each beam and their reactions transferred to the supporting beams and girders. Further, it displays the column axial load map, which shows the column's load in each level down to the foundation. It is important to note that to obtain correct results, one must define a correct analytical structure model in Revit before launching the tool.

Reports can be generated from the **Load Takedown** tool and can be verified using hand or spreadsheet calculations.

Beam Analysis

After understanding the load path in the structure, the beam analysis tool can be used to perform static analysis of beams to determine the internal forces and diagrams (i.e., axial force, shear force, and bending moment diagrams). This can be achieved by first selecting the beam in the BIM model and then invoking the **Beam** tool from the **Extensions** tab under the **Analyze** tab. It is also necessary before instantiating the **Beam** tool to define the loads on the beam. Currently, the tool would not determine the load on the beam automatically from the general loading on the BIM model. It must be explicitly defined. A further limitation of the tool is the application of point loads.

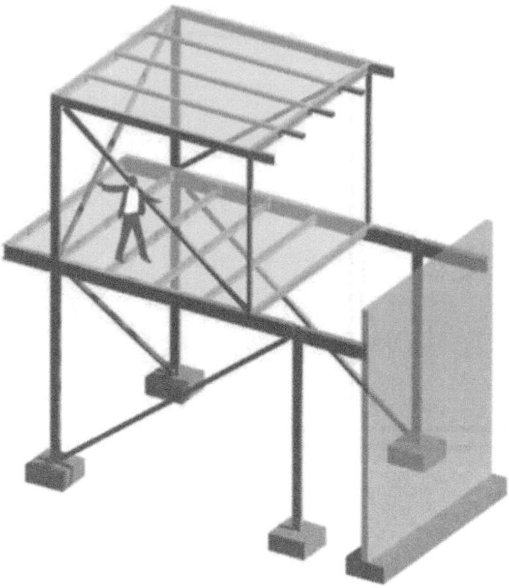

(a)

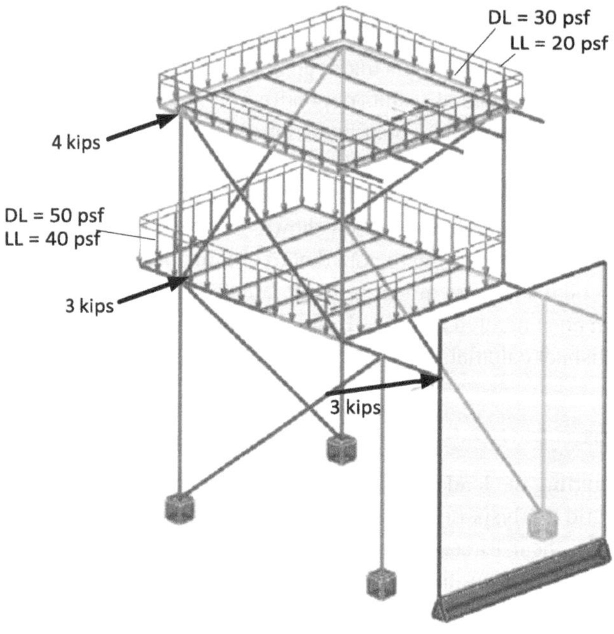

(b)

FIGURE 6.7 Load path using **Load Takedown** tool: (a) BIM model; (b) analytical model showing gravity; (c) **Load Takedown** results with reactions and lateral loads; (d) axial column and foundation loads; (e) north framing elevation.

(Continued)

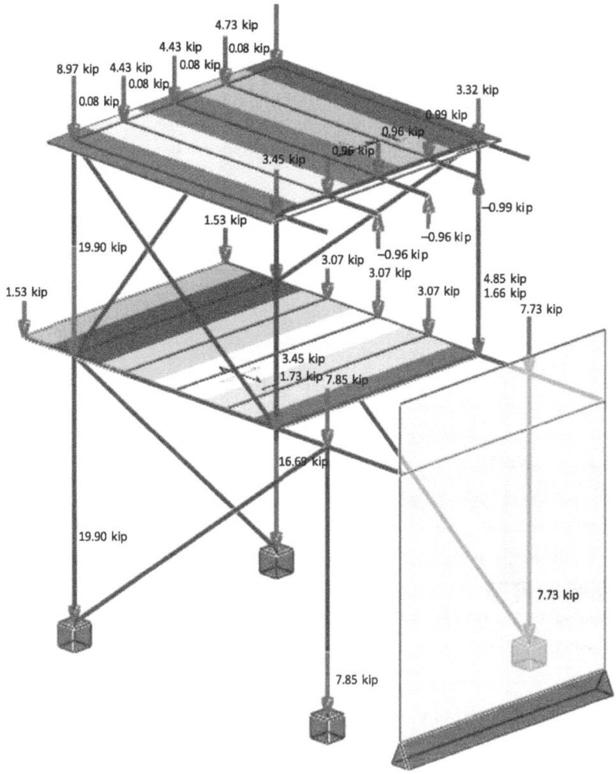

(c)

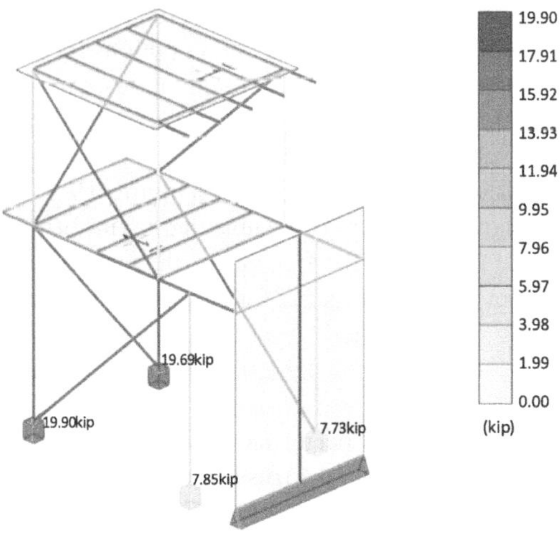

(d)

FIGURE 6.7 (*Continued*) Load path using **Load Takedown** tool: (a) BIM model; (b) analytical model showing gravity; (c) **Load Takedown** results with reactions and lateral loads; (d) axial column and foundation loads; (e) north framing elevation.

(Continued)

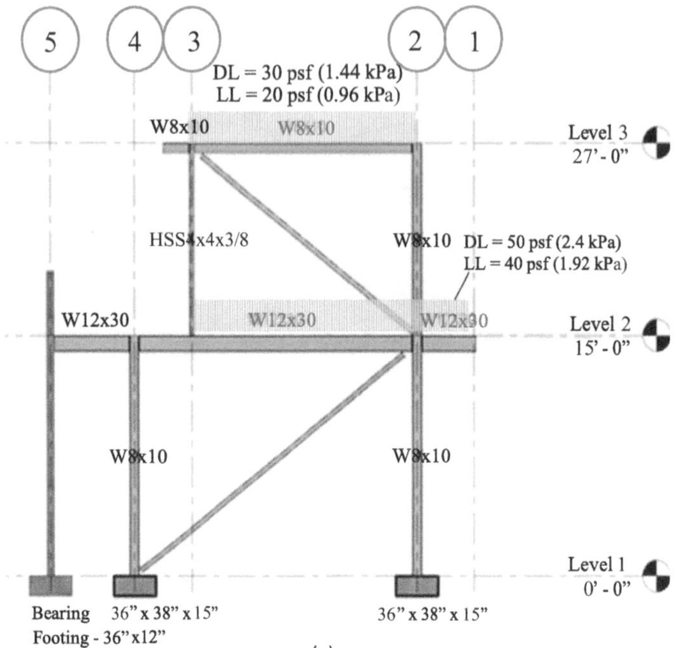

(e)

FIGURE 6.7 (*Continued*) Load path using **Load Takedown** tool: (a) BIM model; (b) analytical model showing gravity; (c) **Load Takedown** results with reactions and lateral loads; (d) axial column and foundation loads; (e) north framing elevation.

Again, similar to the **Load Takedown** dialog, the interface is divided into three main parts as shown in Figure 6.8. Part 1 displays options for selecting items used to define beam parameters, perform analysis computations, and receive parameters of result. Part 2 is a graphical viewer of the beam under consideration. In part 3, there is a table that depends on the tab selected in part 1 of the dialog.

The application of the **Beam** tool is depicted in Figure 6.9. Results obtained are for the cantilever beam shown in Figure 6.10 subjected to uniform dead load of 0.5 k/ft and a point live load of 2.0 kips at the free end of the cantilever beam. These results can be utilized to select preliminary sizes for beams and girders.

Truss Analysis

The current **Truss Analysis** tool performs static analysis of trusses subjected to nodal forces only. You need to select the truss in the Revit model before invoking the **Truss Analysis** tool from the **Extensions** tab. Nodal loads must be defined before instantiating the tool from the **Extensions** tab.

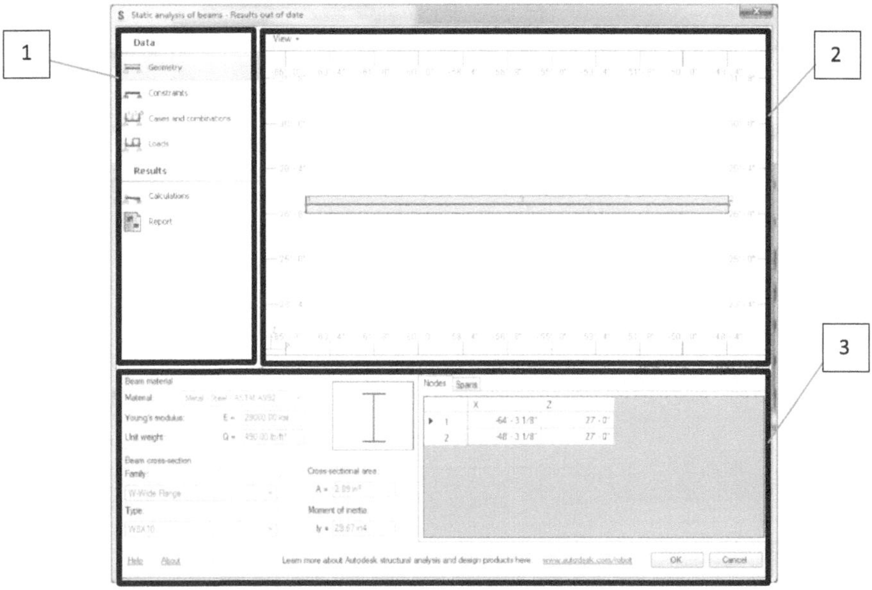

FIGURE 6.8 The **Beam** tool.

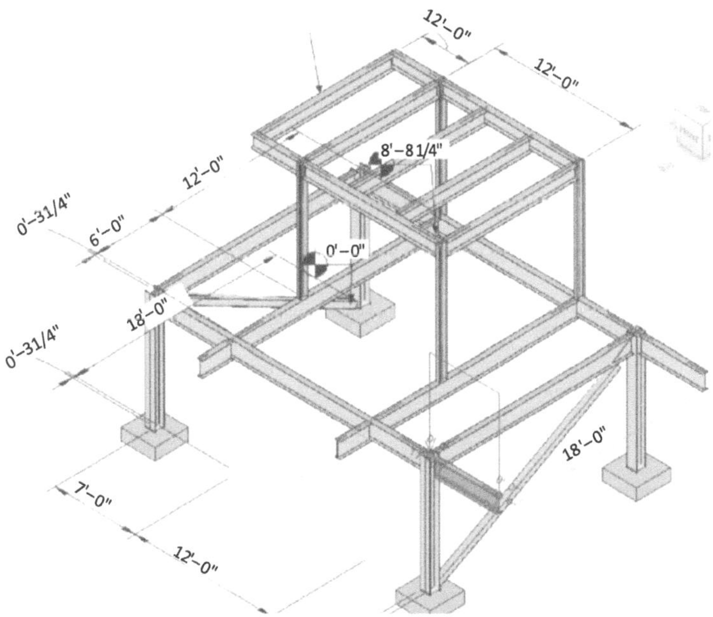

FIGURE 6.9 BIM model showing beam selected for analysis using the **Beam** tool.

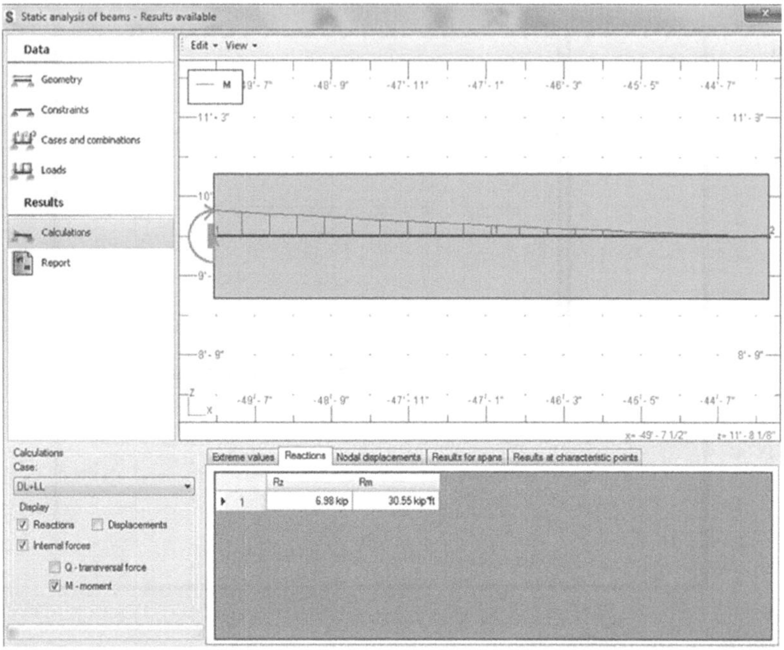

(a)

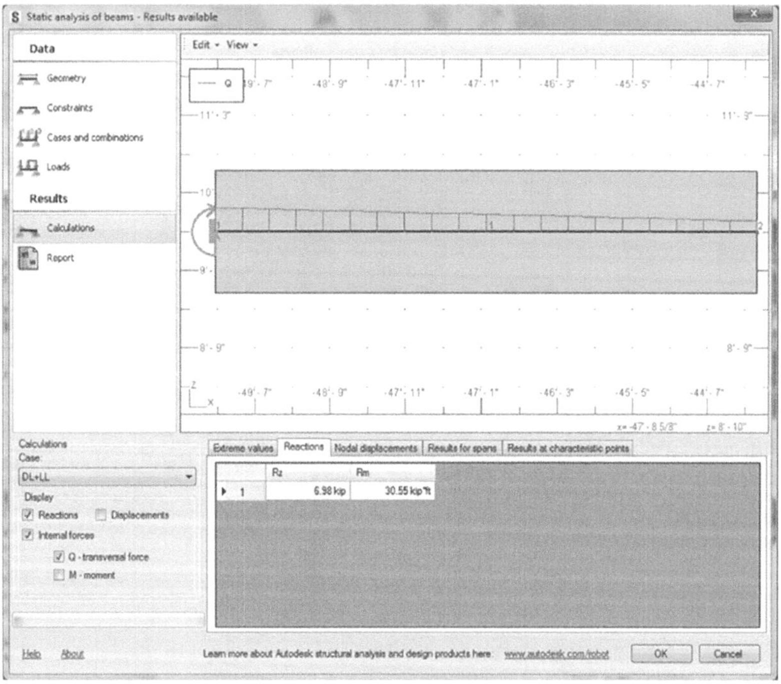

(b)

FIGURE 6.10 (a) Bending moment diagram and reactions. (b) Shear force diagram.

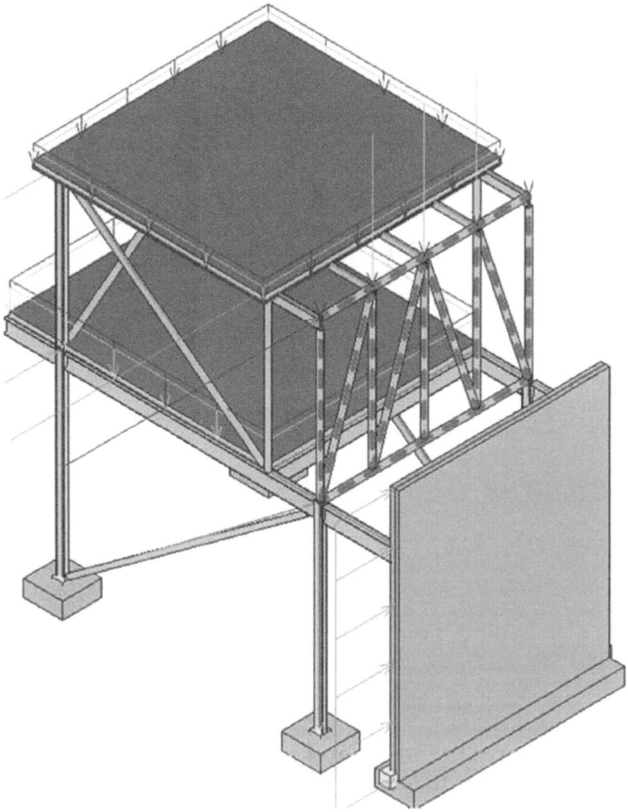

FIGURE 6.11 Truss selected for analysis.

Figure 6.11 shows the truss selected in a BIM model before invoking the truss tool. After launching the tool, the dialog interface shown in Figure 6.12 will appear on the screen. It has similar divisions as the previously discussed **Extensions** tools. That is, it has three main parts. Part 1 (top left) shows options for selecting components used to define truss data and analysis results; part 2 (top right) is a graphic viewer with the selected truss. Part 3 (bottom) deals with specifications of parameters corresponding to the tab selected.

Before running the analysis, it is recommended to check the support conditions and applied loads and load cases. Then, select the load case from the drop-down, Figure 6.10a and b shows the results of the analysis in part 3 of the screen, and press calculate to perform static analysis. You can also specify the type of results you are interested in (i.e., reactions, displacements, and normal forces). In part 3 of the dialog box, result tables are displayed for each type of result (reactions, displacements, and normal forces). Each displays minimum/maximum values, the bar, and the load case for which this value was obtained (Figure 6.13c).

Furthermore, the tool offers a full report of results that can be exported to a Microsoft Excel® or Microsoft Word® document (see Figure 6.14).

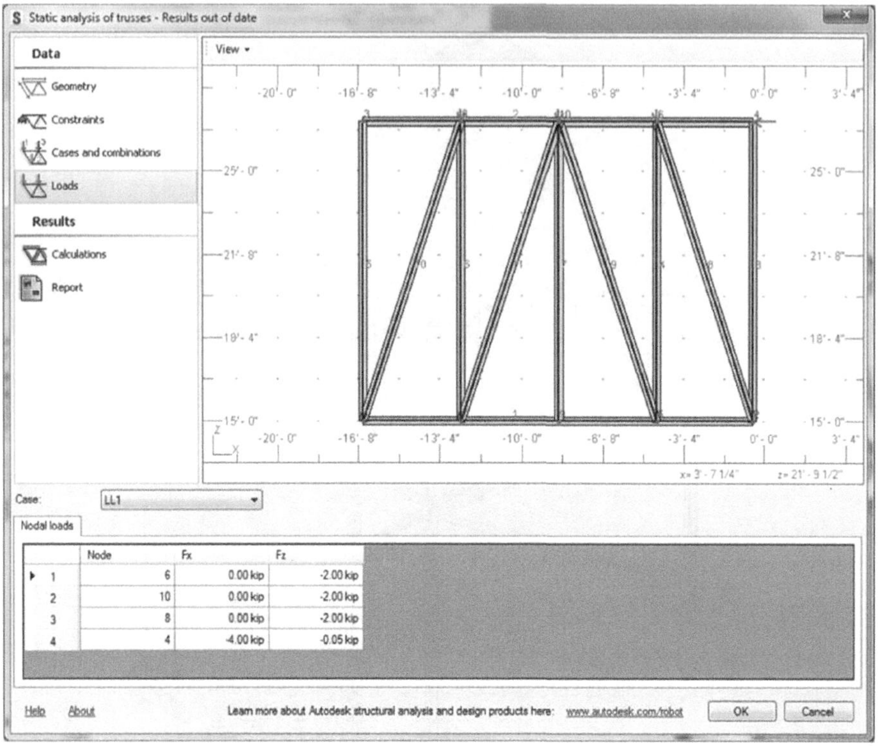

FIGURE 6.12 The **Truss Analysis** tool.

Frame Analysis

With the **Frame Analysis** tool, one can analyze any 2D subframe of the BIM model. The tool performs static analysis of 2D frames. It is important to recognize that the extension analyzes this subframe as a 2D frame. For loads applied to the frame, this **Extensions** tool computes elements acting in the plane of the frame and applies these values to the frame (elements that are perpendicular to the plane of the frame are ignored). The tool replaces with supports all elements adjoining the frame (e.g., edge girders) that lie in a plane different from the plane of the frame.

Before instantiating the **Frame Analysis** tool, one should select elements of the frame (see Figure 6.15). The tool generally ignores offsets defined in a Revit model during calculations (it is assumed that all elements of the frame are positioned at their centers of gravity).

Next, invoke the static analysis of frames tool from the **Extensions** tab under the **Analysis** tab. After launching the tool, the dialog depicted in Figure 6.16 will appear on the screen. It is similar to the previously discussed tools (has three main parts). Part 1 (top left) shows options for selecting components used to define the frame data and analysis results, part 2 (top right) is a graphic viewer with the selected subframe. Part 3 (bottom) deals with specifications of parameters corresponding to the tab selected.

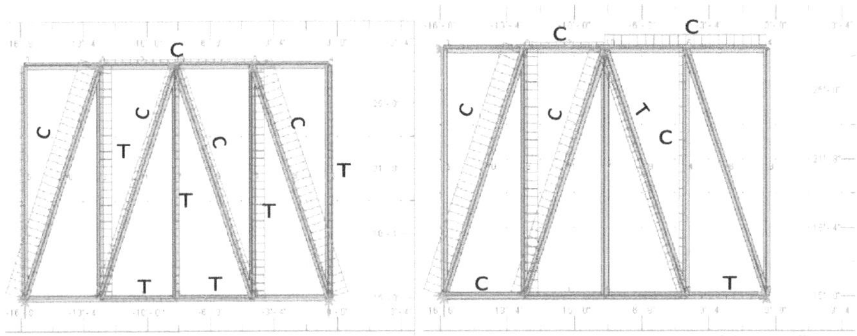

(a) (b)

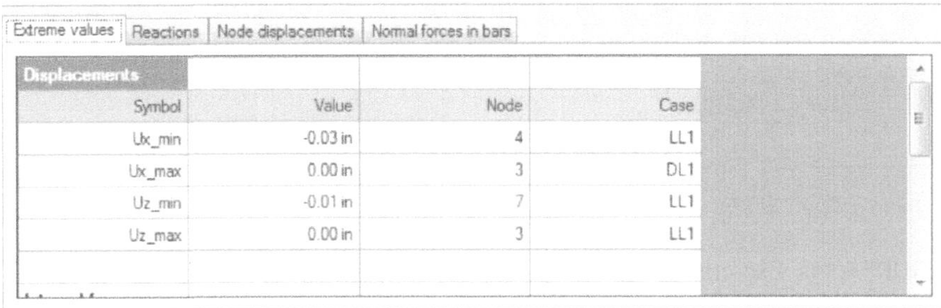

Displacements			
Symbol	Value	Node	Case
Ux_min	-0.03 in	4	LL1
Ux_max	0.00 in	3	DL1
Uz_min	-0.01 in	7	LL1
Uz_max	0.00 in	3	LL1

(c)

FIGURE 6.13 Normal force results using the truss tool: (a) normal forces caused by dead load; (b) normal forces caused by live load; (c) maximum and minimum deformation of the truss members.

The frame shown in Figure 6.15 is subjected to a uniform dead load of 5.0 k/ft and a live load of 8.0 k/ft on the first level. At the roof level, it carries a uniform dead load of 2.0 k/ft and a live load of 5.0 k/ft.

This tool has rich settings options for display, analysis, and reporting. For example, the settings for the graphic display can be instantiated from the drop-down of the **View** menu that shows in the top left of the graphic viewer. This settings dialog is shown in Figure 6.17.

Results obtained for a defined frame model are displayed on the **Calculations** tab in graphical (diagrams of a selected quantity) and tabular form. In addition, data can be presented in an HTML report, saved to a file, or sent to a Microsoft Excel or Microsoft Word document.

Using the options on the **Calculations** tab (Figure 6.18), one can select a load case and a quantity for which results are to be presented in the graphic viewer and in the results table (**Reactions**, **Displacements**, **Internal forces**). The results table in this part of the interface offers values for extreme data, reactions, displacements of nodes or elements, internal forces, and reactions; each part displays minimum/maximum (min/max) values, the identifier of an element (node or bar) at which the extreme value was achieved, and the load case for which this value was obtained.

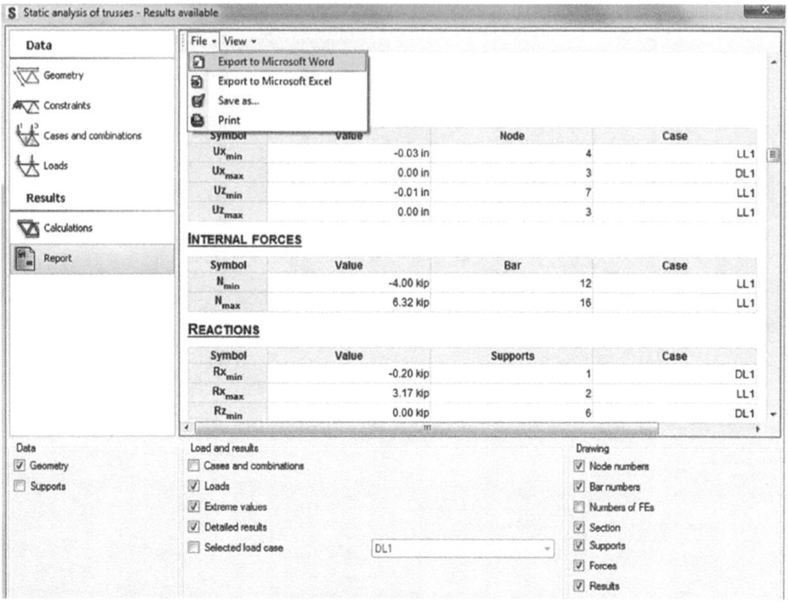

FIGURE 6.14 Exporting results to an MS word document.

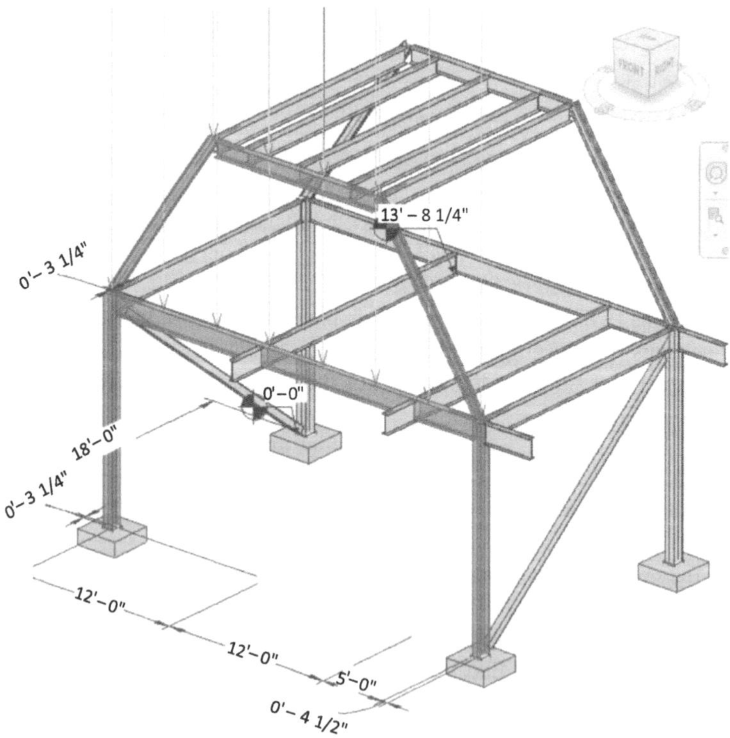

FIGURE 6.15 Subframe selected for analysis.

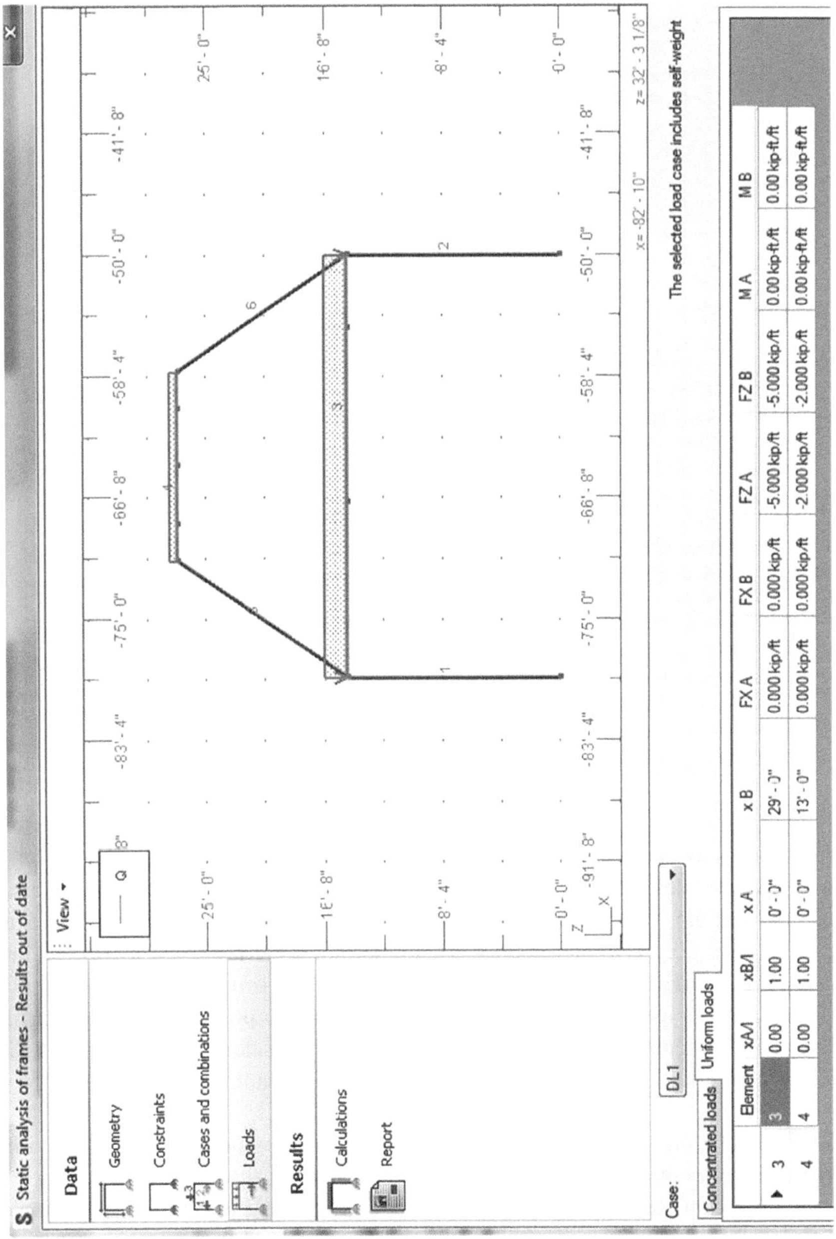

FIGURE 6.16 Static analysis of frames

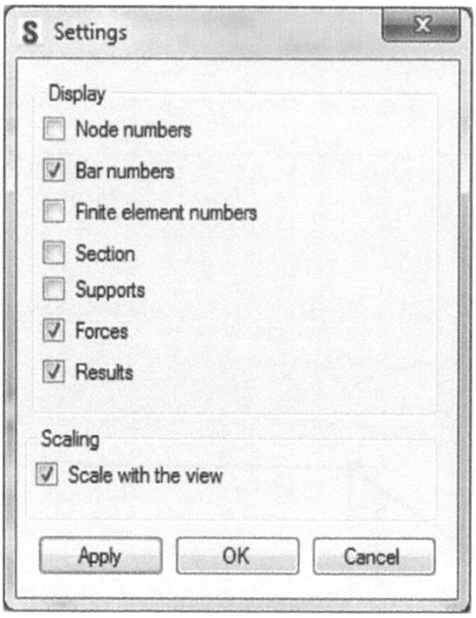

FIGURE 6.17 Setting options for the frame analysis tool.

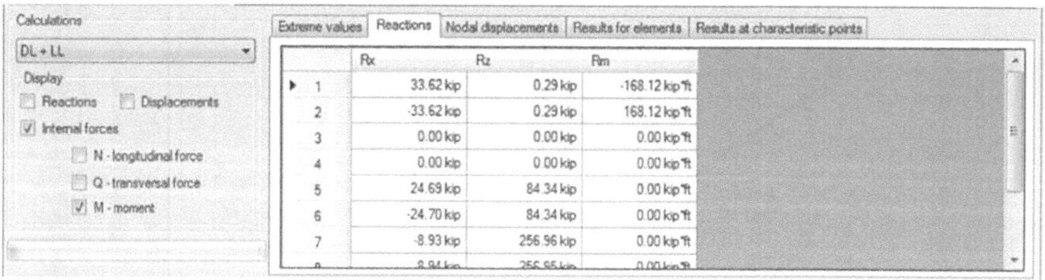

FIGURE 6.18 Calculation settings and tabular results of the **Frame Analysis** tool.

Furthermore, tabular results include nodal displacements (presented in the global coordinate system) and results for individual bars of the frame (min/max values for bars); deflections and displacements of bars are presented in the local coordinate system (the x axis runs along the bar length, and the z axis is perpendicular to the x axis), and results are at characteristic points of a frame. One can define a **Characteristic Point** at any point of the frame. This is extremely helpful as it gives users the flexibility to add calculation points as needed within the frame. Users can define a **Characteristic Point** by selecting it from the **Edit** drop-down menu item in the top left corner of the graphic viewer of the tool (Figure 6.19).

Alternatively, one can define a **Characteristic Point** by defining the x/l ratio in the table (x is a coordinate of a point, and l is a length of the frame element) or by defining an x value (a coordinate of a **Characteristic Point** in the local coordinate system).

To correctly interpret the results output of the static analysis of frame tool, the sign convention for internal forces must be understood. Figure 6.20 illustrates the internal forces sign convention using the local coordinate system.

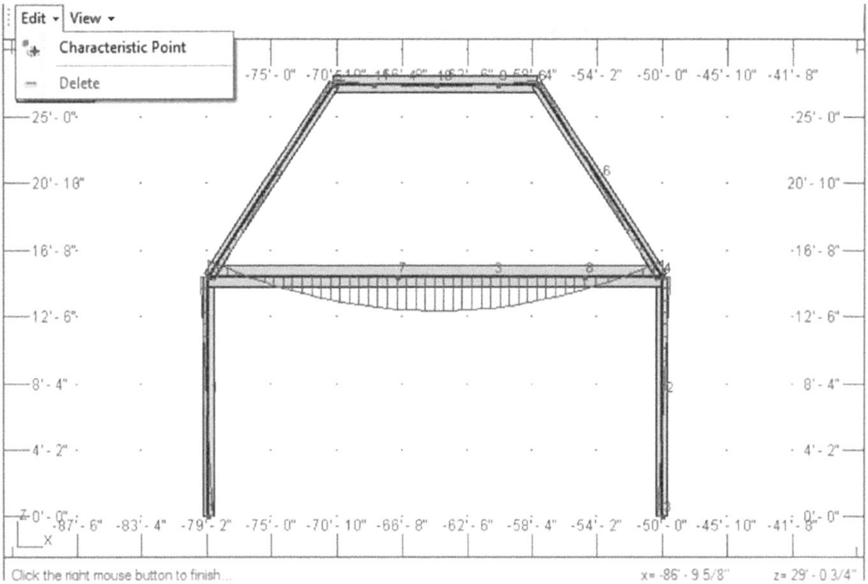

FIGURE 6.19 Defining a characteristic point.

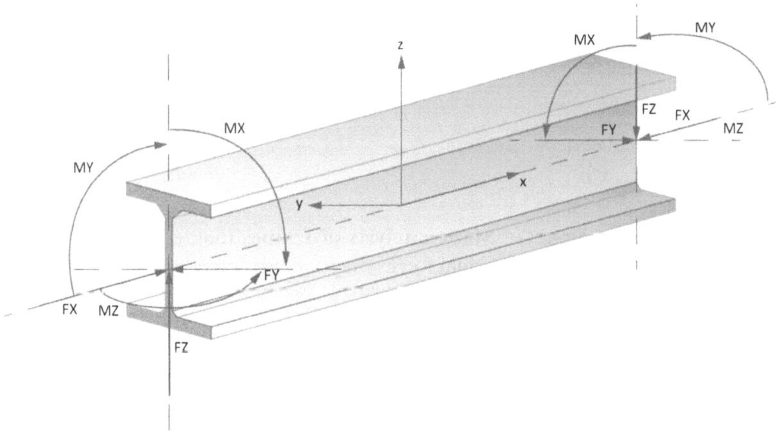

FIGURE 6.20 Local coordinate system and sign convention for internal forces.

For the frame selected, results for a particular quantity (diagrams of shear force, bending moment, internal forces, displacements, and reactions) are shown in the graphic viewer for a selected load case (Figure 6.21). Notice also the possibility of hovering with the mouse over any point in the frame to display results at that point.

Slab Analysis

The **Static analysis of slabs** tool can be utilized to analyze horizontal slabs defined in Revit irrespective of the type of slab. One would invoke the slab analysis tool

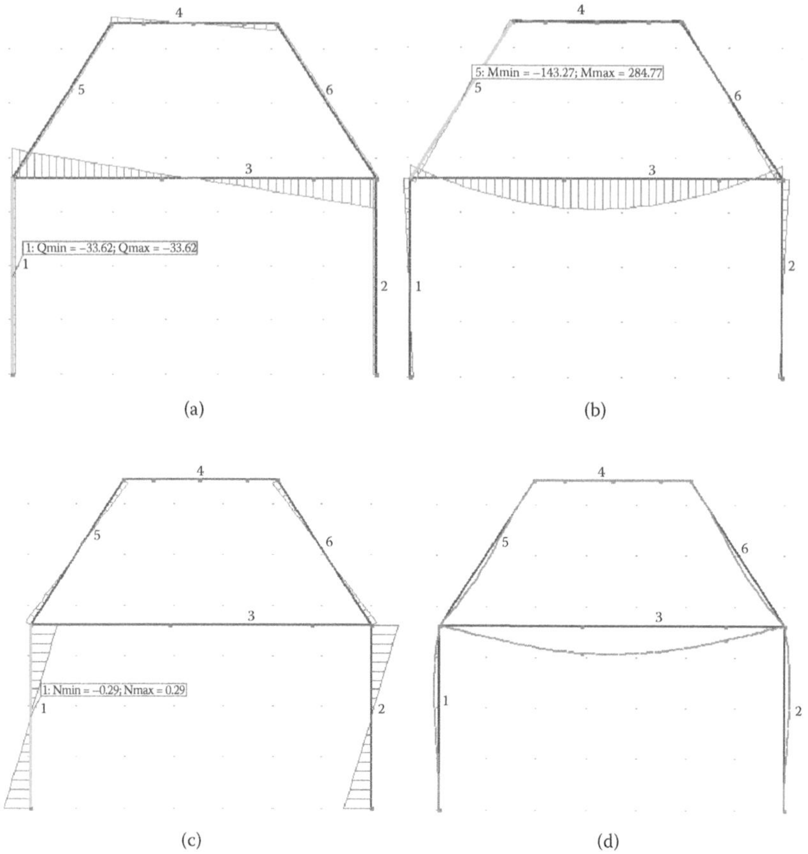

FIGURE 6.21 Results from the **Static analysis of frames** tool: (a) shear force diagram; (b) bending moment diagram; (c) normal forces diagram; (d) frame deformed shape.

similar to the other **Extensions** tools discussed. Before launching the tool, one must select a particular slab in the BIM model (see Figure 6.22). Once the tool is launched, you will see a similar dialog as the previously introduced **Extensions** tools. Again, the dialog box has three main parts, part 1 (top left) displays options for selecting components used to define slab parameters and parameters of result display, part 2 (top right) is a graphic viewer for the slab undergoing analysis, and part 3 (bottom) is a table that depends on the tab selected in the dialog (Figure 6.23).

The **Geometry** tab of the slab analysis tool displays general information about the slab loaded from the BIM model, such as material name, modulus of elasticity, passion ratio, and unit weight. Furthermore, at the bottom right of the user interface is the element table with tabs for each type, such as data for contours, openings, and segments that respond to the current selection. Each tab contains coordinates and identifiers of nodes (Figure 6.23).

The **Constraints** tab allows you to define the type of supports of the slab (Figure 6.24). Revit Extensions recognizes two types of slab supports (Figure 6.25): supports that are defined directly in Revit and supports that result from the slab being

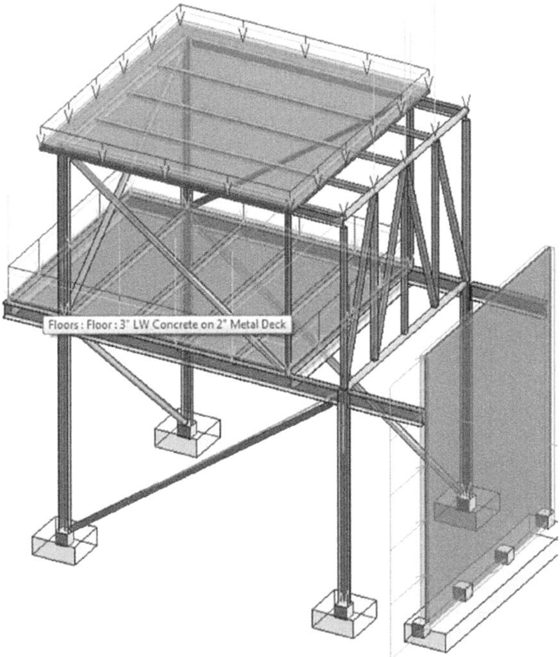

FIGURE 6.22 Slab selected for the analysis.

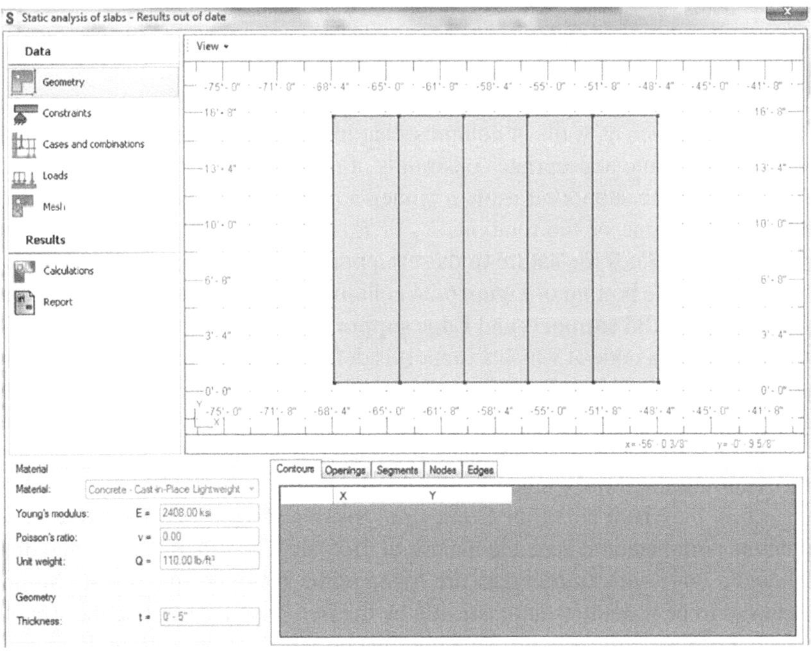

FIGURE 6.23 Static analysis of slabs tool.

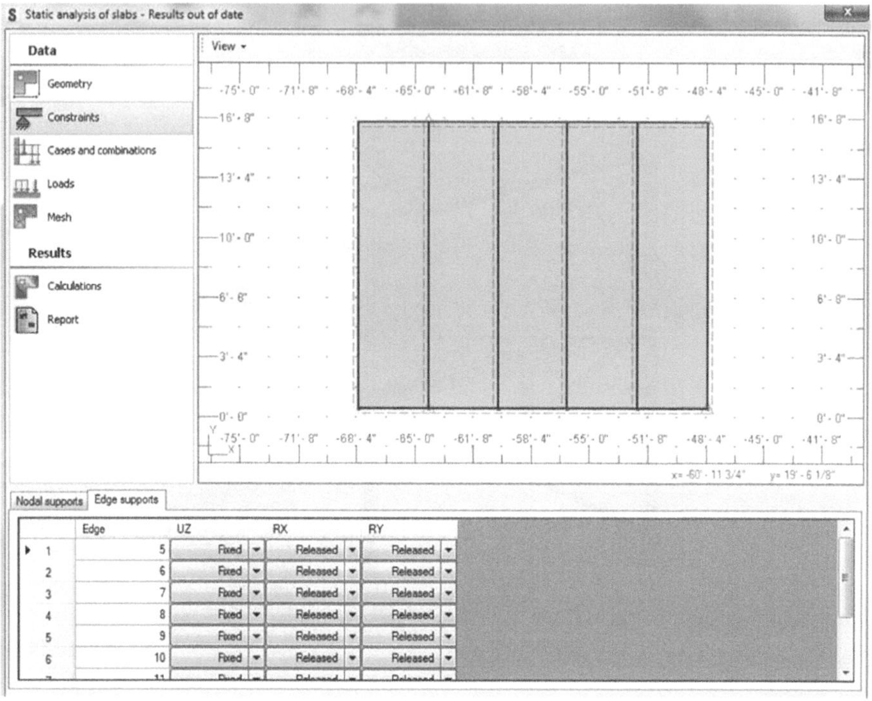

	Edge	UZ	RX	RY
1	5	Fixed	Released	Released
2	6	Fixed	Released	Released
3	7	Fixed	Released	Released
4	8	Fixed	Released	Released
5	9	Fixed	Released	Released
6	10	Fixed	Released	Released

FIGURE 6.24 Type of supports for structural slabs.

supported by other elements in a structure model (walls, columns, beams, etc.). It is important to note that only the second type can be modified in the **Extensions** (vertical restraint in the direction of the z axis). Further, if the BIM model includes structural elements (such as walls or columns) that intersect with the slab, those elements will be changed into appropriate constraints of the selected slab.

The slab can be supported using a pinned support (only the Z direction is fixed), a fixed support (one of the rotations R_X or R_Y is fixed), or a combination of both. Figure 6.25 illustrates the nature of these supports.

The table at the bottom of Figure 6.24 contains the data for the two basic types of supports (i.e., **Nodal supports** and **Edge supports**). These include numbers of a support, numbers of a node at which a support is defined, numbers of an edge on which a support is defined, and released/fixed degrees of freedom (displacement or rotations of a support) as shown in Figure 6.24.

Another important step in the slab analysis is to set up the meshing parameters before running the finite element computation (Figure 6.25c). This can be achieved by selecting the **Mesh** tab of the **Static analysis of slabs** tool. On the **Mesh** tab, one can define parameters to generate a mesh of the slab. The general parameters for the mesh and a table with emitter data are found at the bottom of the screen. Mesh can be selected to be generated automatically by the tool or you can define the size of the element. The finer the mesh is, the better the computed results will be. However, for larger slabs, refined mesh can slow the computation and the responsiveness of the tool.

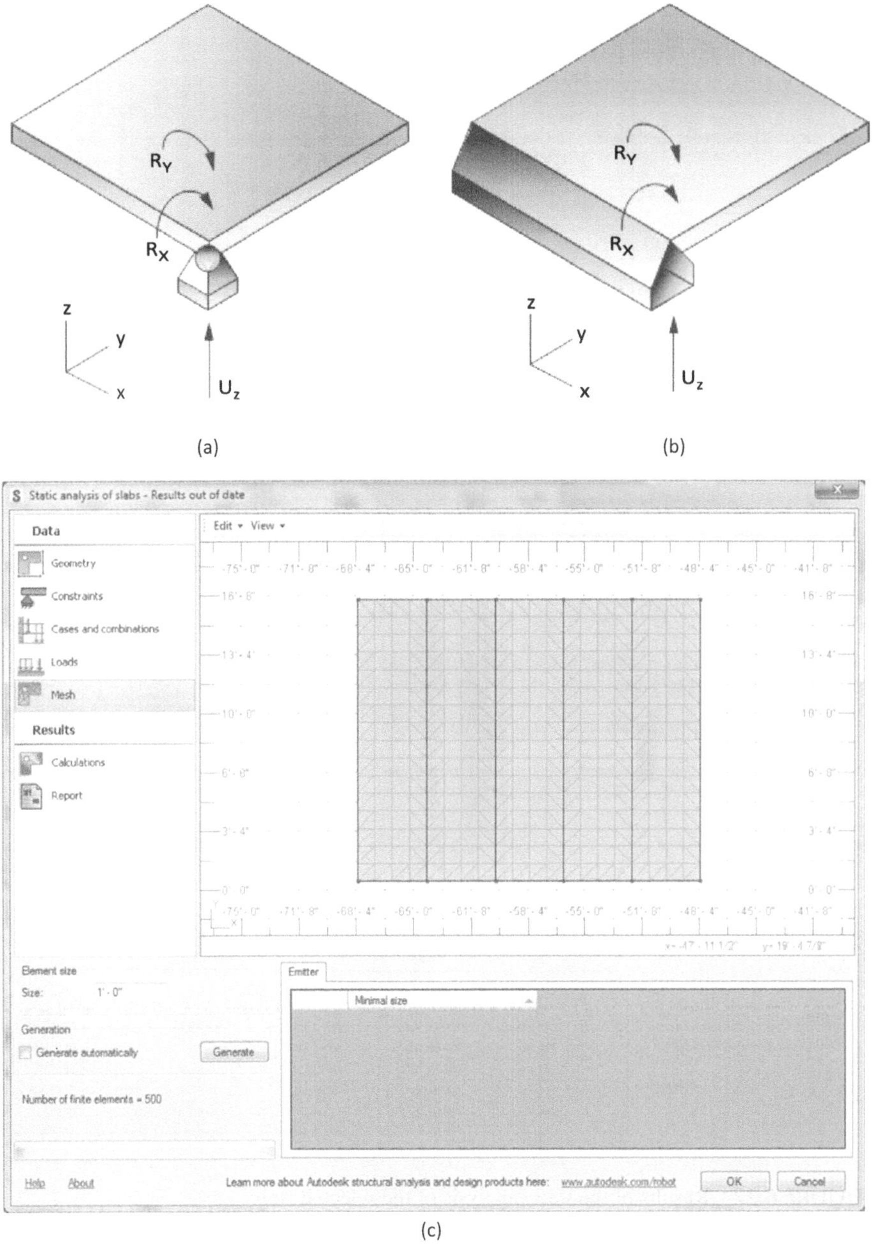

FIGURE 6.25 Reaction forces of a structural slab: (a) nodal support for the slab; (b) edge support for the slab; (c) define mesh parameters for the slab.

The bottom screen of the slab analysis tool also displays an **Emitter** table. Emitters are the nodes in the neighborhood of which the mesh will be refined. You can define emitters by selecting **Emitter** from the **Edit** drop-down menu in the

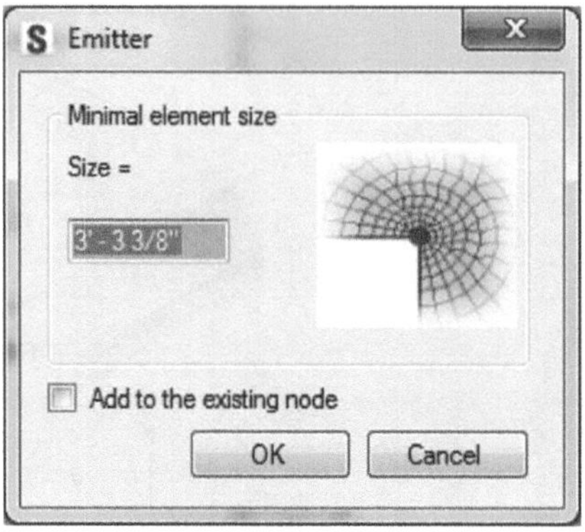

FIGURE 6.26 Defining emitters for meshing the slab.

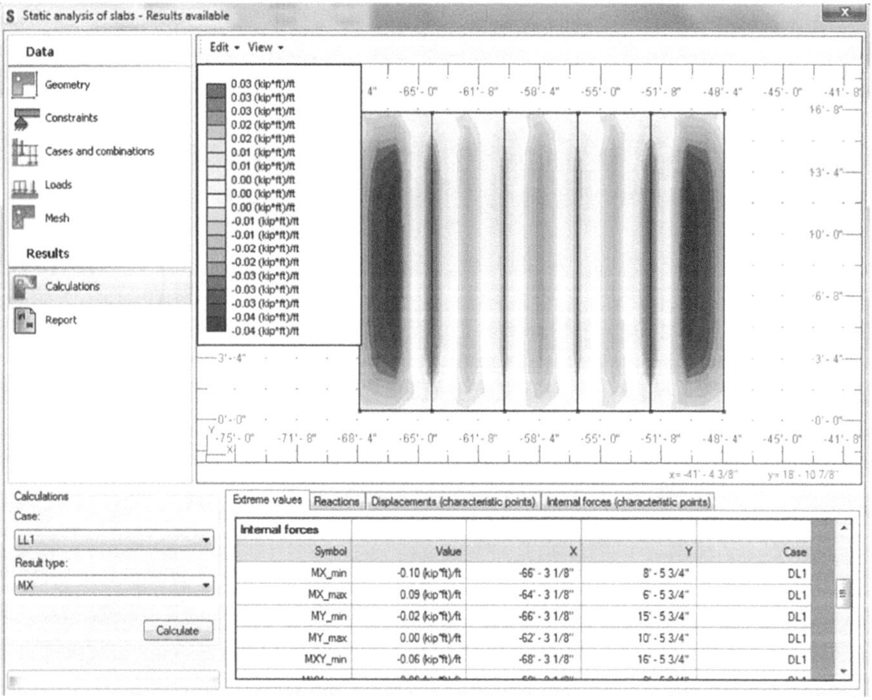

FIGURE 6.27 Results of the static analysis of the selected slab.

graphic viewer (Figure 6.25c). The table will display data for emitters that you have defined (Figure 6.26).

Analysis can be performed by selecting the **Calculations** tab. Before invoking the calculation, one needs to choose a **Load Case** and **Quantity** for which results are to be presented in the graphic viewer and in the results table at the bottom. Results obtained include displacements, reactions, bending moments, and shear forces (Figures 6.27 and 6.28). The table of results shows four types of data:

(1) **Extreme values**, which displays min/max values, coordinates of the point at which the extreme value was achieved, and the load case for which this value was obtained; (2) **Displacements**, which shows results at user-defined characteristic points; (3) **Internal forces**, such as bending moments and shear forces (results at user-defined characteristic points); (4) **Reactions**, which shows reaction forces for edge supports and reaction values at characteristic points of the mesh are presented.

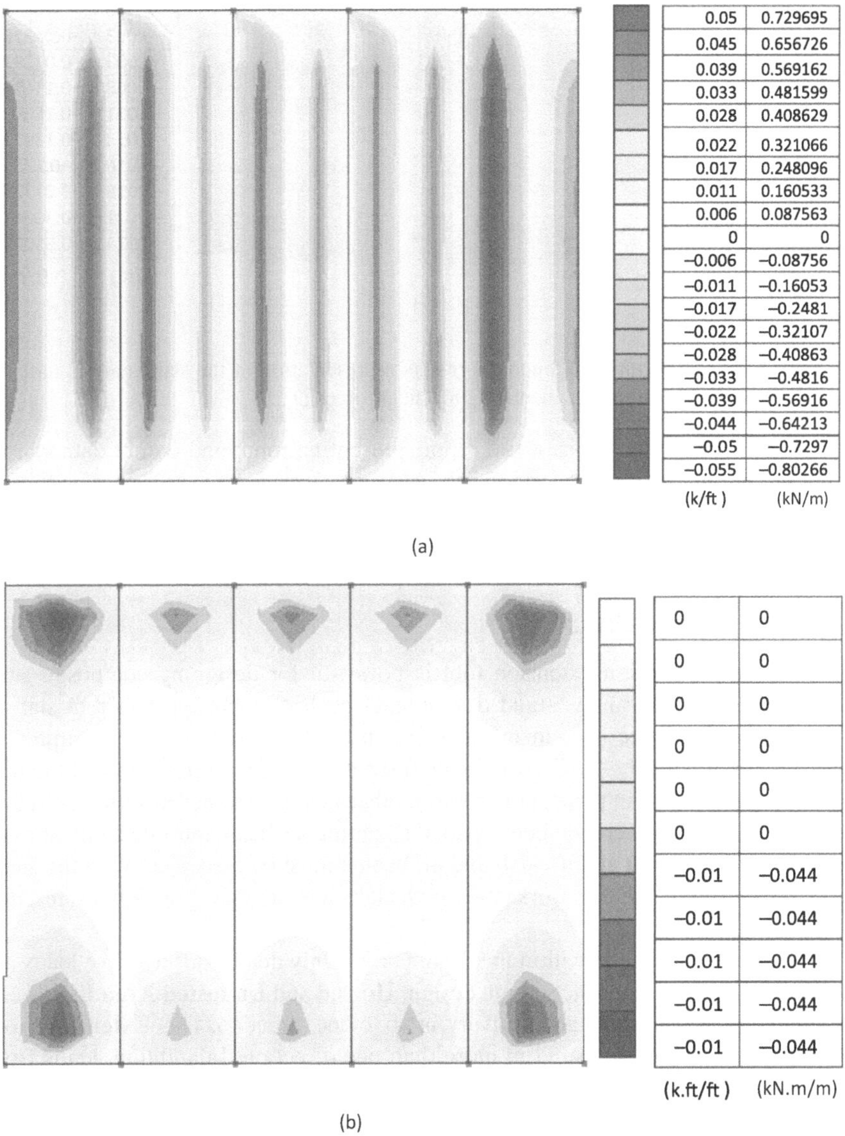

FIGURE 6.28 Graphical results representation of the slab analysis tool: (a) shear forces Q_x; (b) bending moments M_y; (c) shear forces Q_y.

(Continued)

(k/ft)	(kN/m)
0.025	0.364848
0.022	0.321066
0.019	0.277284
0.016	0.233502
0.014	0.204315
0.011	0.160533
0.008	0.116751
0.005	0.07297
0.003	0.043782
0	0
−0.003	−0.04378
−0.005	−0.07297
−0.008	−0.11675
−0.011	−0.16053
−0.013	−0.18972
−0.016	−0.2335
−0.019	−0.27728
−0.021	−0.30647
−0.024	−0.35025

(c)

FIGURE 6.28 (*Continued*) Graphical results representation of the slab analysis tool: (a) shear forces Q_x; (b) bending moments M_y; (c) shear forces Q_y.

Similar to the other tools, a full report presenting input and output data can be generated by customizing the report tab. Also, reports can be exported to and saved as Microsoft Excel or Microsoft Word documents (see the companion website to this book at http://www.crcpress.com/product/isbn/9781482240436).

Composite Section Design

The **Composite Design** extension tool is powerful for designing composite steel beams. The tool can analyze and design steel beams supporting concrete slab or slabs that are concrete on a metal deck (Figure 6.29). The tool is now limited to beams that are linear, AISC (American Institute of Steel Construction) Standard Wide Flange shapes, in plane, not crossing other beams, and not forming a circular bearing chain with other members. Also, the continuous beam must be colinear to be considered a "moment chain"; slab and all beams must be associated with the same level. Furthermore, the tool works only with slabs that are one-directional slabs, and they must be planar.

Loads must be applied within the design area. Only dead loads and live loads are considered for the composite section design. **Hosted** and **Unhosted Area Loads** can be applied to any slab area and can vary in up to one direction. This **Extensions** tool does not support loads tapered in more than one direction. In addition, loads from columns are not taken into account. In other words, point loads must be defined by users. Only the component of a load normal to the plane of the design area will be included in the design computation.

The composite section design tool works similarly to the static analysis of beams covered previously. The user interface for this tool has additional tabs to perform

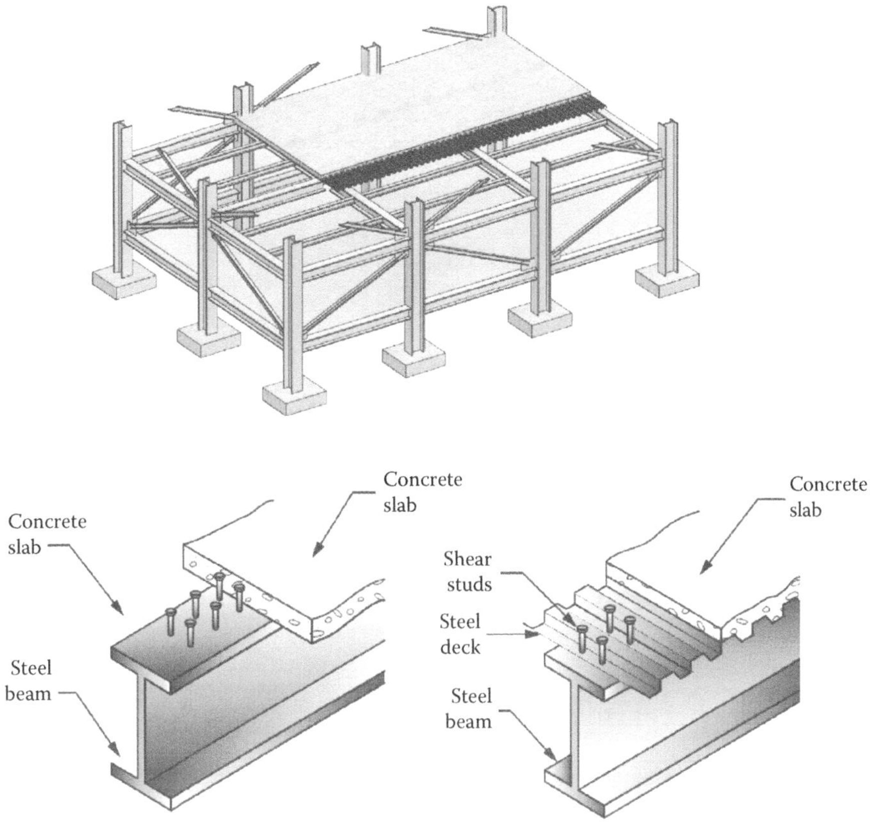

FIGURE 6.29 Composite steel beams.

steel beam design according to the AISC 2005 standard. One selects the beam in the BIM model (Figure 6.30) and then invokes the **Composite Design** tool from the **Analysis** tab under the **Extensions** tab.

As depicted in Figure 6.31, the beam selected is **W6X12**. The load case chosen for the analysis is the factored load **1.2 Dead + 1.6 Live** (1.2DL + 1.6LL). The results of the analysis are shown in Figure 6.32 and Table 6.1. These results include details about the reactions at the supports, shear forces in the beam, and bending moments. Deformation can also be determined using service loads as required by ASCE (American Society of Civil Engineers) 7 (2010).

The **Design** tab offers a number of options for designing the composite steel beam (Figure 6.33). Under **Design Procedure** options, one can choose to design the selected beam or let the tool find the best (optimum) section for the given loading conditions. In addition, you can choose to design the beam as composite or noncomposite. Furthermore, design can be performed for the selected beam or for all beams supporting the slab on that level. At the bottom corner of the tool dialog, there is an icon that indicates the success or failure of the designed beam by showing a red circle with an *X* or a green OK symbol.

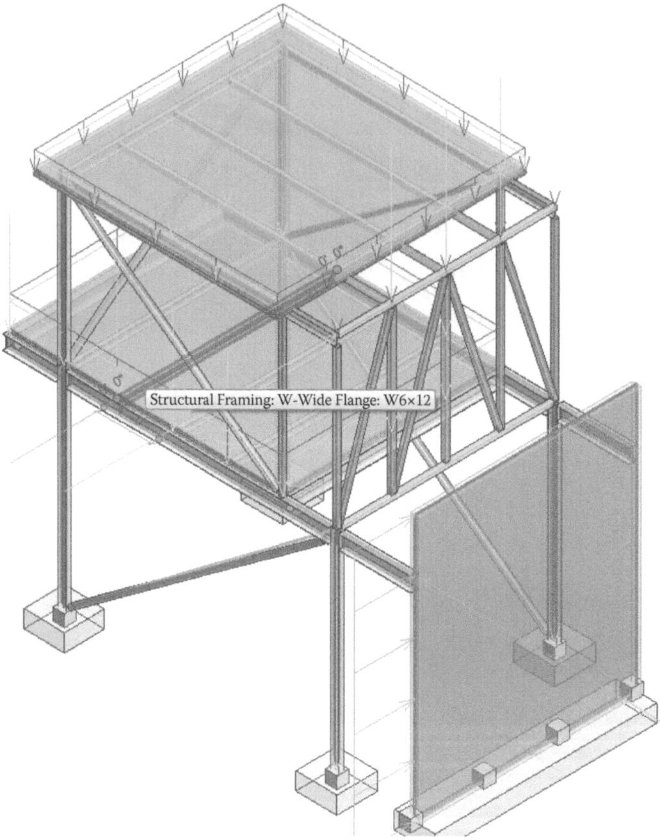

FIGURE 6.30 Beam selected for the composite design tool.

If the beam is designed as a composite, then additional parameters for the camber and studs can be defined. Also, the camber can be manipulated directly within values specified in the **Composite Settings** section of the dialog screen under the **Edit** menu in the graphic viewer (Figure 6.34). During the design of a composite section, the tool will compute the number of studs required to develop the moment capacity, deflection stiffness, or minimum composite action as required by the AISC code. For beams with no point loads, the studs are uniformly distributed across the beam. For girders or other beams with point loads, the studs are distributed in segments to ensure that there are enough studs between every point load and its nearest support to develop the appropriate moment capacity of the composite section.

Results after performing the design for **W6X12** indicated that the beam failed and it was not adequate. Changing the design procedure to **Pick Best Section** and rerunning the design check yielded a beam section **W10X12** for the given load case (see Figure 6.35).

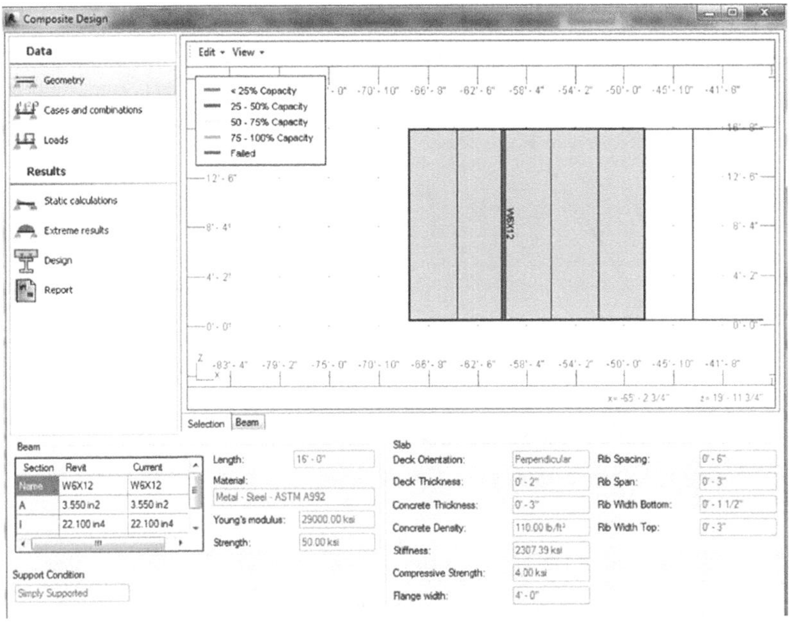

FIGURE 6.31 Composite design tool user interface.

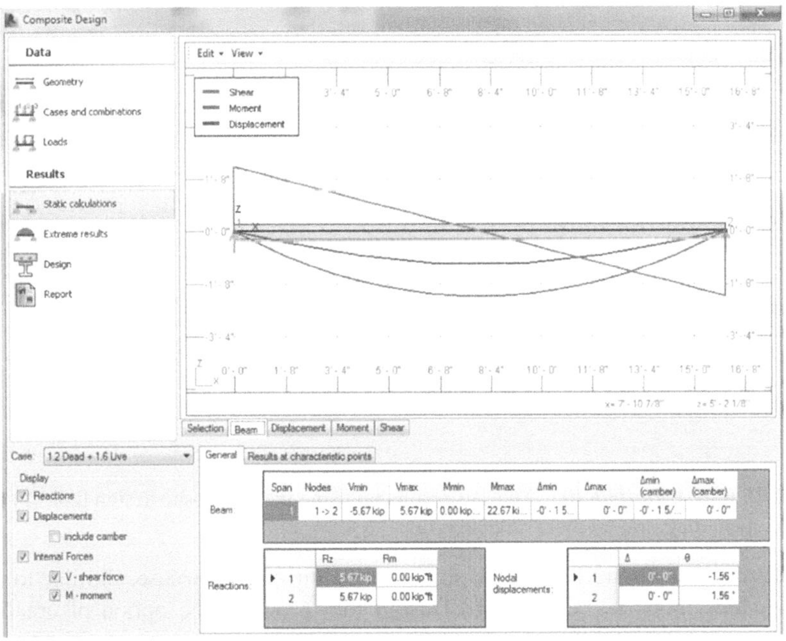

(a)

FIGURE 6.32 Analysis results (a) using the composite design tool; (b) bending moment diagram; (c) shear force diagram.

(Continued)

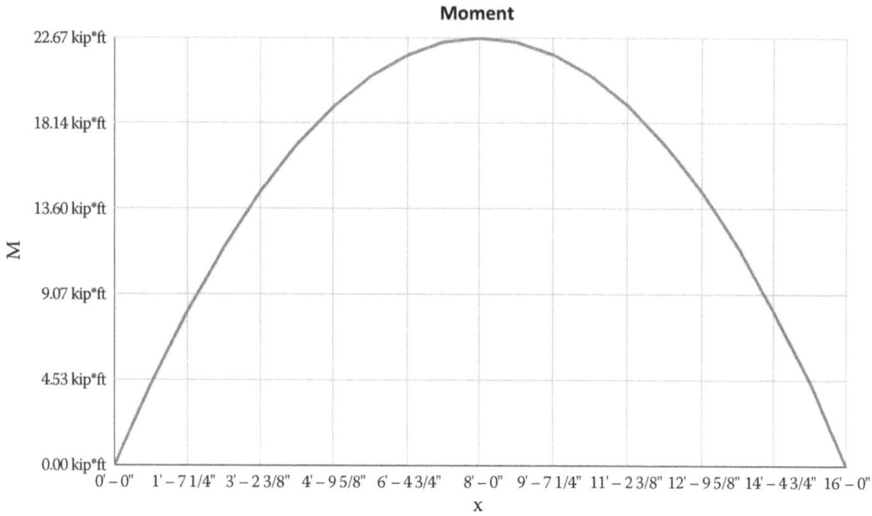

(b)

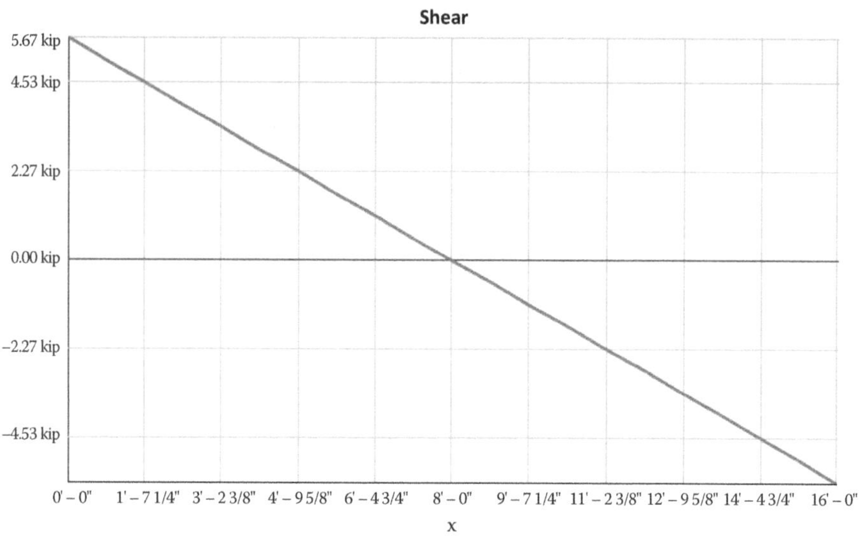

(c)

FIGURE 6.32 (*Continued*) Analysis results (a) using the composite design tool; (b) bending moment diagram; (c) shear force diagram.

Results also showed that nine studs are required in this case. Similar to other **Extensions** tools, the **Composite Design** tool provides the option of creating a full report of detailed design results. These reports can be exported to a Microsoft Word document or Microsoft Excel spreadsheet. Examples of these documents are given in the companion website to this book (http://www.crcpress.com/product/isbn/9781482240436).

TABLE 6.1
Critical Values

Beam Displacement

Symbol	Value (in.)	Value (mm)	*X* (ft)	*X* (m)	Case
Δ min	−0′ 1¼″	−31.75	8′ 0″	2.5	Unfactored
Δ max	0′ 0″	0	0′ 0″	0	Construction Dead
Δ Camber_min	−0′ 1¼″	−31.75	8′ 0″	2.5	Unfactored
Δ Camber_max	0′ 0″	0	0′ 0″	0	Dead

Internal Forces

Symbol	Value	Value (SI)	*X* (ft)	*X* (m)	Case
V_min	−5.67 kip	−25.22 KN	16′–0″	5	1.2 Dead + 1.6 Live
V_max	5.67 kip	25.22 KN	0′–0″	0	1.2 Dead + 1.6 Live
M_min	0.00 kip.ft	0 KN.m	0′–0″	0	Construction Dead
M_max	22.67 kip.ft	30.73 KN.m	8′–0″	2.5	1.2 Dead + 1.6 Live

Reaction

Symbol	Value	Value (SI)	Node	Case
Rz_min	0.00 kip	0 KN	1	Construction Dead
Rz_max	5.67 kip	25.22 KN	1	1.2 Dead + 1.6 Live
Rm_min	0.00 kip.ft	0 KN.m	1	Construction Dead
Rm_max	0.00 kip.ft	0 KN.m	1	Construction Dead

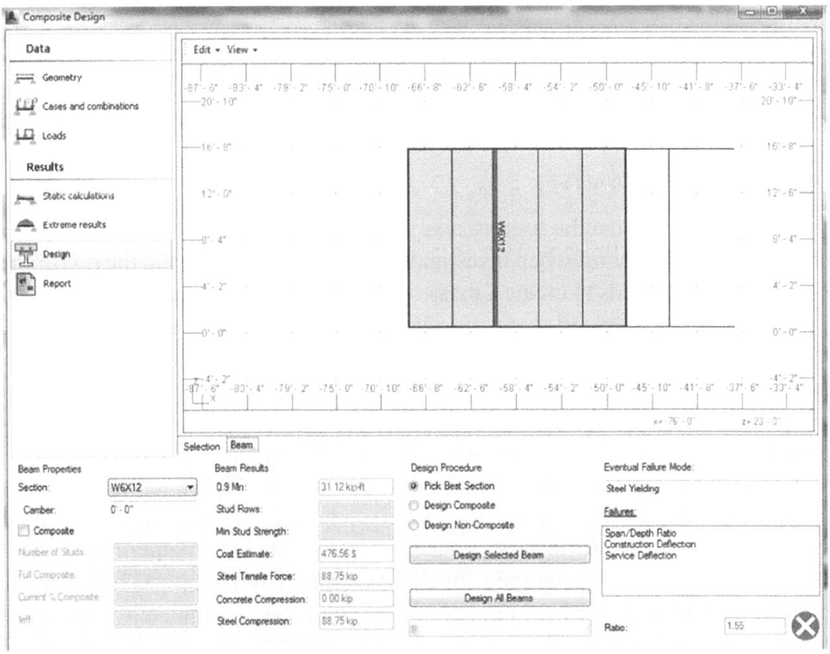

FIGURE 6.33 **Design** tab options.

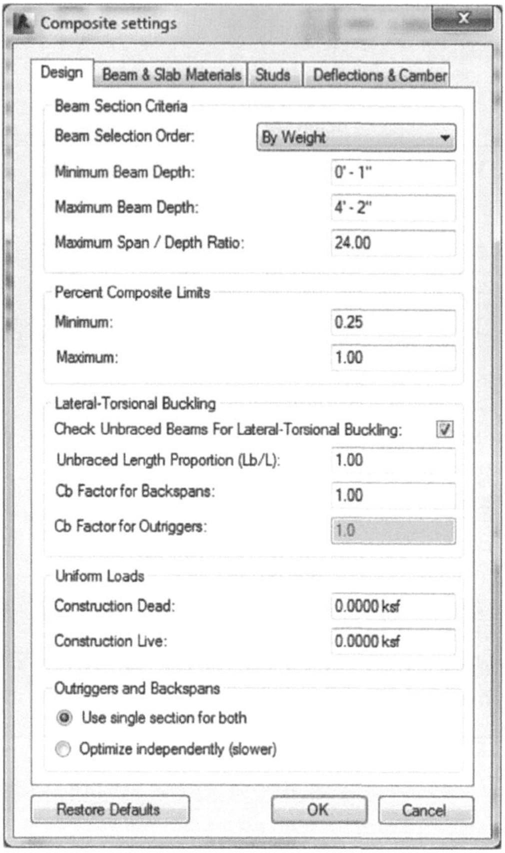

FIGURE 6.34 Composite beam settings.

Conceptual Form Analysis

This **Extensions** tool can be used to analyze any conceptual mass form using static analysis. The tool is useful when investigating a creative form during the preliminary design phase. One needs to create a mass solid using the conceptual mass family editor in Revit and then upload that mass into a Revit project. The solid mass surfaces must be divided into a rectangular pattern (meshing pattern) before using the form in a project (Figure 6.36).

The next step is to select the mass solid created and invoke the **Conceptual form simulation** tool from the **Extensions** tab under **Analysis**. The tool has a similar interface as the previously discussed **Extension** tools (see Figure 6.37a).

The solid mass shown in Figure 6.36 is selected for the analysis using the **Conceptual form simulation** tool. Once the tool is launched, supports (pinned or fixed) for the mass can be defined. In this example, fixed supports are defined at a number of points as shown in Figure 6.36. The mass is subjected to dead, live, and wind loads of 1.0 ksf (47.88 kPa), 0.5 ksf (23.94 kPa), and 0.05 ksf (2.394 kPa),

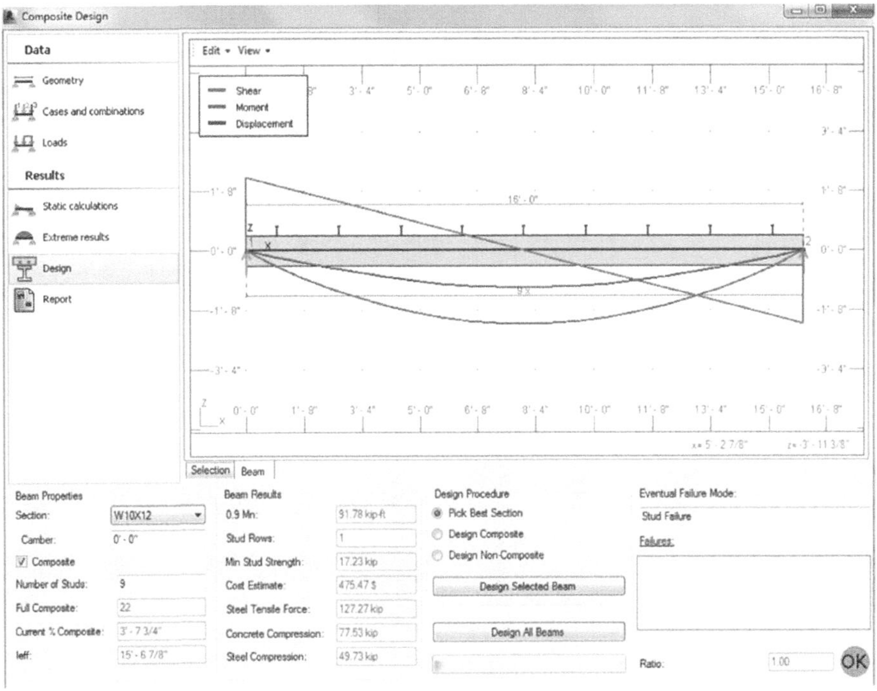

FIGURE 6.35 Design results for the composite section.

respectively. Parts of the analysis computation results are shown in Figure 6.37b–e for the various load cases. The tool can display the following results: (1) internal forces (M_{xx}, M_{yy}, M_{xy}, N_{xx}, N_{yy}, N_{xy}); (2) deformation (U_x, U_y, U_z). Internal forces (bending moment, shear forces, and axial forces) are presented in the local —coordinate system, where the *X* axis is always set according to the *Z* axis of the global coordinate system. The only exceptions are horizontal elements (where the *X* axis is set according to the *Y* axis of the global coordinate system). Displacements are presented in the global coordinate system. The tabular results also show extreme values (maximum and minimum) for the calculated internal forces and reactions (Table 6.2). A full report showing the internal forces and reactions can be produced and exported to a Microsoft Word or a Microsoft Excel document.

A few important notes about this tool include the definition of loads and supports. First, not all loads defined in Revit are exported to the **Conceptual form simulation** tool automatically. Only point loads applied at nodes are considered in tool calculations. Second, changes made by adding supports or loads using the tool are not saved in the original Revit file. Furthermore, automatically generated wind loads act as horizontal forces applied to surfaces, which are visible from the windward side, and then applied proportionally to nodes. Area loads defined in this tool are automatically converted to projected, concentrated loads that are applied at mesh nodes.

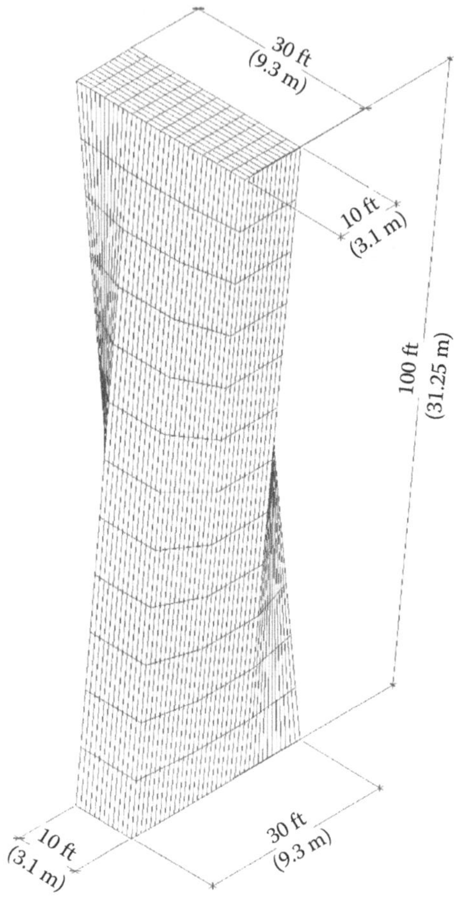

FIGURE 6.36 Conceptual mass with meshing pattern.

TABLE 6.2
Extreme Values Caused by Live Load

Extreme values

Symbol	Minimum value	Maximum value	Case
Mxx	-566.30 kip-ft/ft (kip*ft)/ft	912.19 kip-ft/ft (kip*ft)/ft	Live
Myy	-1789.41 kip-ft/ft (kip*ft)/ft	1036.47 kip-ft/ft (kip*ft)/ft	Live
Mxy	-581.34 kip-ft/ft (kip*ft)/ft	544.75 kip-ft/ft (kip*ft)/ft	Live
Nxx	-832.759 kip/ft kip/ft	21.973 kip/ft kip/ft	Live
Nyy	-269.109 kip/ft kip/ft	90.644 kip/ft kip/ft	Live
Nxy	-533.827 kip/ft kip/ft	348.268 kip/ft kip/ft	Live
Ux	-0' - 7 31/32"	0' - 8 3/16"	Live
Uy	-0' - 1 13/16"	0' - 2 1/16"	Live
Uz	-1' - 6 5/32"	0' - 0"	Live

New Load Takedown Tool

Since **2023**, the **Load Takedown (LTD) tool** has been exclusively available in the **Autodesk Robot Structural Analysis (RSA) platform** rather than **Revit**. This section provides an in-depth overview of the **LTD tool**, including its features and functionality, accompanied by practical examples.

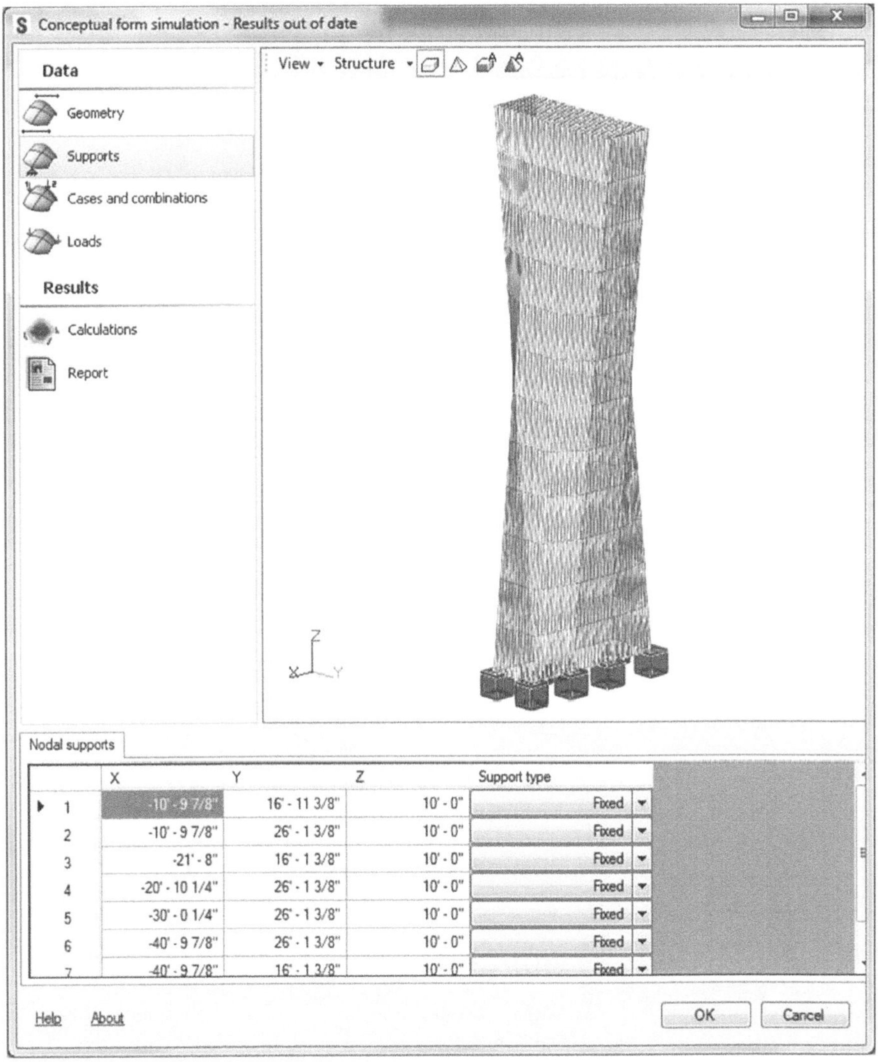

(a)

FIGURE 6.37 (a) Example of results from the **Conceptual form simulation** tool; (b) M_{xx} due to dead load; (c) M_{xx} due to wind load; (d) N_{xx} due to wind load; (e) lateral deformation U_x due to wind load.

(Continued)

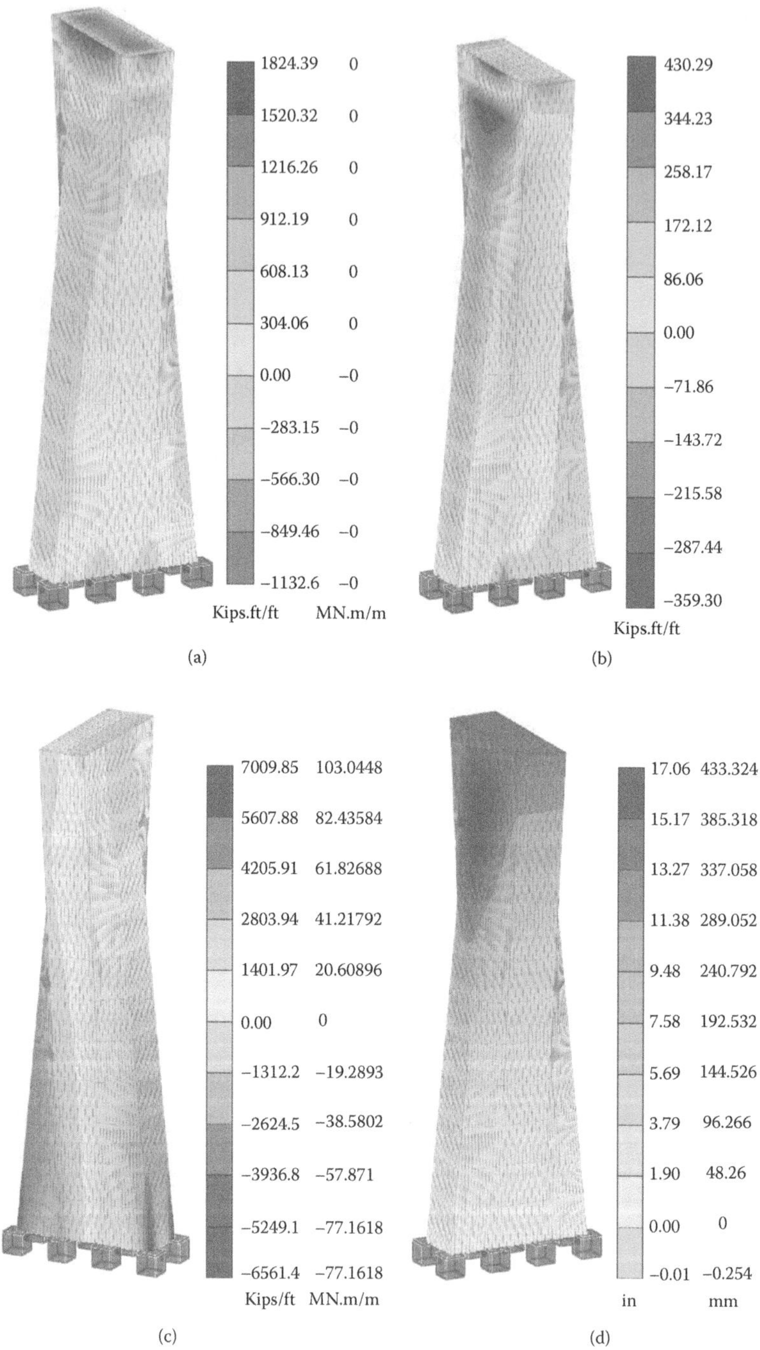

FIGURE 6.37 (*Continued*) (a) Example of results from the **Conceptual form simulation** tool; (b) M_{xx} due to dead load; (c) M_{xx} due to wind load; (d) N_{xx} due to wind load; (e) lateral deformation U_x due to wind load.

The **LTD** systematically distributes **floor loads** onto **beams, columns, and walls**, ensuring accurate load transfer from the upper stories down **to** the foundation. This method provides a fast and insightful analysis, making it particularly valuable during **early design stages**, where understanding **gravity load distribution** is crucial for informed building design decisions. The three primary objectives for this new capability are to enable engineers to:

- Estimate foundation reactions at the early design stages, even while work on the upper structure is still in progress.
- Analyze building behavior to optimize column layout and load distribution for structural efficiency.
- Transfer loads to individual structural elements, enabling object-based code design using traditional engineering tools or spreadsheets.

The outcomes of the LTD analysis are

i. The resulting loads on individual structural objects such as beams, columns, and walls coming from the floor load distribution, which can be used as a load input in external design applications.
ii. Internal forces in beams and columns based on obtained resulting loads which can be consumed in internal design modules.
iii. Reactions on supports.

To start the LTD tool in Autodesk RSA platform, one begins with a new project as shown in Figure 6.38.

After that selection, a simple structure will be defined as shown in Figure 6.39. It consists of 6 in (12 cm) concrete slab, 18 × 18 in (45 × 45 cm) concrete columns, and

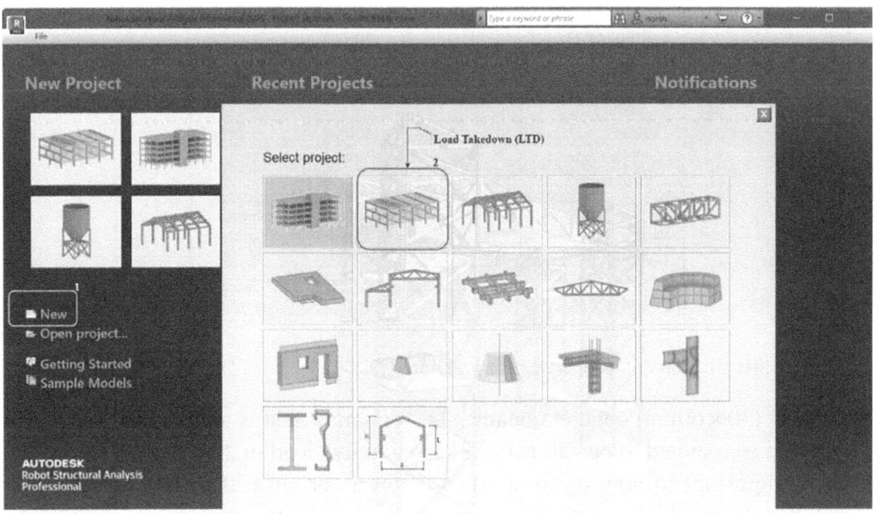

FIGURE 6.38 Starting the LTD tool in RSA platform.

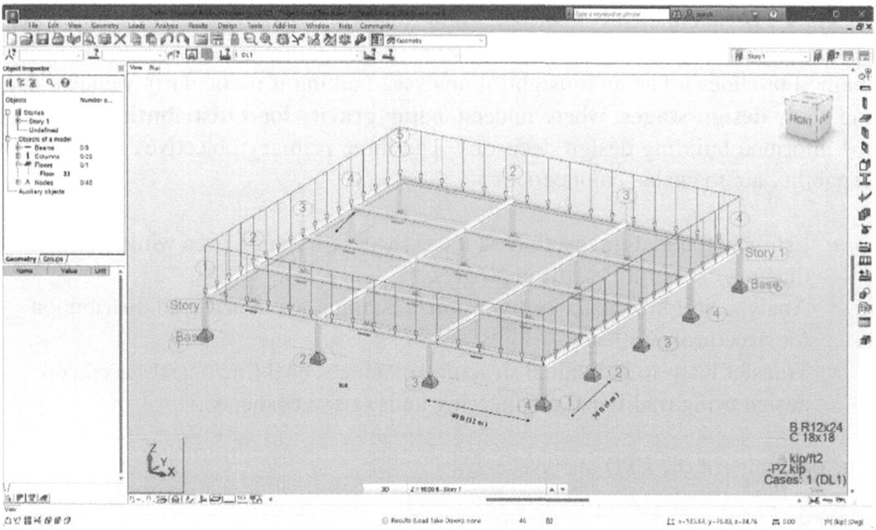

FIGURE 6.39 Illustration of the LTD tool in RSA platform.

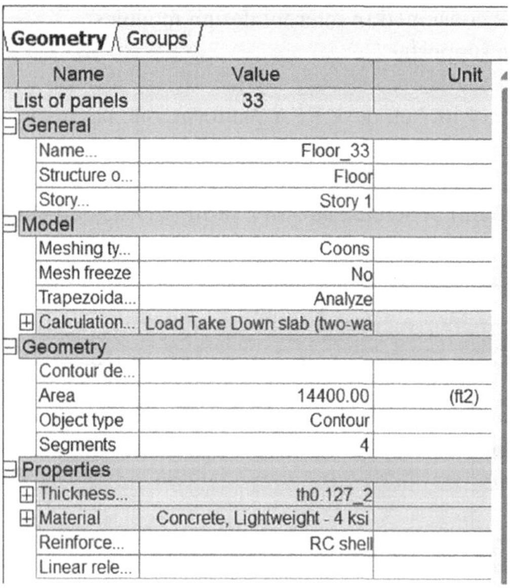

Geometry / Groups

Name	Value	Unit
List of panels	33	
General		
Name...	Floor_33	
Structure o...	Floor	
Story...	Story 1	
Model		
Meshing ty...	Coons	
Mesh freeze	No	
Trapezoida...	Analyze	
Calculation...	Load Take Down slab (two-wa	
Geometry		
Contour de...		
Area	14400.00	(ft2)
Object type	Contour	
Segments	4	
Properties		
Thickness...	th0.127_2	
Material	Concrete, Lightweight - 4 ksi	
Reinforce...	RC shell	
Linear rele...		

FIGURE 6.40 Floor slab properties for LTD analysis.

12×24 in (30×60cm) concrete beams. The columns grid is at 40×30ft (12×9m). Dead load is assumed to be 150 psf (7.2kPa) and live load of 20 psf (1.0kPa).

It is important to note the properties of the floor since this will influence the Load Takedown (LTD) analysis. Figure 6.40 depicts the properties of the floor slab. The important parameter here is the "Calculation Method", which should be "Load Take Down".

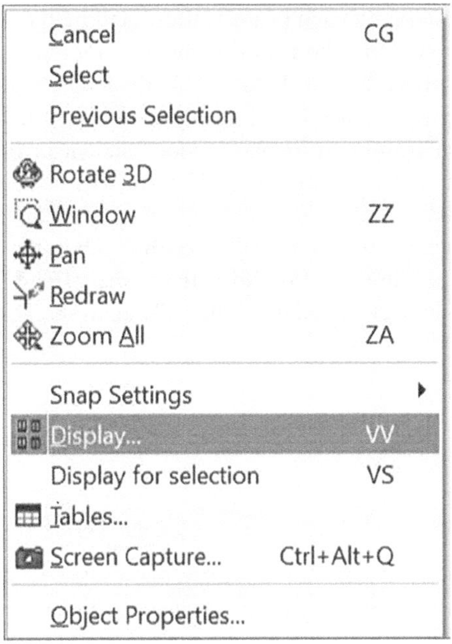

FIGURE 6.41 Invoking the display options.

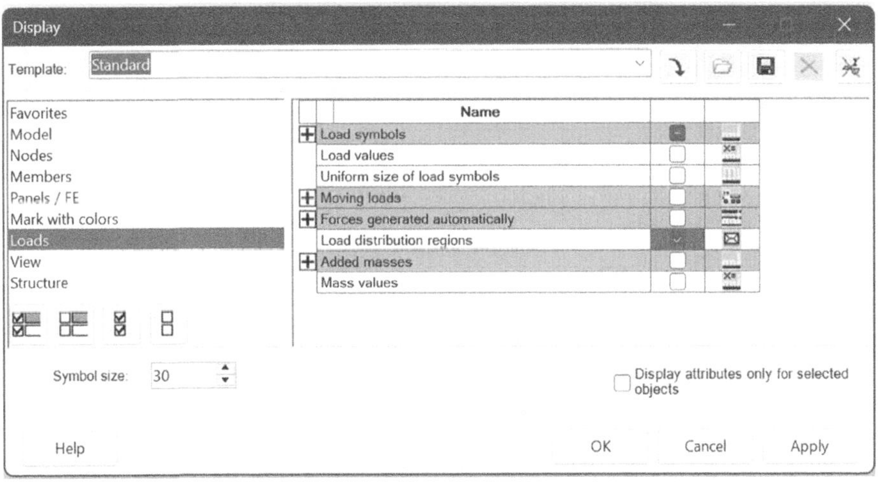

FIGURE 6.42 Display options.

You can now perform the analysis in RSA software by clicking the "Calculator" icon located in the top toolbar. Once the analysis is complete, access the display options by right-clicking anywhere within the drawing area and selecting "Display" from the context menu (refer to Figure 6.41).

The display options are illustrated in Figure 6.42. One needs to select "Loads" from the lefthand side menu, and then click on "Load distribution regions" and then click on the "Apply" button. The results of these actions are shown in Figure 6.100f. The load tributary areas distribution on each beam is illustrated by the different regions in the case of two-way slab system. The load distribution along the shorter span follows a triangular pattern, whereas the distribution along the longer span is trapezoidal.

To see the load acting on each beam, one needs to activate the "Forces generated automatically" option from load display options shown in Figure 6.42. Once that step is completed, the forces acting on each beam will be displayed as depicted in Figures 6.43 and 6.44.

Alternatively, you can navigate to the **"Results"** tab in the top menu of the RSA software platform (see Figure 6.45) and activate the **"Load Take Down Tables"** option to display the numerical results of the LTD analysis. This will present various

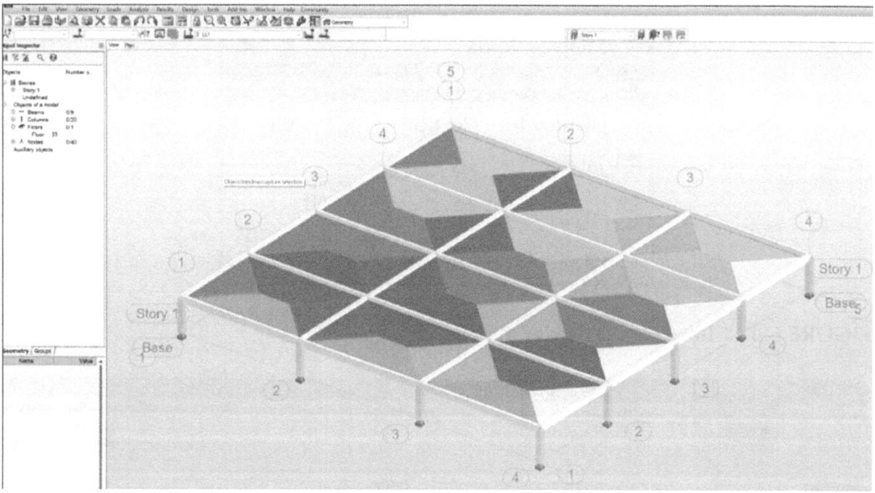

FIGURE 6.43 LTD tributary areas diagram.

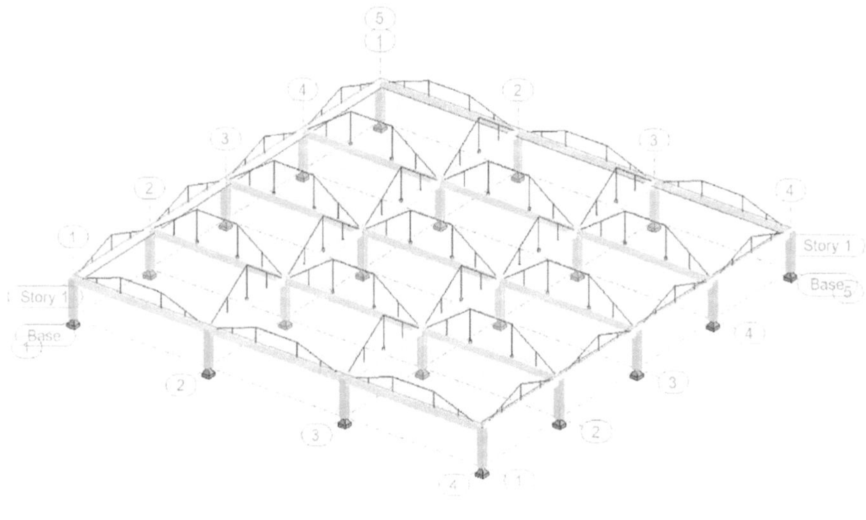

FIGURE 6.44 Force distribution using LTD analysis.

options at the bottom of the screen—such as column and beam loads, as well as tributary areas—to view detailed numerical data (refer to Figure 6.46).

Also, one can select "Load Take Down Diagrams" option from the "Results" tab (Figure 6.46) to display further numerical values of the LTD analysis as illustrated in Figure 6.47. For example, the results for columns and beams are shown in Figure 6.48.

As for the internal forces analysis such as finding the shear force, reactions, stresses, deformation and bending moment diagrams on beams, columns, or slabs, one can use the "Results" option from the dropdown menu. Figure 6.49 illustrates the different internal forces diagrams options while Figure 6.50 presents some examples. You can also get detailed analysis for each single structural elements by selecting

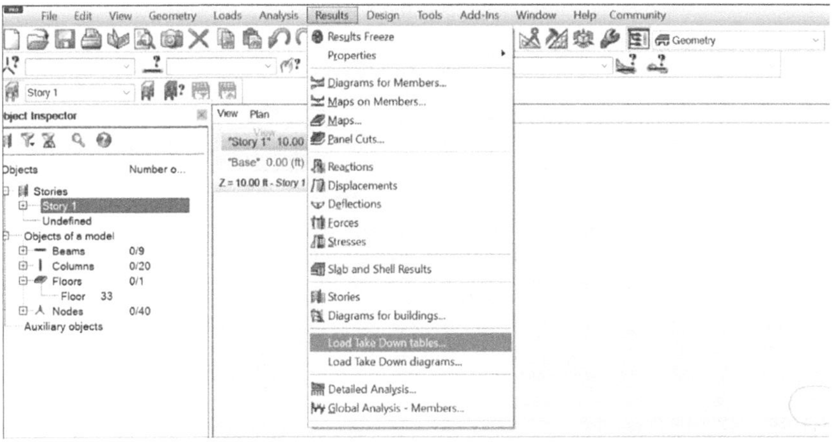

FIGURE 6.45 Load take down tables.

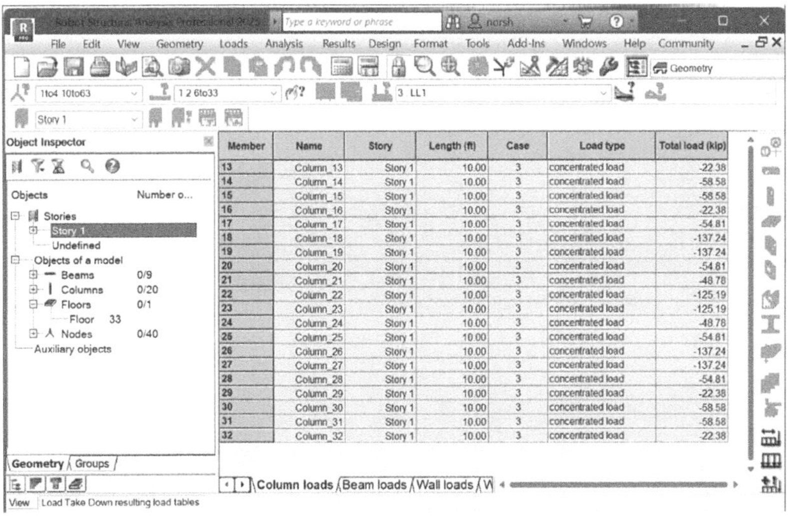

FIGURE 6.46 LTD tables.

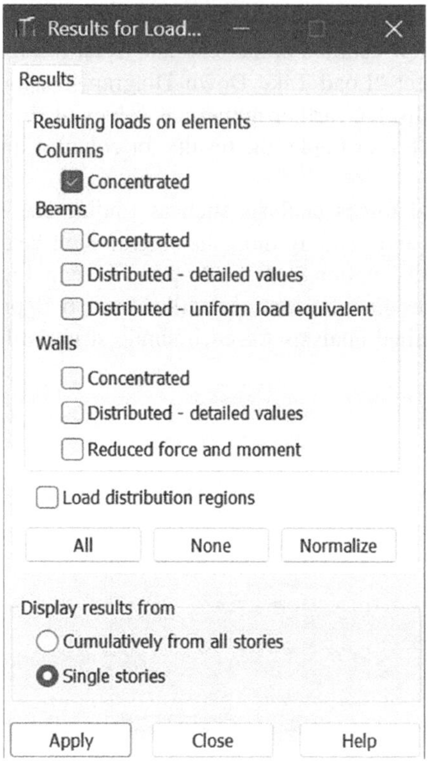

FIGURE 6.47 LTD results.

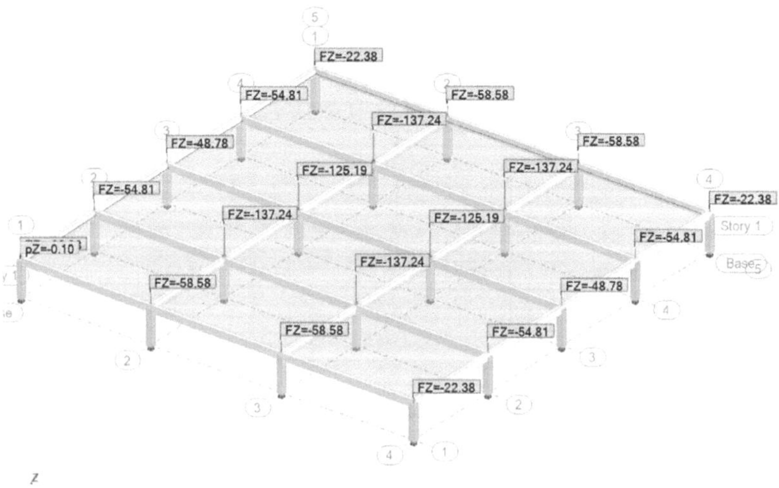

(a)

FIGURE 6.48 (a) Columns forces due to dead load using the LTD analysis. (b) Forces on beams as determined by the LTD analysis.

(Continued)

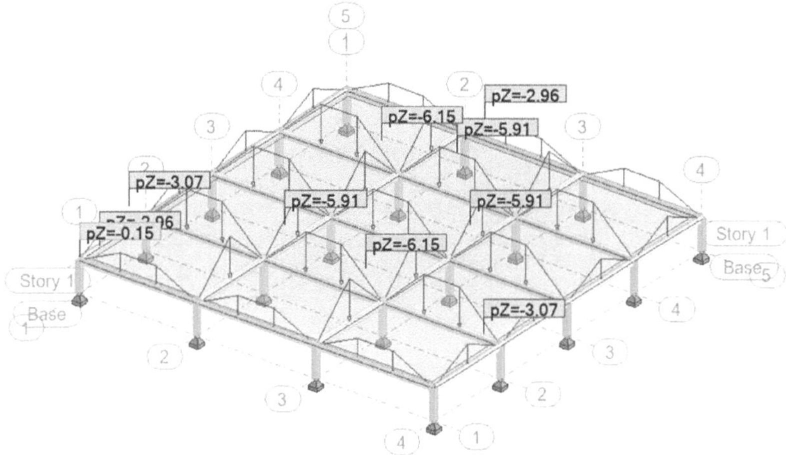

(b)

FIGURE 6.48 (*Continued*) (a) Columns forces due to dead load using the LTD analysis. (b) Forces on beams as determined by the LTD analysis.

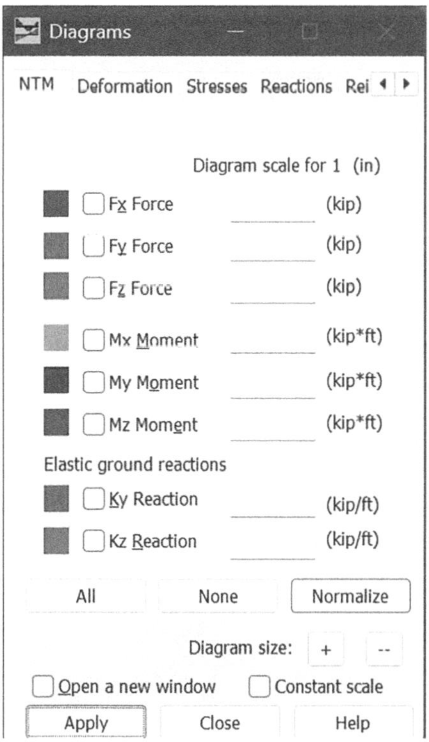

FIGURE 6.49 Diagrams options for the internal forces and deformation.

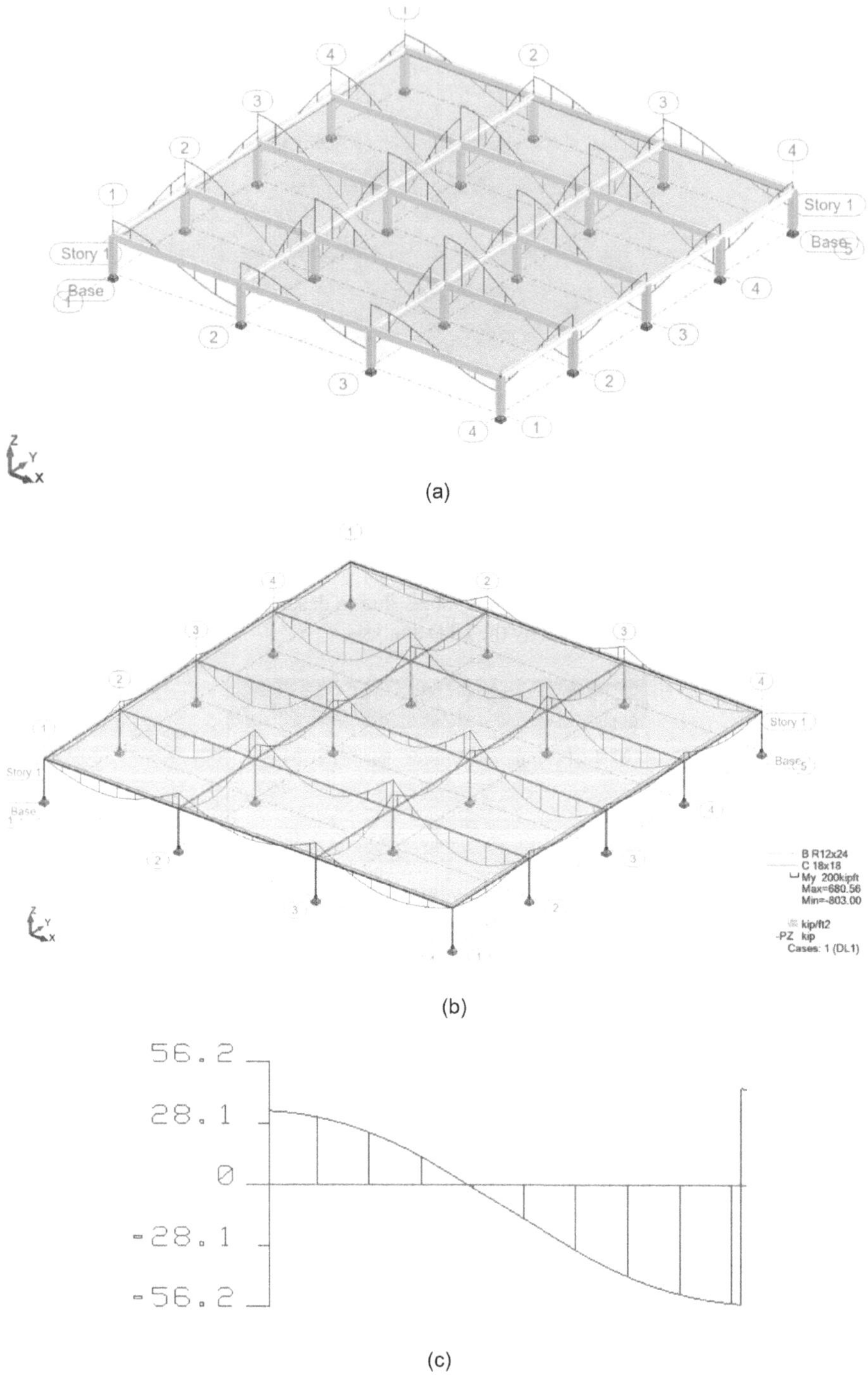

FIGURE 6.50 (a) Shear force distribution due to dead load. (b) Bending moment diagrams due to dead load. (c) Shear force distribution for an interior beam due to dead load.

the "Detailed Analysis" option. For instance, Figure 6.50c depicts the shear force distribution of an interior beam.

Next, let's add four additional stories to the initial structure and perform the Load Take Down (LTD) analysis. For multi-story buildings, LTD analysis is particularly useful for quickly estimating preliminary member sizes. It can also serve as a tool to validate the results of more detailed structural analyses.

To add stories in the RSA software, right-click on **"Stories"** in the **Object Inspector** located on the left side of the screen, and select the **"Copy Story"** command (see Figure 6.51).

Now, let's run the LTD analysis again for the 5-story structure and repeat the same previous steps for the entire structure or you can select one specific story. Figure 6.52 depicts the results for the tributary areas for each beam in the entire structure. Furthermore, one can utilize the "Results" command to navigate different results (Figure 6.53).

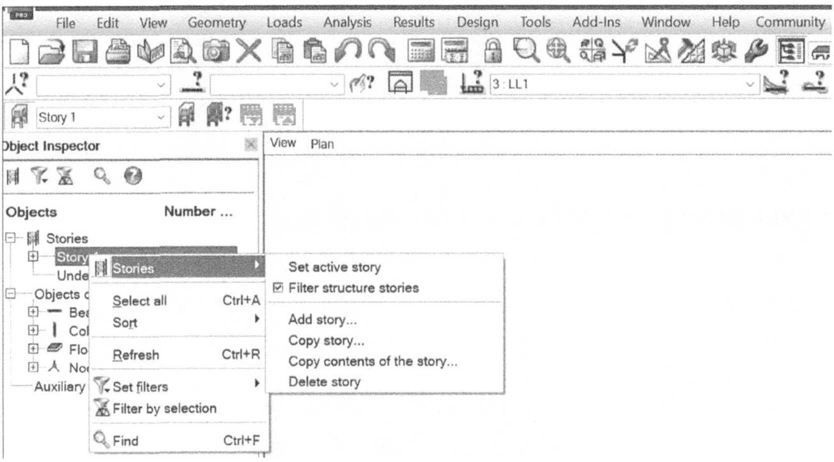

(a)

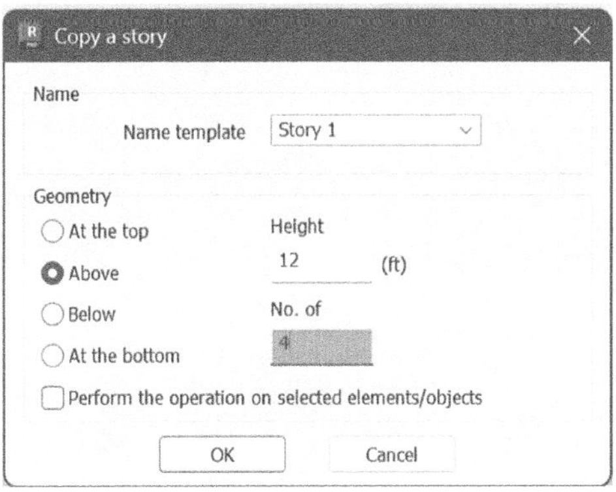

(b)

FIGURE 6.51 Adding typical stories to the initial structure.

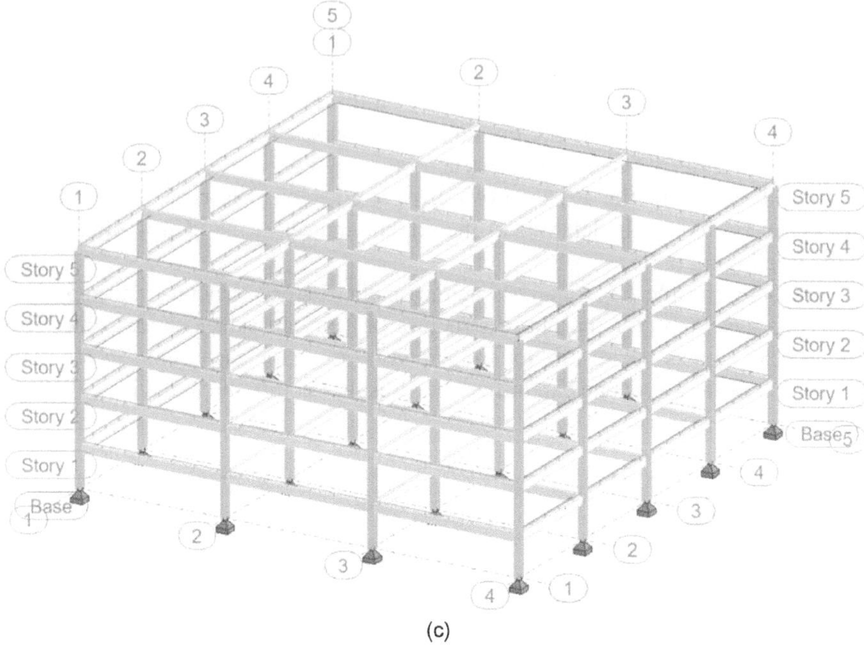

FIGURE 6.51 Adding typical stories to the initial structure.

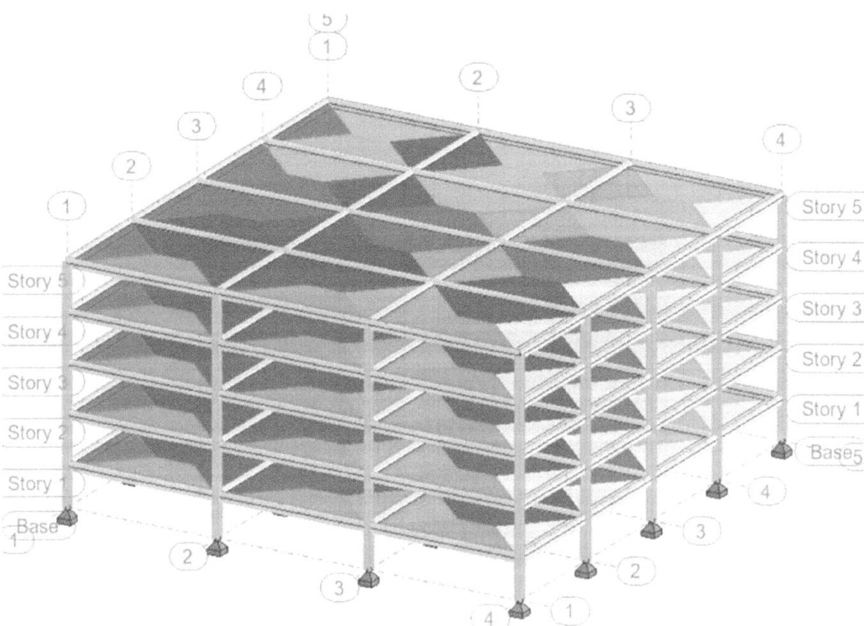

FIGURE 6.52 Load tributary areas for each beam.

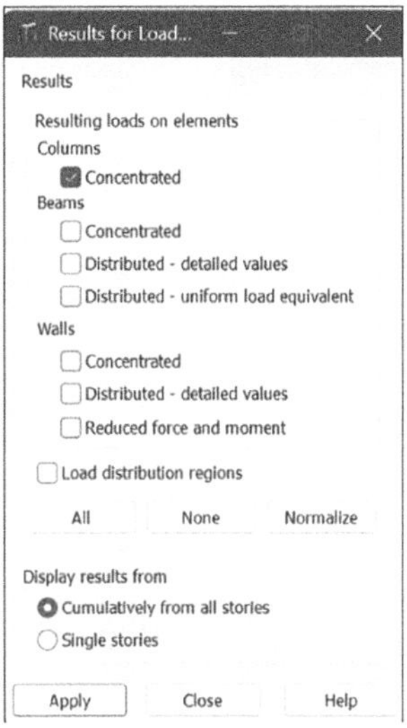

(a)

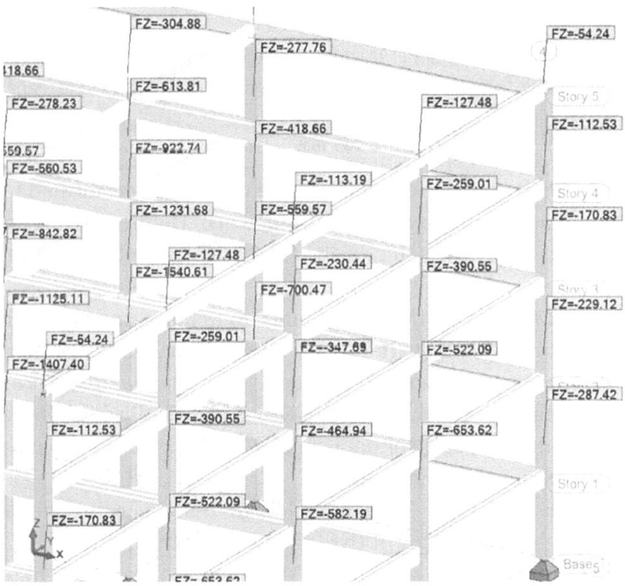

(b)

FIGURE 6.53 (a) LTD Results dialog options, (b) Columns forces due to dead load using the cumulative forces option, (c) Forces on beams due to dead load using LTD analysis.

(*Continued*)

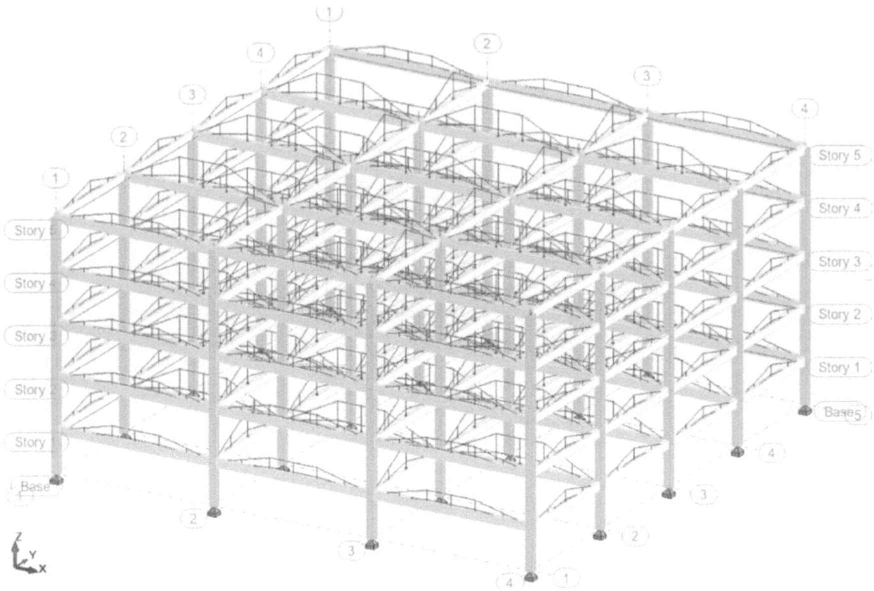

(c)

FIGURE 6.53 (*Continued*) (a) LTD Results dialog options, (b) Columns forces due to dead load using the cumulative forces option, (c) Forces on beams due to dead load using LTD analysis.

ADVANCED STRUCTURAL DESIGN

In the following sections, detailed analysis and design are performed using the BIM models developed according to the SAS framework. Before starting this process, it would be beneficial to review some of the SAS concepts mentioned in the previous chapters.

One of the key aspects of BIM is the ability to manage and communicate project data among collaborating trades. In broader perspective, this refers to interoperability of technology to enable efficiency in practice. In other words, a project team can freely exchange data across different applications and platforms. BIM as such serves as a shared knowledge resource for data about a facility, thus forming a reliable basis for design and construction decisions.

The interoperability of the BIM model with other general-purpose structural design software is the main focus of the following sections.

The first interoperability method for BIM models is the direct link of data transfer. With this direct link method of interoperability, there is typically a menu item in the BIM authoring platform that references the other structural software program. For example, in Revit, there is a menu that has both **Import from** and **Export to** RISA-3D Structural software under the **Add-Ins** tab. When the **Export To RISA** is selected, a dialog box pops up prompting a choice between **RISA Floor** or **RISA-3D** along with other export settings (see Figure 6.54).

This is normally done through APIs (application programming interfaces), which allow BIM model data to be directly accessed. Sometimes, these APIs are proprietary,

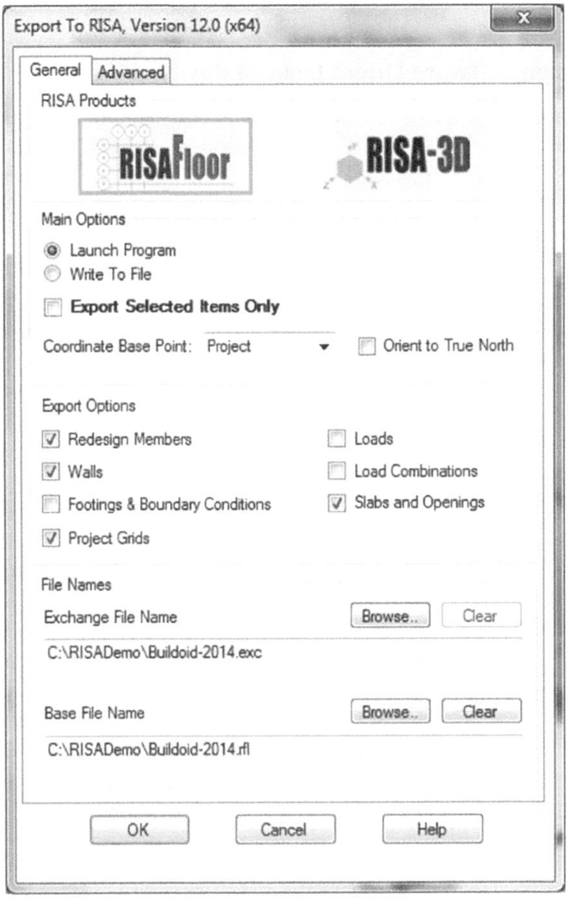

FIGURE 6.54 Direct link to export Revit models to RISA.

and often they are freely available. For instance, many structural engineering software companies (e.g., Computers & Structures) are making these direct links available through their websites (see Table 6.3).

Another method of interoperability of data is using a file conversion method, often with a neutral data format such as the IFC (Industry Foundation Classes). Then, this neutral data file can be imported into the structural design software program. Results of the analysis are converted again to a neutral data format (IFC), then imported into the BIM authoring platform.

This method still works well, but it can be like the game "Telephone." When you played Telephone as a kid, you usually sat in a circle, and one person whispered a sentence to the next person, who then whispered to the next, and so on down the line. Sometimes, the sentence sounded the same as the original when it was whispered to the last person, but sometimes a word here or there was different. Model transfer can be similar: As the model is converted multiple times, it may lose some of its integrity. In general, this is an effective method; however, there is more likelihood of model degradation.

TABLE 6.3
Structural Design Software Direct Links to Revit

Structural Analysis and Design SOFTWARE	Direct Link to BIM Platform (Autodesk Revit)
ADAPT-Builder	Bidirectional http://www.adaptsoft.com/revitstructure/
CSC Fastrack	Bidirectional http://www.cscworld.com/Regional/USA/Products/CSC-Integrator.aspx
ETABS	Bidirectional https://www.csiamerica.com/building-information-modeling
RAM Structural System	Bidirectional http://www.bentley.com/en-US/Free+Software/rss+revit+link.htm
S-FRAME	Bidirectional https://s-frame.com/BIMLinks.htm
RISA-3D/RISAFloor	Bidirectional http://www.risa.com/partners/prt_revit.html
Robot Structural Analysis Professional	Bidirectional https://apps.exchange.autodesk.com/RVT/en/Detail/Index?id=appstore. exchange.autodesk.com%3astructuralanalysisandcodecheckingtoolkit2014%3aen
SAP2000	Bidirectional https://www.csiamerica.com/building-information-modeling
STAAD.Pro	One way (SI) http://www.bentley.com/en-US/Free+Software/bentley+iware.htm

Wood Systems

Example 6.1

In this section, the design of wood structures is demonstrated utilizing the direct link between Revit and RISAFloor and RISA-3D. The RISA software used in the example that follows is the free demo version of the software. It is important to recognize the size limitation of the model when using this demo version because it has limitations in the number of structural objects that it can handle.

At the beginning of the structural computations, it is customary to distinguish between the two main supporting systems:

1. **Vertical (Gravity) Resisting System**: This includes beams, columns, walls, slabs, roofs, trusses, and foundations.
2. **Lateral Resisting System**: Examples of lateral support elements include bracing, shear walls, slabs, and fixed connections (rigid frames).

The organization of these systems is essential for the overall stability of the structure. Revit will allow users to set parameters to define the role of the member in the resisting structures. The example used to demonstrate these steps along with the detailed analysis and design are depicted in Figure 6.55. The BIM physical and analytical models are shown in Figures 6.54 and 6.56, respectively. Loads are applied to the first floor and the roof as given in Figure 6.40b. The vertical and lateral resisting elements are specified in the analytical model (see Figure 6.40b). Loads applied include a dead load of 30 psf (1.44 kPa) and a live load of 40 psf (1.92 kPa) for the floor. The roof receives a dead load of 10 psf (0.48 kPa) and a live load of 20 psf (0.96 kPa). Wind and seismic loads are computed using RISA software, as illustrated in Figure 6.56.

Initially, columns used in the first floor are southern pine wood columns of 8×8 in. (191×191 mm actual). Girders in that floor are 4×12 in. (89×286 mm actual), and beams are 2×10 in. (38×235 mm actual). The roof is supported by four columns that are 4×4 in. (89×89 mm actual), and roof beams are 2×10 in. (38×235 mm actual). Brace elements are 6×6 in. (140×140 mm actual) southern pine wood, grade 1.

Using the direct link between Revit and RISA, the model is exported to RISAFloor first (Figure 6.54) and then from RISAFloor to RISA-3D to perform lateral load analysis. One invokes this link from the **Add-Ins** tab in Revit main user interface. The results for the gravity load analysis based on the US National Design Specification (NDS) for Wood Structures (2005; Allowable Stress Design, ASD).

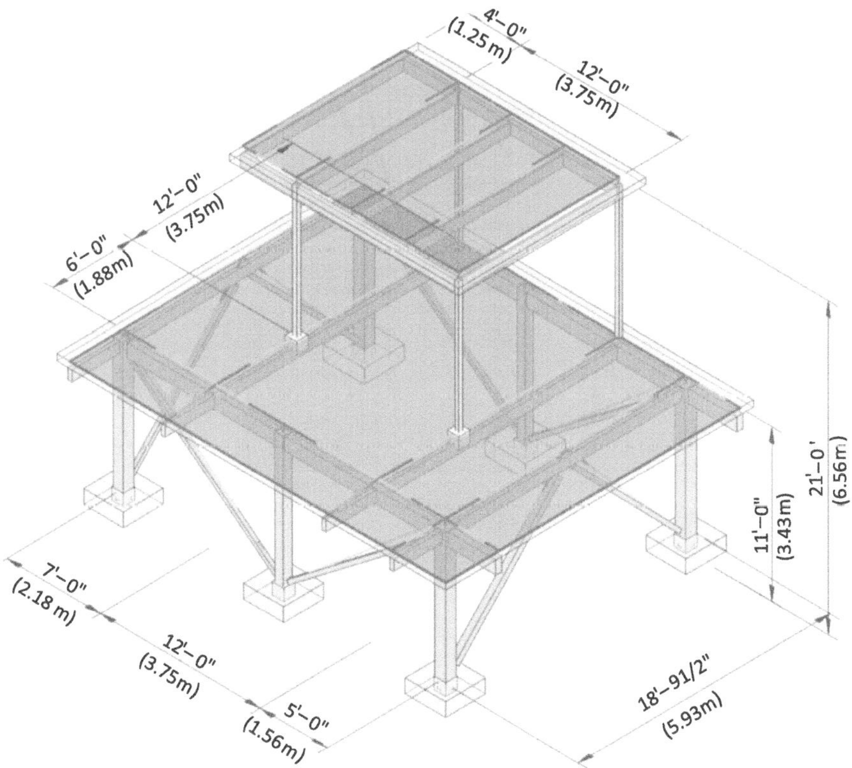

(a)

FIGURE 6.55 (a) BIM model for wood structure. (b) Analytical model along with end releases and resisting role.

(Continued)

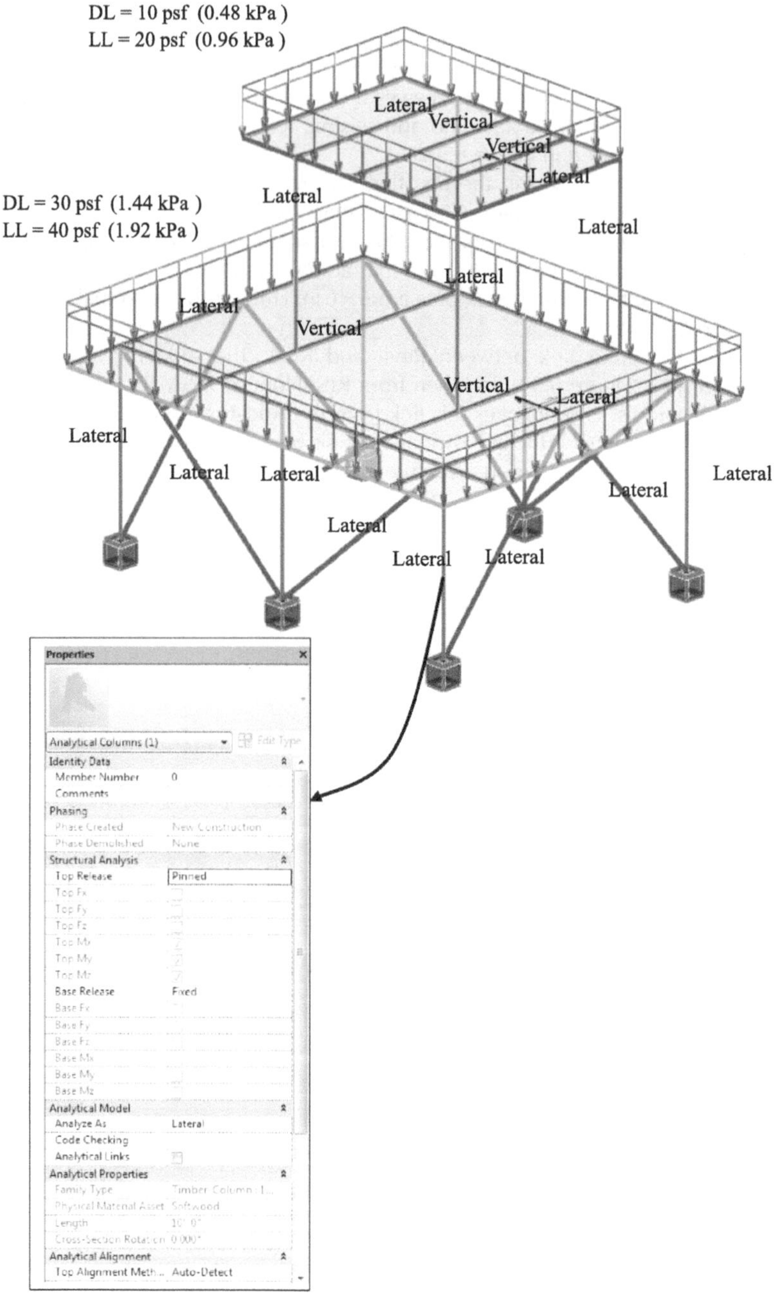

(b)

FIGURE 6.55 (*Continued*) (a) BIM model for wood structure. (b) Analytical model along with end releases and resisting role.

The load combinations are produced by RISA **Load Combination Generator** using the 2012 International Building Code (IBC) ASD (Figure 6.56).

After generating the load combinations, design computation for gravity loads can be carried out by selecting the **Solve** menu in RISAFloor. Results are summarized in Figure 6.57. The framing plan shown in Figure 6.42 also depicts end reactions for each member. Notice also that all the original sizes are now changed to support gravity loads.

In addition, you can select one element from the framing plan to display detailed analysis and design computation checks. An example of the design check details for the roof beam at Grid 7 is given in the companion website to this book (http://www.crcpress.com/product/isbn/9781482240436). Part of that report is given in Figure 6.58. Students are encouraged to verify such results by carrying out hand computations.

The next step is to design members to resist lateral loads. RISAFloor has a direct link to RISA-3D that shows up at the top right corner of the RISA-Floor screen. Selecting that link will export the model to RISA-3D, where wind load and seismic load parameters can be defined for automatic lateral load generation. The dialog boxes that appear once you initiate the RISA-3D link are for the wind and seismic load parameters (see Figure 6.59).

These loads are then generated according to the ASD ASCE 7–10 standard. One needs to generate the load combinations before conducting the design computations. RISA-3D has a load combination generator that includes various later loads, such as wind and seismic forces and building codes.

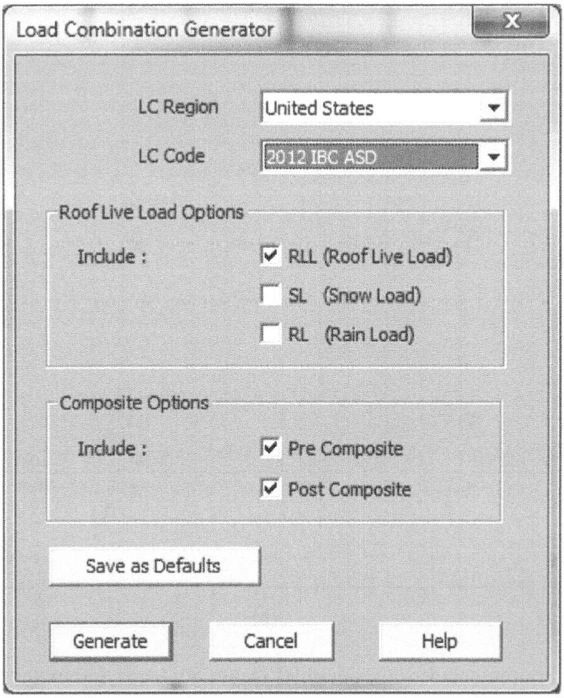

FIGURE 6.56 RISA load combination generator.

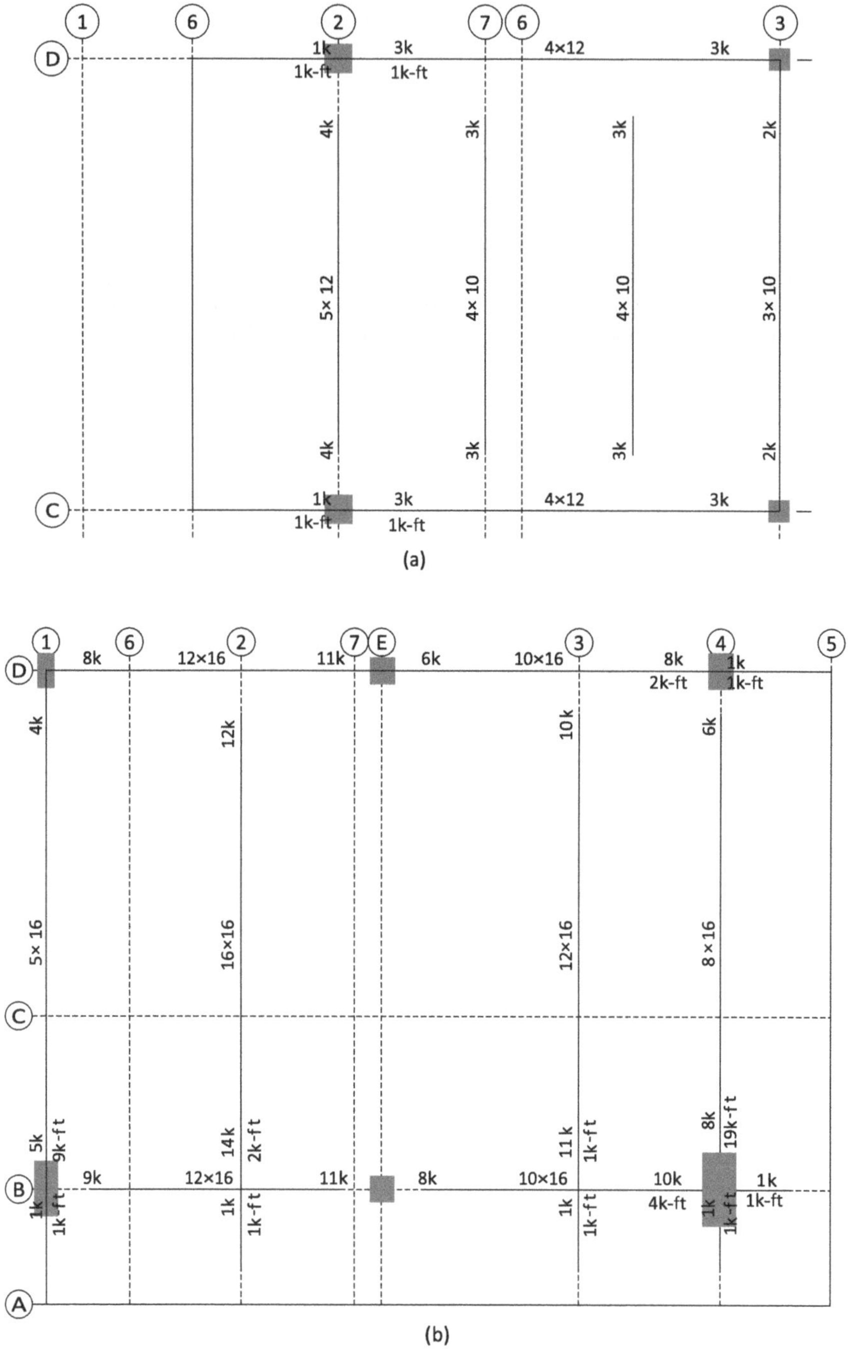

FIGURE 6.57 Framing plan as a result of RISA gravity loads design: (a) roof; (b) floor.

Shear: 51.4% Capacity at 0 ft for LC 1 (Comb. 1 -DL)
V = 1.94 k at 0 ft fv = .09 ksi Fv' = 0.21875 ksi

Wood Shear Factors
Fv = .175 ksi CD = 1.25 Cm = 1 Ct = 1

Bending: 99.3% Capacity at 6 ft for LC 6 (IBC 16-10 (a) Post)
M = 7.261 k-ft Rb = 10.4276 le-bend Top = 12 ft le-bend Bot = 12 ft fb = 1.746 ksi
Fb' = 1.75721 ksi

Wood Bending Factors
Fb = 1.3 ksi CD = 1.25 Cm = 1 Ct = 1 Cfu = 1.1 CL = 0.983053 CF = 1.1 Cr = 1

Deflections: 79.9% Capacity at 6 ft. (Camber = 0 in)

	PreDL	DL	LL	DL+LL	None	None
Deflection (in):	**.337**	**.384**	**.095**	**.48**	**0**	**0**
Span Ratio	**427**	**375**	**1514**	**300**	**10000**	**10000**

FIGURE 6.58 An extract from the RISA floor report for the roof beam at grid 7.

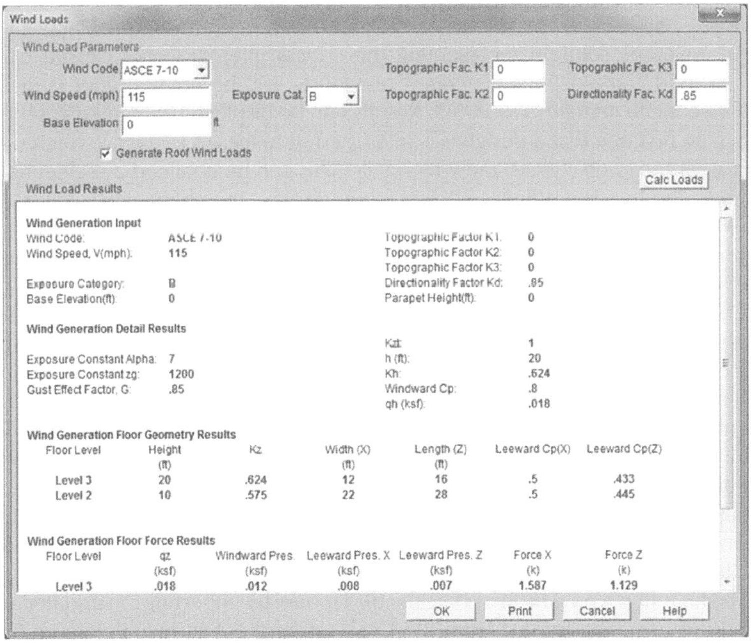

(a)

FIGURE 6.59 Setting parameters for wind and seismic load generation: (a) wind load parameters; (b) seismic load parameters.

(Continued)

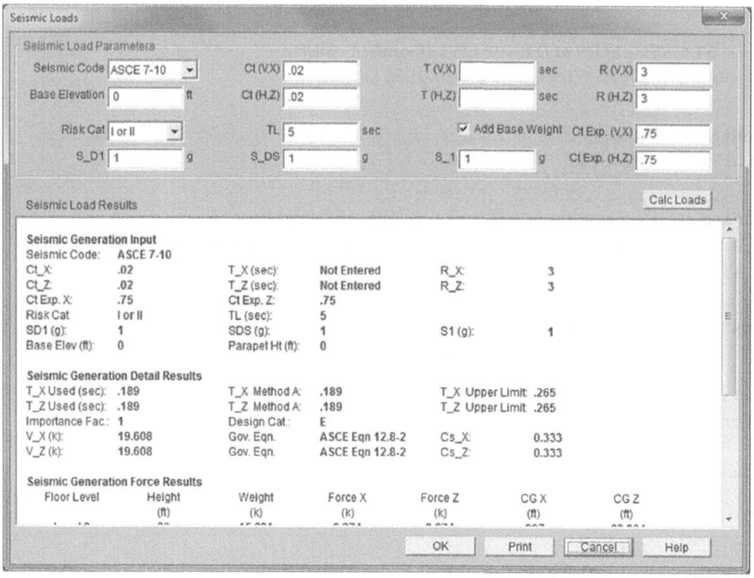

(b)

FIGURE 6.59 (*Continued*) Setting parameters for wind and seismic load generation: (a) wind load parameters; (b) seismic load parameters.

An envelope solution representing the critical internal forces caused by the various load combinations generated previously will be emitted after launching the **Solve** menu item from RISA-3D. Results can be displayed in many different formats. The first one that a designer is interested in relates to knowing which member passes the code checks and which fails. This can be obtained by selecting, for instance, the bending criteria. RISA-3D then displays members with a color code showing which members pass or fail (Figure 6.60). A number is indicated also for each member indicating the ratio of strength required/strength provided. A value less than one means the member passes.

Changing from grade 1 to Better and Selected along with modifying the sizes of the member failed resulted in a structural solution by which all members pass the code checks (Figure 6.60b). Other structural analysis results can be obtained, such as bending moment, shear force diagrams, and axial forces (see Figure 6.61). The ability to examine the entire behavior of the structure under the various load conditions in such a manner will enhance students' understanding of the system and develop a sense of structural performance and solutions.

Similar to the gravity load analysis in RISA Floor, RISA-3D allows for detailed results for each structural element under the critical load case. Figure 6.62 depicts detailed analysis and design checks for a column at grid intersection B-1.

The final step would be to update the BIM model by importing the updated sizes from RISA using the **Import from RISA** tool under the **Add-Ins** tab (Figure 6.63). Member sizes in the BIM model are now different from initially assumed. For example, columns in the first floor include 10 × 10 in. (241 × 241 mm actual), 8 × 14 in. (191 × 343 mm actual), and 8 × 16 in. (191 × 394 mm actual); girders in the same floor include 12 × 16 in. (292 × 394 mm actual) and 10 × 16 in. (241 × 394 mm actual). Floor beams are updated to 6 × 12 in. (140 × 292 mm actual), while all

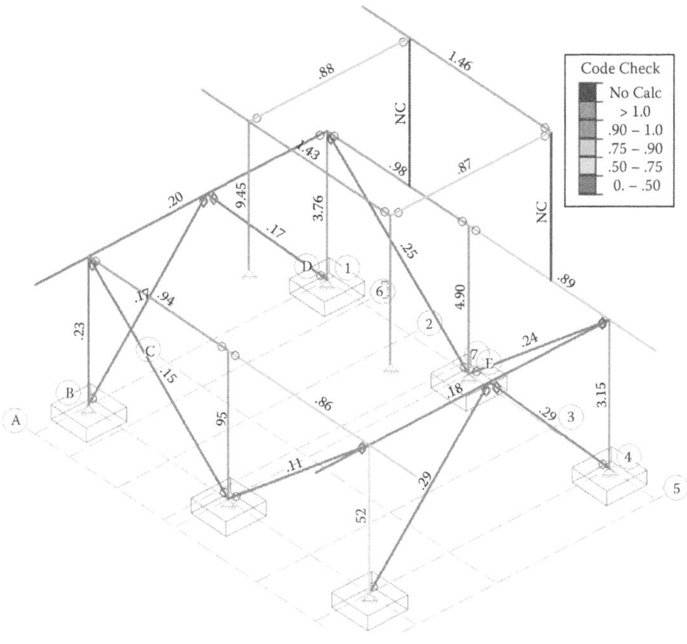

(a)

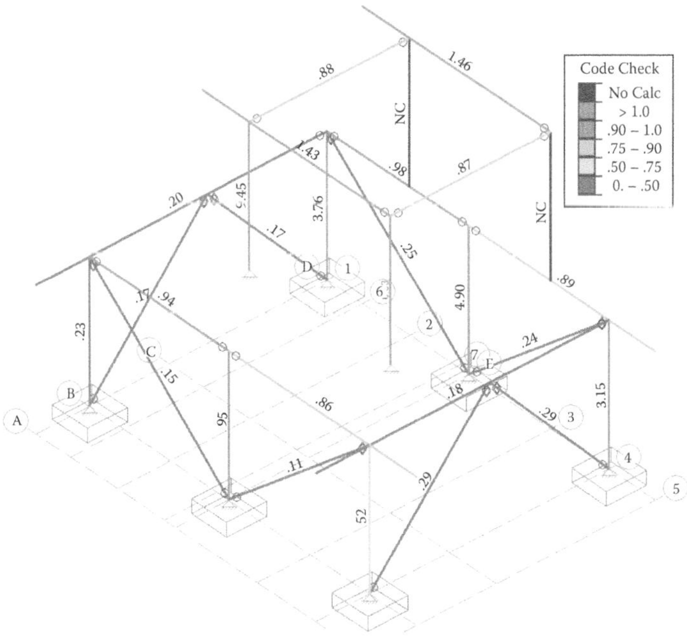

(a)

FIGURE 6.60 (a) Initial results for bending strength code checks for structural members. (b) Final results for bending strength code checks for structural members.

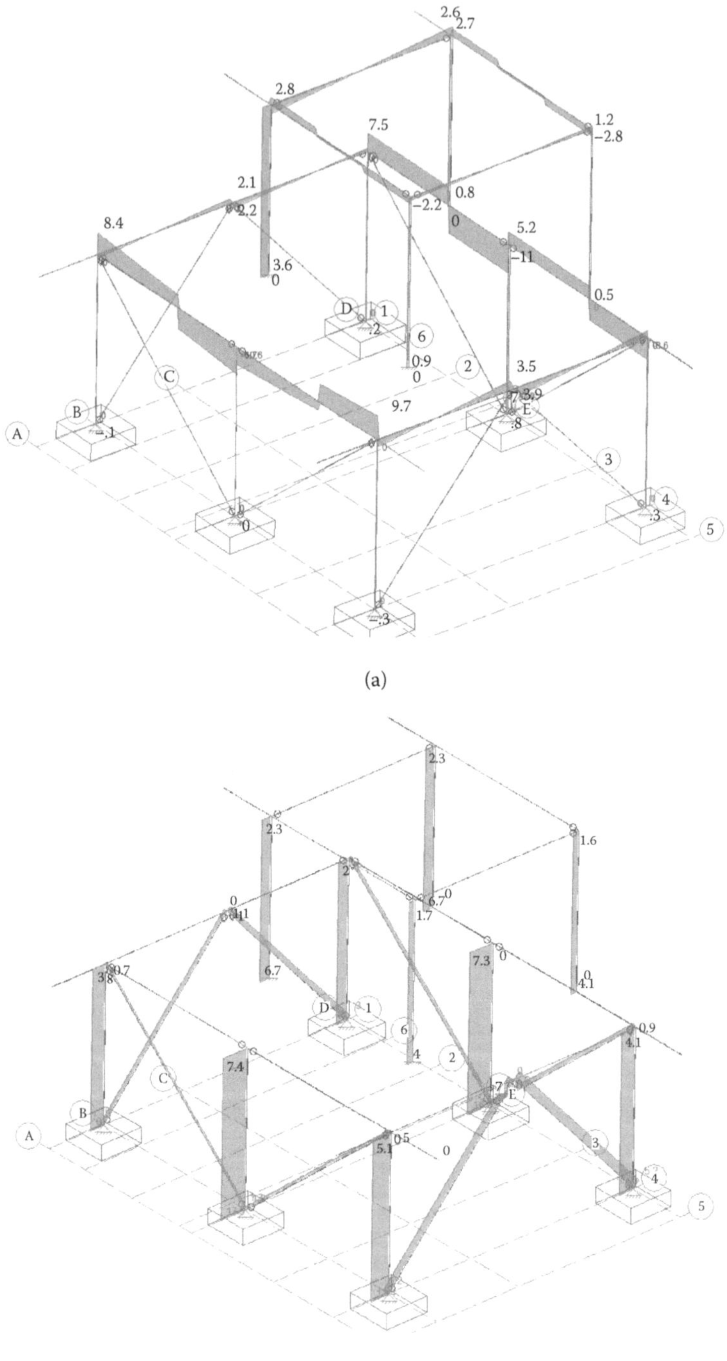

FIGURE 6.61 (a) Examples of RISA-3D analysis results: (a) shear force diagram. (b) Examples of RISA-3D analysis results: (b) axial forces.

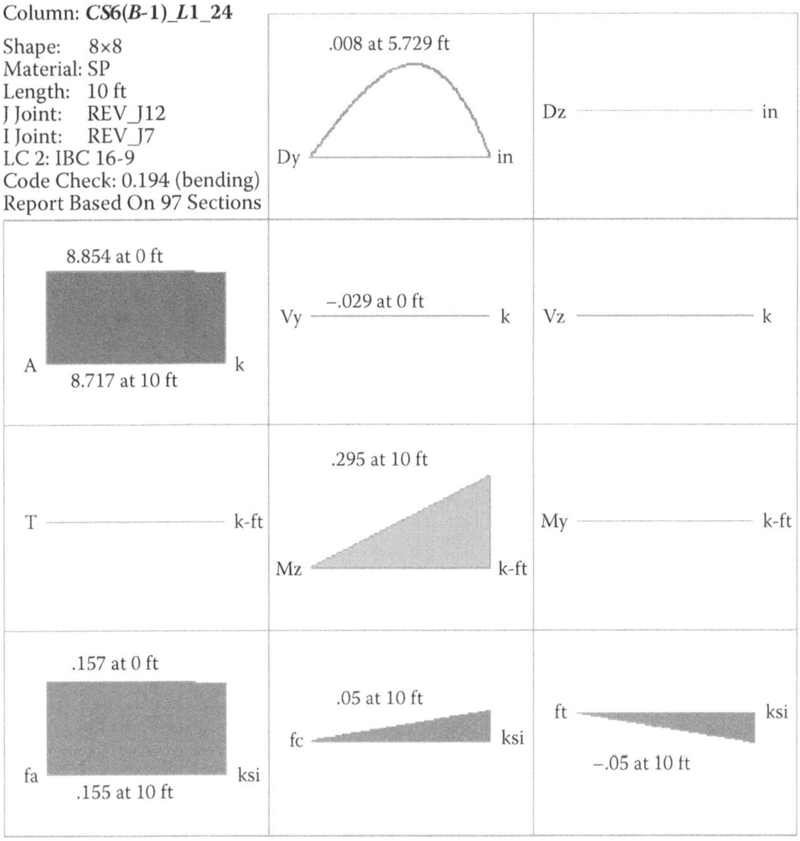

AF&PA NDS-05/08: ASD Code Check

Max Bending Check	0.194	Max Shear Check	0.005 (y)
Location	0 ft	Location	0 ft
Equation	3.6.3	Max Defl Ratio	L/10000

CD 1	RB 4	CL 1
Cr 1	Cfu 1	CP .854

	(ksi)	Cm	Ct	CF
Fc'	.811	1	1	1
Ft'	1	1	1	1
Fb1'	1.5	1	1	1
Fb2'	1.5	1	1	1
Ev'	.165	1	1	
E'	1500	1	1	

	y-y	z-z
Lb	10 ft	10 ft
le/d	16	16
Sway	No	No

Le-Bending Top 10 ft
Le-Bending Bot 10 ft

FIGURE 6.62 Detailed analysis and design report for column 1B.

brace elements remain 6 × 6 in. (140 × 140 mm actual). In the companion website (http://www.crcpress.com/product/isbn/9781482240436) to this book, readers can download both the starting BIM model file and the updated Revit file.

Another example of analyzing wood system will focus on using Autodesk Revit and RSA 2025. Figure 6.64 depicts a wood structure that will be analyzed.

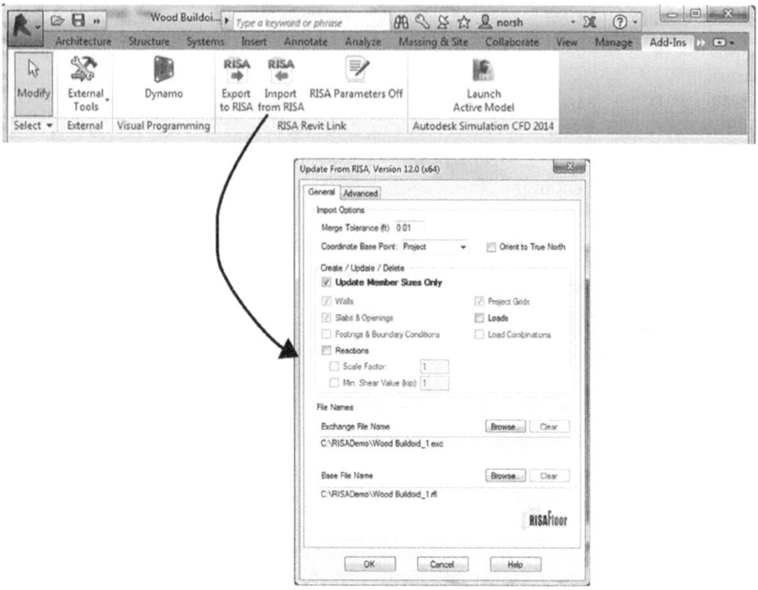

FIGURE 6.63 Import data from RISA into Revit.

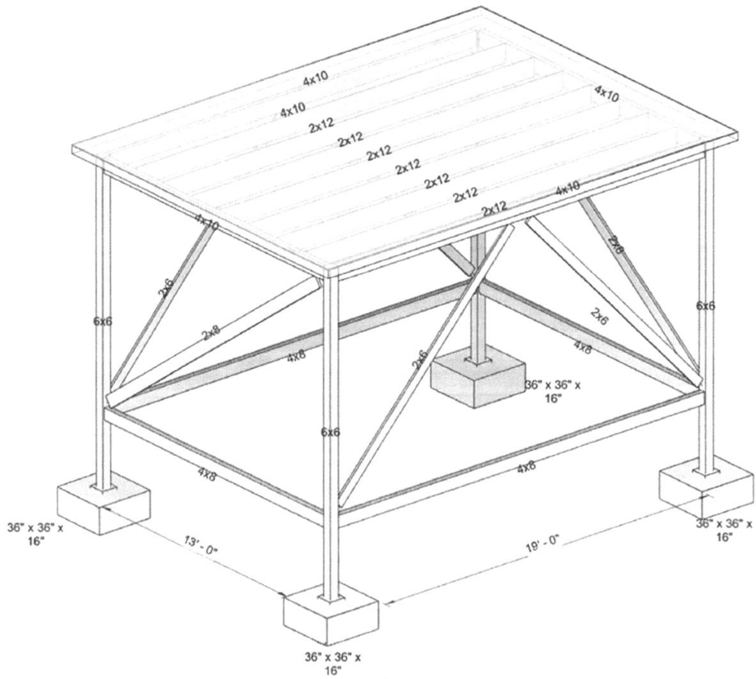

(a)

FIGURE 6.64 (a) Wood structural system. (b) Wind load pressure on the structure.

(Continued)

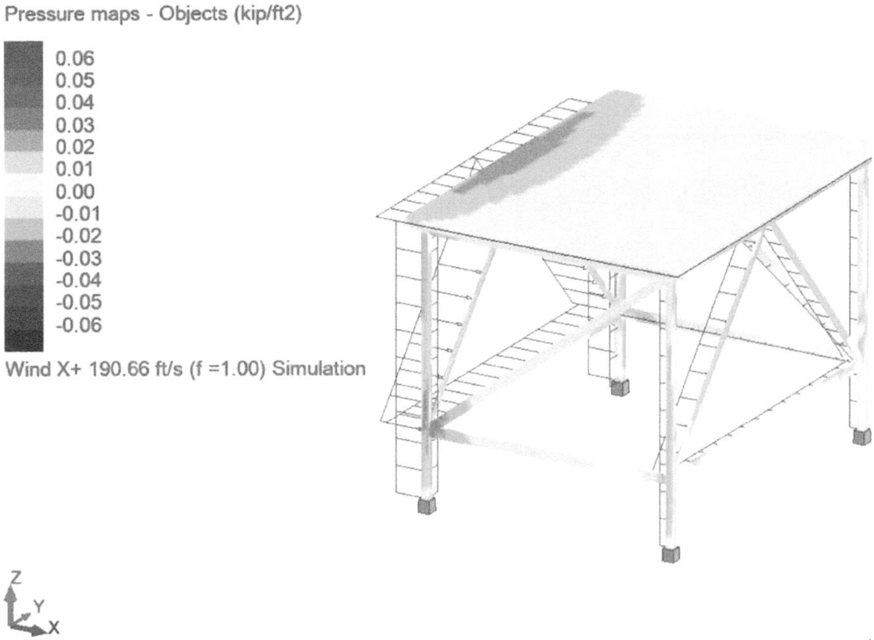

(b)

FIGURE 6.64 (*Continued*) (a) Wood structural system. (b) Wind load pressure on the structure.

Example 6.2

In this example, the wood structure will be analyzed using the direct integration between Revit and RSA software. As shown in Figure 6.64a, a simple wood frame model is used to illustrate the analysis and design process for wood structural systems. The roof is subjected to 15 psf (718 Pa) dead load and 20 psf (958 Pa) live load. A wind load of 130 mph is assumed. The Wind simulation tool within RSA software is used to calculate the wind pressure on the structural elements (see Figure 6.64b).

The analysis is performed in accordance with the LRFD provisions of ASCE 7–22. Examples of shear force and bending moment distributions on a typical floor beam and a girder are shown in Figure 6.65. After running the computations, you can select the "Timber Design" option from the "Design" tab of the RSA software to begin designing the wood members (refer to Figure 6.66).

The design results are shown in Table 6.4 and Figure 6.67. The brace element 2×8 was not sufficient to resist the wind load and has been replaced by 4×4 southern pine, No.1 section. Detailed design reports for each member are depicted in Figure 6.67.

Steel Systems

The analysis and design of a structural steel system is demonstrated using the direct link between Revit and Robot Structural Analysis software. The example for this analysis is the buildoid shown in Figure 6.68. The structure is subjected to dead, live,

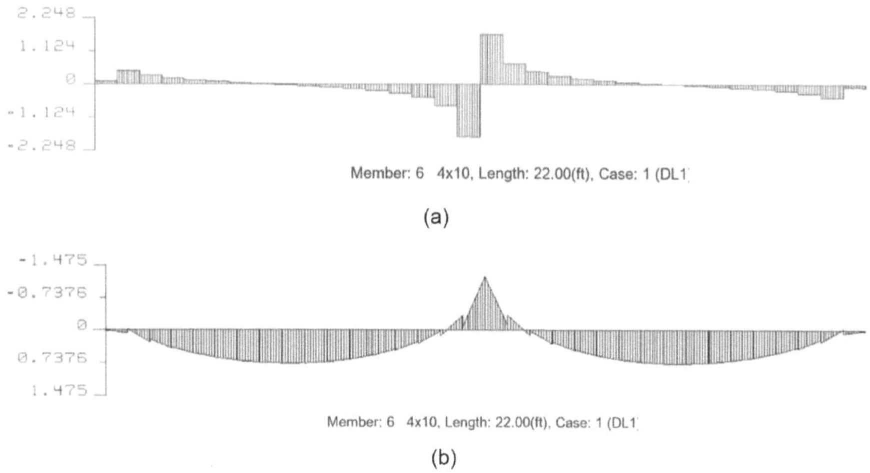

FIGURE 6.65 (a) Shear force diagram on a typical girder. (b) Bending moment diagram on a typical girder.

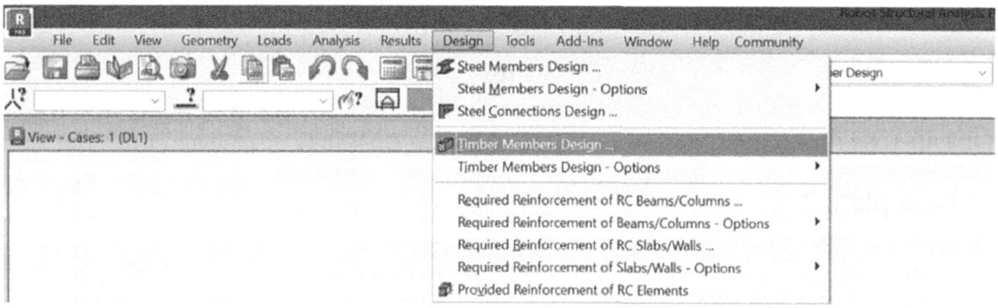

FIGURE 6.66 Timber members design.

TABLE 6.4
Design Results

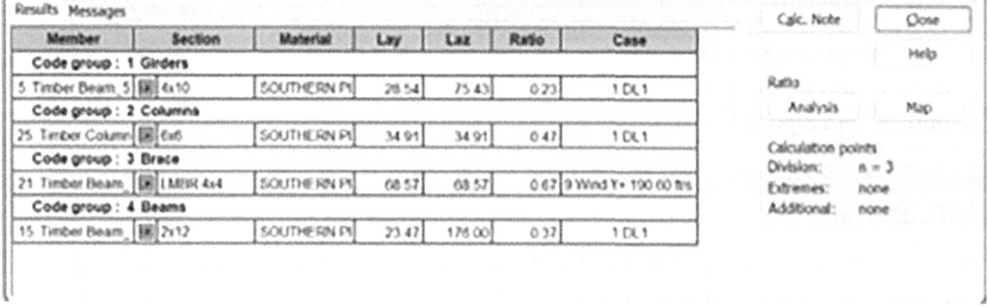

ANSI/AWC NDS-2012 LRFD - Code Group Verification (ULS) 1to4

Results Messages

Member	Section	Material	Lay	Laz	Ratio	Case
Code group : 1 Girders						
5 Timber Beam_5	4x10	SOUTHERN PI	28.54	75.43	0.23	1 DL1
Code group : 2 Columns						
25 Timber Column	6x6	SOUTHERN PI	34.91	34.91	0.47	1 DL1
Code group : 3 Brace						
21 Timber Beam_	LMBR 4x4	SOUTHERN PI	68.57	68.57	0.67	9 Wind Y+ 190.00 fts
Code group : 4 Beams						
15 Timber Beam_	2x12	SOUTHERN PI	23.47	178.00	0.37	1 DL1

Calc. Note
Close
Help
Ratio
Analysis
Map
Calculation points
Division: n = 3
Extremes: none
Additional: none

and wind loads as depicted in Figure 6.68b. Sizes for the steel member are initially assumed to be for columns in the first floor, W8X10; all beams in the first floor are W6X12. Beams supporting the roof are HSS4X4X1/2, while roof girders and columns are W8X10.

The direct link to Robot is found in the drop-down-menus for **Structural Analysis** under the **Analyze** tab. Once launched, the screen shown in Figure 6.69 will pop up to give the options for the export of the model to Robot. After sending the model, the

(a)

FIGURE 6.67 (a) Calculation report for a typical wood girder. (b) Calculation report for a typical column. (c) Calculation report for a brace element.

(Continued)

C:\Users\nnawa\OneDrive\Documents\Autodesk\Output\rd_res.rtf

File Edit View Insert Format Tools Window Help

TIMBER STRUCTURE CALCULATIONS

CODE: *ANSI/AWC NDS-2012 LRFD*
ANALYSIS TYPE: Code Group Verification

CODE GROUP: 2 Columns
MEMBER: 25 Timber Column_25 **POINT:** 1 **COORDINATE:** x = 0.00 L = 0.00 ft

---

LOADS:
Governing Load Case: 1 DL1

MATERIAL: SOUTHERN PINE No.1 2-4
Visually Graded Southern Pine Dimension Lumber - Tab.4B

Fb=1.50 ksi	Ft=1.00 ksi	Fv=0.18 ksi	Fcp=0.57 ksi
Fc=1.65 ksi	E=1600.00 ksi	Emin=580.00 ksi	

SECTION PARAMETERS: 6x6

d=5.50 in			
b=5.50 in	Ay=20.167 in2	Az=20.167 in2	A=30.250 in2
	Iy=76.260 in4	Iz=76.260 in4	Ix=128.654 in4
	Sy=27.731 in3	Sz=27.731 in3	

MEMBER PARAMETERS:

BUCKLING Y	BUCKLING Z	LT BUCKLING
Key = 1.00	Kez = 1.00	lu = 16.00 ft
ley = 16.00 ft	lez = 16.00 ft	leb = 29.44 ft
ley/d = 34.91	lez/b = 34.91	RB = 8.01
CPy = 0.16	CPz = 0.16	CL = 0.99
FcEy = 0.59 ksi	FcEz = 0.59 ksi	FbE = 16.21 ksi

INTERNAL FORCES AND ACTUAL STRESSES:

N = 8.25 kip	My = -0.18 kip*ft	Mz = -0.12 kip*ft	Vy = 0.03 kip	Vz = -0.11 kip
fc = 0.27 ksi	fby = -0.08 ksi	fbz = -0.05 ksi	fvy = 0.00 ksi	fvz = -0.01 ksi
Mx = -0.00 kip*ft			fvty = 0.00 ksi	fvtz = 0.00 ksi

DESIGN WOOD STRENGHTS:
Fc' = Fc(1.65)*CM(1.00)*Ct(1.00)*CF(1.00)*CP(0.16)*KF(2.40)*Fi(0.90)*Lam(1.00) = 0.56 ksi
Fb' = Fb(1.50)*CM(1.00)*Ct(1.00)*CL(0.99)*CF(1.00)*KF(2.54)*Fi(0.85)*Lam(1.00) = 3.20 ksi
Fv' = Fv(0.18)*CM(1.00)*Ct(1.00)*KF(2.88)*Fi(0.75)*Lam(1.00) = 0.38 ksi
Emin' = Emin(580.00)*CM(1.00)*Ct(1.00)*KF(1.76)*Fi(0.85) = 867.68 ksi

---

RESULTS:
(fc/Fc')^2 + fby/(Fb'*(1-fc/FcEy) + fbz/(Fb'*(1-fc/FcEz-(fby/FbE)^2) = 0.31 < 1.00 [3.9-3] OK!
fc/FcEz + (fby/FbE)^2 = 0.47 < 1.00 [3.9-4] OK!, fc/FcEy = 0.47 < 1.00 [3.9.2] OK!,
(fvy + 3/2*fvty)/Fv' = 0.00 < 1.00 [3.4.1] OK!, (fvz + 3/2*fvtz)/Fv' = 0.01 < 1.00 [3.4.1] OK!
ley/d = 34.91 < 50.00 STABLE, lez/b = 34.91 < 50.00 STABLE, Rb = 8.01 < 50.00 STABLE

Section OK !!!

(b)

FIGURE 6.67 (*Continued*) (a) Calculation report for a typical wood girder. (b) Calculation report for a typical column. (c) Calculation report for a brace element.

(*Continued*)

C:\Users\nnawa\OneDrive\Documents\Autodesk\Output\rd_res.rtf

File Edit View Insert Format Tools Window Help

TIMBER STRUCTURE CALCULATIONS

CODE: *ANSI/AWC NDS-2012 LRFD*
ANALYSIS TYPE: Code Group Verification

CODE GROUP: 3 Brace
MEMBER: 21 Timber Beam_21 **POINT:** 2 **COORDINATE:** x = 0.50 L = 10.00 ft

LOADS:
Governing Load Case: 9 Wind Y+ 190.60 ft/s (f =1.00) Simulation

MATERIAL: SOUTHERN PINE No.1 2-4
Visually Graded Southern Pine Dimension Lumber - Tab.4B

Fb=1.50 ksi	Ft=1.00 ksi	Fv=0.18 ksi	Fcp=0.57 ksi
Fc=1.65 ksi	E=1600.00 ksi	Emin=580.00 ksi	

SECTION PARAMETERS: LMBR 4x4

d=3.50 in			
b=3.50 in	Ay=8.167 in2	Az=8.167 in2	A=12.250 in2
	Iy=12.510 in4	Iz=12.510 in4	Ix=21.098 in4
	Sy=7.149 in3	Sz=7.149 in3	

MEMBER PARAMETERS:
BUCKLING Y **BUCKLING Z** **LT BUCKLING**
lu = 20.00 ft
leb = 33.47 ft
RB = 10.71
CL = 0.97
FbE = 9.07 ksi

INTERNAL FORCES AND ACTUAL STRESSES:

N = -0.07 kip	My = -0.00 kip*ft	Mz = 1.40 kip*ft
ft = -0.01 ksi	fby = -0.00 ksi	fbz = 2.36 ksi

DESIGN WOOD STRENGHTS:
Ft' = Ft(1.00)*CM(1.00)*Ct(1.00)*CF(1.00)*KF(2.70)*Fi(0.80)*Lam(1.00) = 2.16 ksi
Fb' = Fb(1.50)*CM(1.00)*Ct(1.00)*CL(0.97)*CF(1.10)*KF(2.54)*Fi(0.85)*Lam(1.00) = 3.46 ksi
Fb* = Fb(1.50)*CM(1.00)*Ct(1.00)*CF(1.10)*KF(2.54)*Fi(0.85)*Lam(1.00) = 3.56 ksi
Fb** = Fb(1.50)*CM(1.00)*Ct(1.00)*CL(0.97)*CF(1.10)*KF(2.54)*Fi(0.85)*Lam(1.00) = 3.46 ksi

RESULTS:
ft/Ft' + fby/Fb* + fbz/Fb' = 0.67 < 1.00 [3.9-1] OK!
(fby-ft)/Fb** + fbz/Fb' = 0.66 < 1.00 [3.9-2] OK!
Rb = 10.71 < 50.00 STABLE

Section OK !!!

(c)

FIGURE 6.67 (*Continued*) (a) Calculation report for a typical wood girder. (b) Calculation report for a typical column. (c) Calculation report for a brace element.

Robot interface will be instantiated and start to display the transferred BIM model. This results in full transmission of the model as shown in Figure 6.70.

The next step is to verify that all members, loads, and supports are exported correctly. Normally, the link works well in sending most of the structural objects to Robot. After that step, it is important to generate load combinations according to the selected standard. In this example, load and resistance factor design (LRFD) ASCE 7–10 is chosen along with the simplified automatic combinations. Analysis computations can be started by selecting the **Calculation** icon from the Robot interface. Any errors in

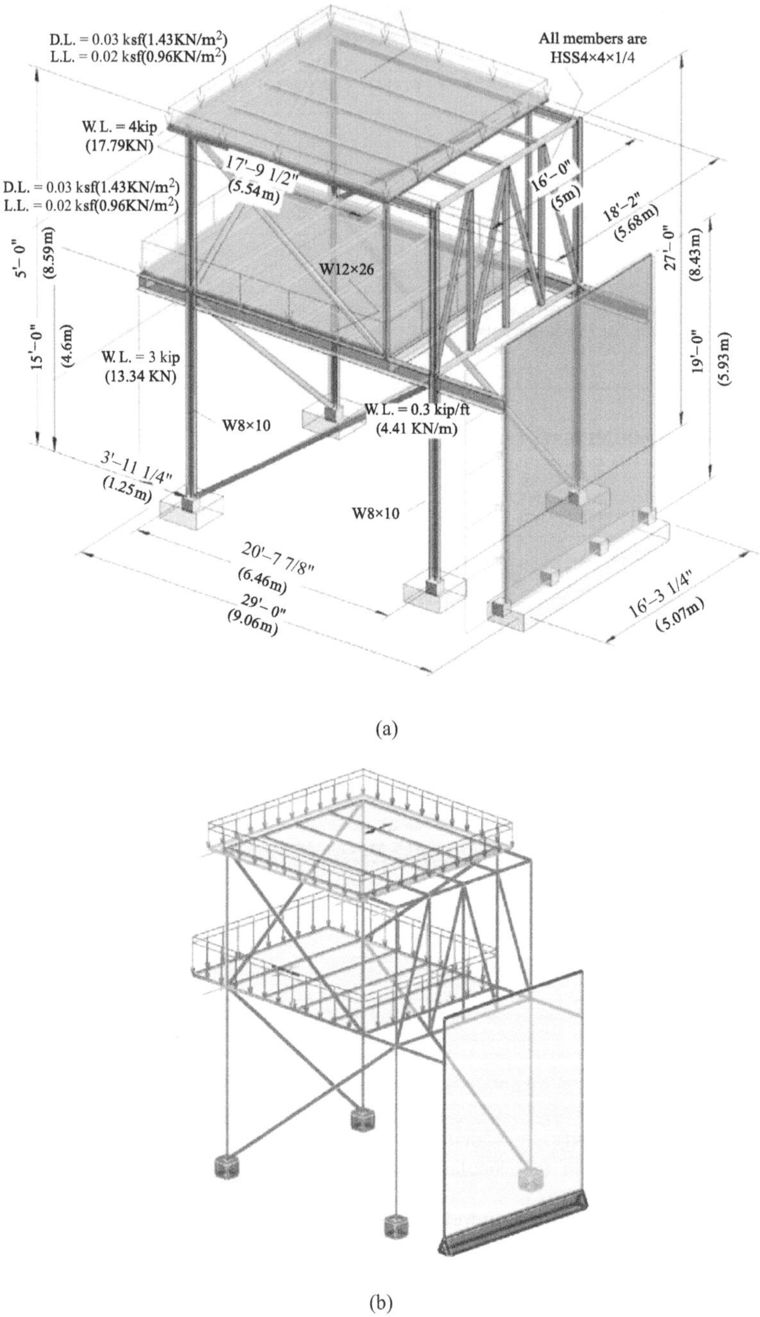

FIGURE 6.68 BIM model for steel system structural design: (a) 3D view of the BIM model; (b) analytical model.

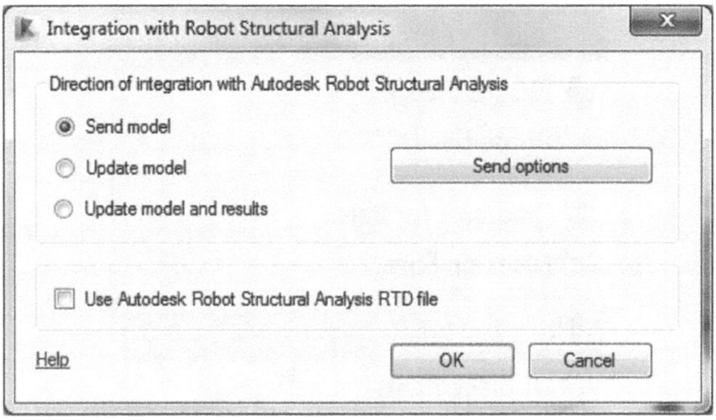

FIGURE 6.69 Export options for the BIM model to robot.

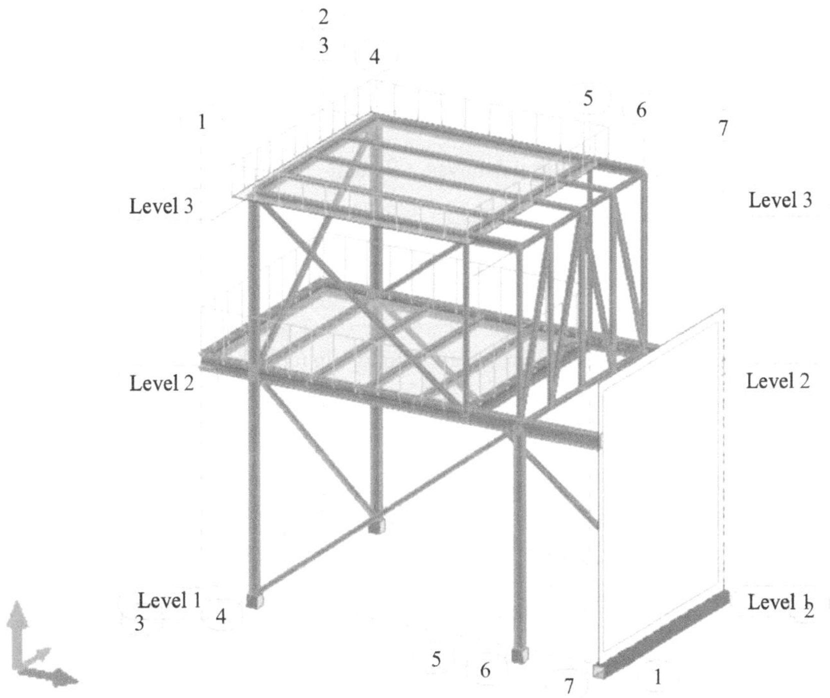

FIGURE 6.70 Exported BIM model in robot.

the models will be reported by Robot, and these errors, if any, should be resolved before proceeding to the analysis of the results and design.

Results of the analysis can be represented in various ways. Robot has rich presentation offerings to display results of the FEM computations (Figure 6.71). These include both graphical and tabular representations of results. For example, the diagrams for the bar option will display the diagrams of the section forces selected (see Figure 6.72). Maps of planar and linear elements can also be created by selecting the

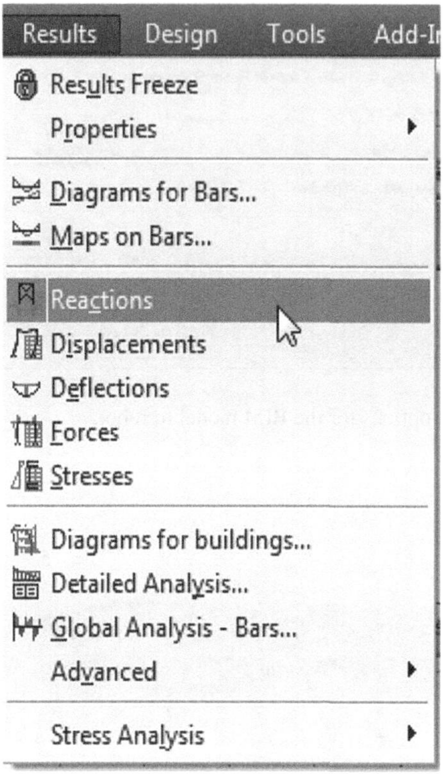

FIGURE 6.71 Analysis result representation options in robot.

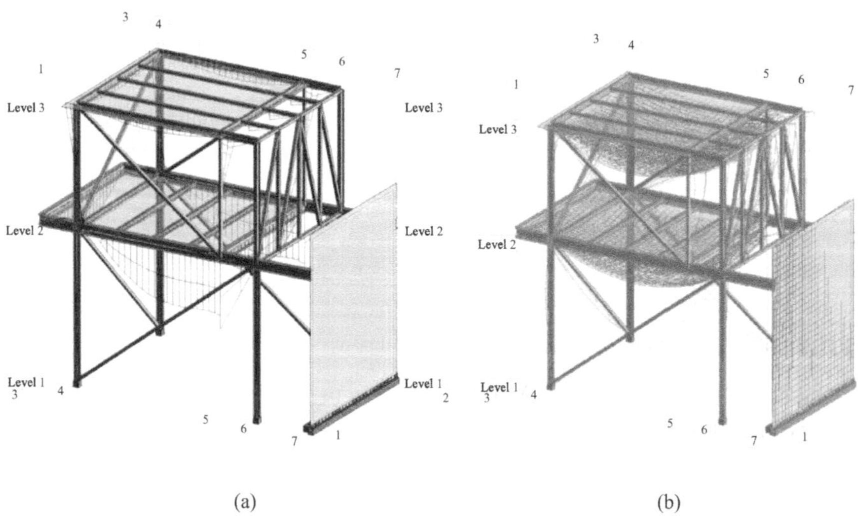

FIGURE 6.72 Examples of robot structural analysis result representations: (a) bar diagrams for bending moments; (b) bar diagrams for deformations.

option **Maps on Bars**. This will open a new dialog box for options to display map results. For example, Figure 6.73 depicts a map result for the vertical displacements of the planar elements.

Tabular displays of results can be exhibited for each type of result by selecting **Tables** from the Robot **View** menu item. A dialog box with many table options will appear (Figure 6.74). Readers can choose any number of tables, and in each table, a filter can be utilized to limit the number of members or nodes that need to be seen in the table. Furthermore, the table of reaction forces automatically displays when the results layout is opened (Table 6.5).

In addition, detailed analysis for each structural object can be performed by selecting **Detailed Analysis** from the results options under the **Results** tab.

This option will give a detailed analysis of a single structural object of the model. Select the object by clicking on it and then make choices about the forces and deformations that are required for the analysis. In the example given in Figure 6.75, detailed results are displayed for bending moment, shear force, and bending stresses of the edge beam W12X26 (). Along with the graphical display, the detailed analysis also shows a table for the numerical values at each point along the span of the beam (Table 6.6) as well as the maximum and minimum values.

After viewing the analysis results, the design of the steel member can be instantiated by selecting the design of steel members from the design menu item. This will cause the steel design calculation options screen to pop up (Figure 6.76). The steel design in this example is performed according to the AISC 2005 code.

An important step before carrying out the calculations is to check the options for optimization. Clicking on the options button will allow setting up the required

FIGURE 6.73 Map results for the vertical displacements of planar elements.

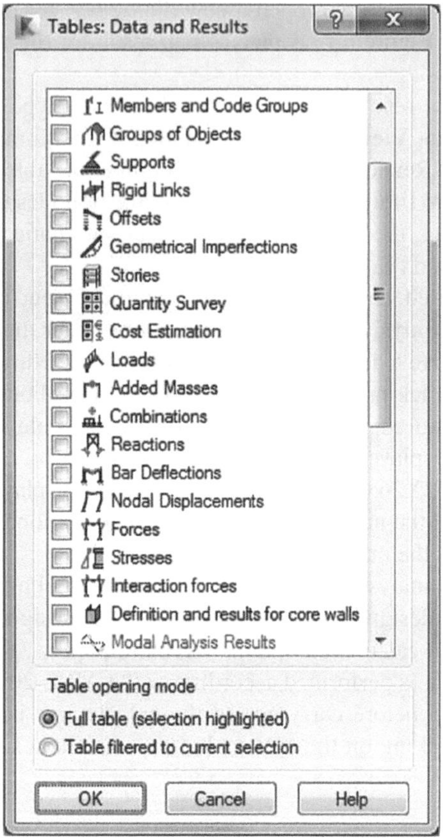

FIGURE 6.74 Tabular data and analysis results.

TABLE 6.5
Reactions at the Supports for the Analyzed Structure

Reactions in the coordinate system: global - Cases: 10 11 13 14

Node/Case	FX (kip)	FY (kip)	FZ (kip)	MX (kip-ft)	MY (kip-ft)	MZ (kip-ft)
1/ ULS+	0.44	-0.04	49.35	0.31	2.47	-0.00
1/ ULS-	0.24	-8.02	20.87	0.15	1.33	-0.00
1/ SLS+	0.37	-0.05	44.71	0.28	2.17	-0.00
1/ SLS-	0.27	-8.01	23.19	0.17	1.48	-0.00
3/ ULS+	0.94	0.09	40.66	-0.19	1.38	-0.00
3/ ULS-	-2.84	0.04	7.32	-0.45	0.60	-0.00
3/ SLS+	0.74	0.07	31.70	-0.22	1.06	-0.00
3/ SLS-	-2.78	0.04	9.84	-0.35	0.68	-0.00
5/ ULS+	-0.00	-0.00	21.09	0.19	-0.08	0.0
5/ ULS-	-0.01	-0.02	12.07	0.01	-0.25	0.0
5/ SLS+	-0.00	-0.00	18.38	0.19	-0.09	0.0
5/ SLS-	-0.01	-0.02	13.41	0.01	-0.24	0.0
7/ ULS+	2.43	-0.00	22.23	0.19	0.61	0.00
7/ ULS-	-1.55	-0.03	10.89	0.01	0.25	0.00

Values / Envelope / Global extremes / Info

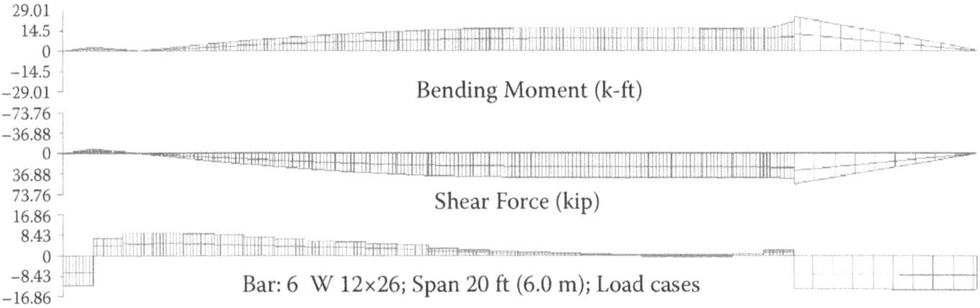

FIGURE 6.75 Detailed analysis for a floor beam in robot.

TABLE 6.6
Results of Robot Detailed Analysis of Bar 6 W12×26 (W310×39)

Bar / Point (ft)	FZ (kip)	MY (kip-ft)	UZ (in)	S max (ksi)
Current value	-12.38	0.0	-0.0837	-0.09
for bar:		6		
in point:		x=0.0 (ft)		
6 / origin	-12.38	0.0	-0.0837	-0.09
6 / auto x=0.67 (-)	-12.40	-8.23	-0.1320	2.90
6 / auto x=0.67 (+)	7.19	-7.64	-0.1323	2.71
6 / auto x=1.33 (-)	7.17	-2.89	-0.1812	1.00
6 / auto x=1.33 (+)	9.20	-3.14	-0.1814	1.10
6 / auto x=2.00 (-)	9.18	2.95	-0.2307	1.03
6 / auto x=2.00 (+)	9.25	2.67	-0.2310	0.93
6 / auto x=2.67 (-)	9.22	8.79	-0.2799	3.13
6 / auto x=2.67 (+)	8.92	8.50	-0.2802	3.03
6 / auto x=3.33 (-)	8.90	14.40	-0.3279	5.15
6 / auto x=3.33 (+)	8.40	14.11	-0.3282	5.04
6 / auto x=4.00 (-)	8.38	19.67	-0.3741	7.04
6 / auto x=4.00 (+)	7.22	19.37	-0.3744	6.94
6 / auto x=4.67 (-)	7.20	24.15	-0.4178	8.65
6 / auto x=4.67 (+)	6.62	23.84	-0.4181	8.54
6 / auto x=5.33 (-)	6.60	28.22	-0.4584	10.11

Values | Local extremes | Global extremes

optimization criterion for the steel design (Figure 6.77). In this example, the **Weight** of the structural member is the optimization parameter chosen.

Design results will be displayed in a tabular format, with colors distinguishing passed or failed sections. Red normally designates a failed section; green means the section is adequate (see Figure 6.78). In this example, the columns in the first floor (W8X10) did not pass the code checks because of instability, and size changes are required. By clicking on any of these failed sections, a report depicting the detailed

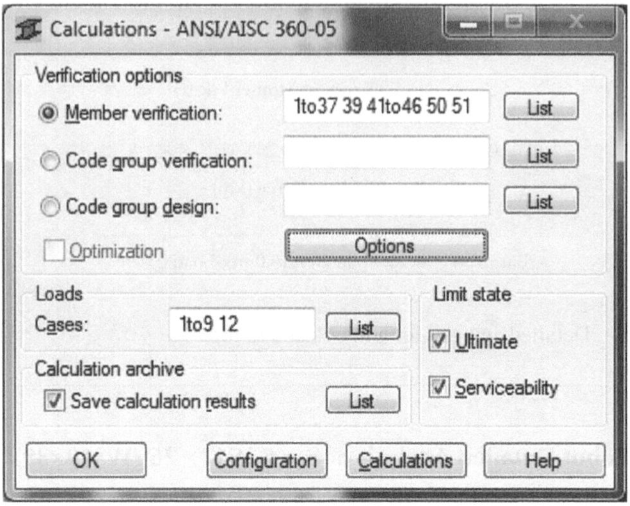

FIGURE 6.76 Steel design computation options.

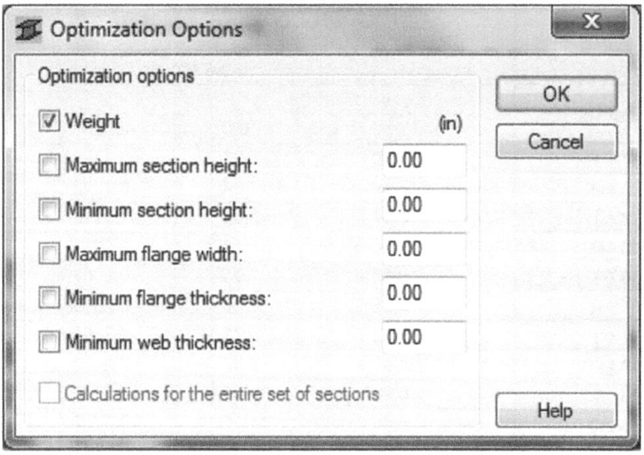

FIGURE 6.77 Steel design optimization options.

design computations according to the AISC 2014 specifications will be created by Robot. A sample of this report is available from the companion website (http://www.crcpress.com/product/isbn/9781482240436).

Modifying the failed structural columns to W8X40 () solved the instability issues, and the design checks for these columns indicated success (see Figure 6.79). The newly updated design results still show some beams that are not adequate according to the AISC 2005 code. These are **Girder_6** (W12X26) and beams (W6X12) on the first floor and girders in the roof (W8X10). They are replaced now by W16X14 for the girder in the first floor and W12X26 for floor beams and roof girders. The final results are depicted in Figure 6.80. Also, a full design calculation report for the entire.

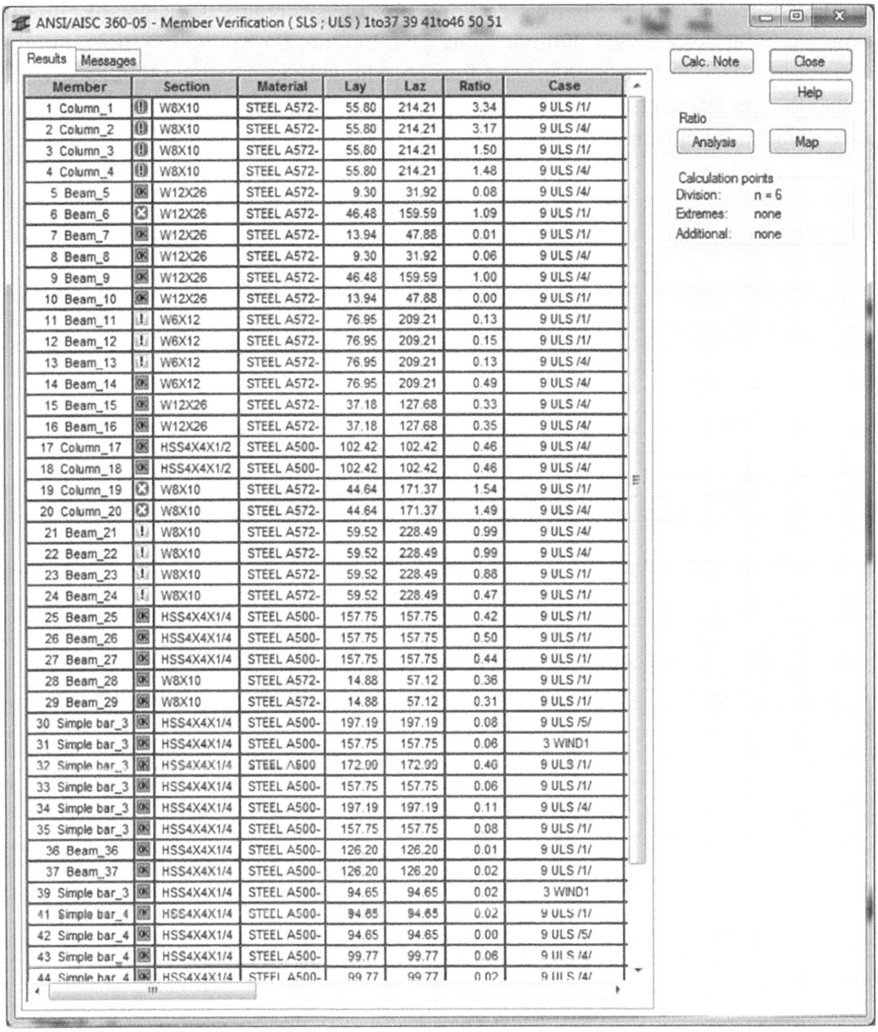
ANSI/AISC 360-05 - Member Verification (SLS ; ULS) 1to37 39 41to46 50 51

Results | Messages

Member	Section	Material	Lay	Laz	Ratio	Case
1 Column_1	W8X10	STEEL A572-	55.80	214.21	3.34	9 ULS /1/
2 Column_2	W8X10	STEEL A572-	55.80	214.21	3.17	9 ULS /4/
3 Column_3	W8X10	STEEL A572-	55.80	214.21	1.50	9 ULS /1/
4 Column_4	W8X10	STEEL A572-	55.80	214.21	1.48	9 ULS /4/
5 Beam_5	W12X26	STEEL A572-	9.30	31.92	0.08	9 ULS /4/
6 Beam_6	W12X26	STEEL A572-	46.48	159.59	1.09	9 ULS /1/
7 Beam_7	W12X26	STEEL A572-	13.94	47.88	0.01	9 ULS /1/
8 Beam_8	W12X26	STEEL A572-	9.30	31.92	0.06	9 ULS /4/
9 Beam_9	W12X26	STEEL A572-	46.48	159.59	1.00	9 ULS /4/
10 Beam_10	W12X26	STEEL A572-	13.94	47.88	0.00	9 ULS /1/
11 Beam_11	W6X12	STEEL A572-	76.95	209.21	0.13	9 ULS /1/
12 Beam_12	W6X12	STEEL A572-	76.95	209.21	0.15	9 ULS /1/
13 Beam_13	W6X12	STEEL A572-	76.95	209.21	0.13	9 ULS /4/
14 Beam_14	W6X12	STEEL A572-	76.95	209.21	0.49	9 ULS /4/
15 Beam_15	W12X26	STEEL A572-	37.18	127.68	0.33	9 ULS /4/
16 Beam_16	W12X26	STEEL A572-	37.18	127.68	0.35	9 ULS /4/
17 Column_17	HSS4X4X1/2	STEEL A500-	102.42	102.42	0.46	9 ULS /4/
18 Column_18	HSS4X4X1/2	STEEL A500-	102.42	102.42	0.46	9 ULS /4/
19 Column_19	W8X10	STEEL A572-	44.64	171.37	1.54	9 ULS /1/
20 Column_20	W8X10	STEEL A572-	44.64	171.37	1.49	9 ULS /4/
21 Beam_21	W8X10	STEEL A572-	59.52	228.49	0.99	9 ULS /4/
22 Beam_22	W8X10	STEEL A572-	59.52	228.49	0.99	9 ULS /4/
23 Beam_23	W8X10	STEEL A572-	59.52	228.49	0.88	9 ULS /1/
24 Beam_24	W8X10	STEEL A572-	59.52	228.49	0.47	9 ULS /1/
25 Beam_25	HSS4X4X1/4	STEEL A500-	157.75	157.75	0.42	9 ULS /1/
26 Beam_26	HSS4X4X1/4	STEEL A500-	157.75	157.75	0.50	9 ULS /1/
27 Beam_27	HSS4X4X1/4	STEEL A500-	157.75	157.75	0.44	9 ULS /1/
28 Beam_28	W8X10	STEEL A572-	14.88	57.12	0.36	9 ULS /1/
29 Beam_29	W8X10	STEEL A572-	14.88	57.12	0.31	9 ULS /1/
30 Simple bar_3	HSS4X4X1/4	STEEL A500-	197.19	197.19	0.08	9 ULS /5/
31 Simple bar_3	HSS4X4X1/4	STEEL A500-	157.75	157.75	0.06	3 WIND1
32 Simple bar_3	HSS4X4X1/4	STEEL A500	172.00	172.99	0.40	9 ULS /1/
33 Simple bar_3	HSS4X4X1/4	STEEL A500-	157.75	157.75	0.06	9 ULS /1/
34 Simple bar_3	HSS4X4X1/4	STEEL A500-	197.19	197.19	0.11	9 ULS /4/
35 Simple bar_3	HSS4X4X1/4	STEEL A500-	157.75	157.75	0.08	9 ULS /1/
36 Beam_36	HSS4X4X1/4	STEEL A500-	126.20	126.20	0.01	9 ULS /1/
37 Beam_37	HSS4X4X1/4	STEEL A500-	126.20	126.20	0.02	9 ULS /1/
39 Simple bar_3	HSS4X4X1/4	STEEL A500-	94.65	94.65	0.02	3 WIND1
41 Simple bar_4	HSS4X4X1/4	STEEL A500-	94.65	94.65	0.02	9 ULS /1/
42 Simple bar_4	HSS4X4X1/4	STEEL A500-	94.65	94.65	0.00	9 ULS /5/
43 Simple bar_4	HSS4X4X1/4	STEEL A500-	99.77	99.77	0.06	9 ULS /4/
44 Simple bar_4	HSS4X4X1/4	STEEL A500-	99.77	99.77	0.02	9 ULS /4/

FIGURE 6.78 Steel design results from robot.

The last phase is to update the BIM model with the steel design results from Robot. This is achieved by launching the direct link to Robot from the **Analyze** tab. This will update all the member sizes in Revit. A report showing these changes is shown in Figure 6.81. The floor-framing plan in Figure 6.82 illustrates this update by contrasting before and after the analysis and design with the direct link to Robot.

CONCRETE SYSTEMS

In this section, a structural reinforced concrete system is analyzed and designed using the direct link between Revit and Robot Structural Analysis software. The buildoid shown in Figure 6.83 is the example used. The structure is subjected to dead, live, and wind loads as depicted in Figure 6.83b Sizes for the structural elements are initially assumed to be as follows: Columns in the first floor are 8×8 in. (203×203 mm); the roof is supported by two 8×8 in. and four 6×6 in. (152×152 mm); all beams in the first floor and the roof are 8×12 in. (203×300 mm); and both floor and roof slabs have a thickness of 6 in. (152 mm).

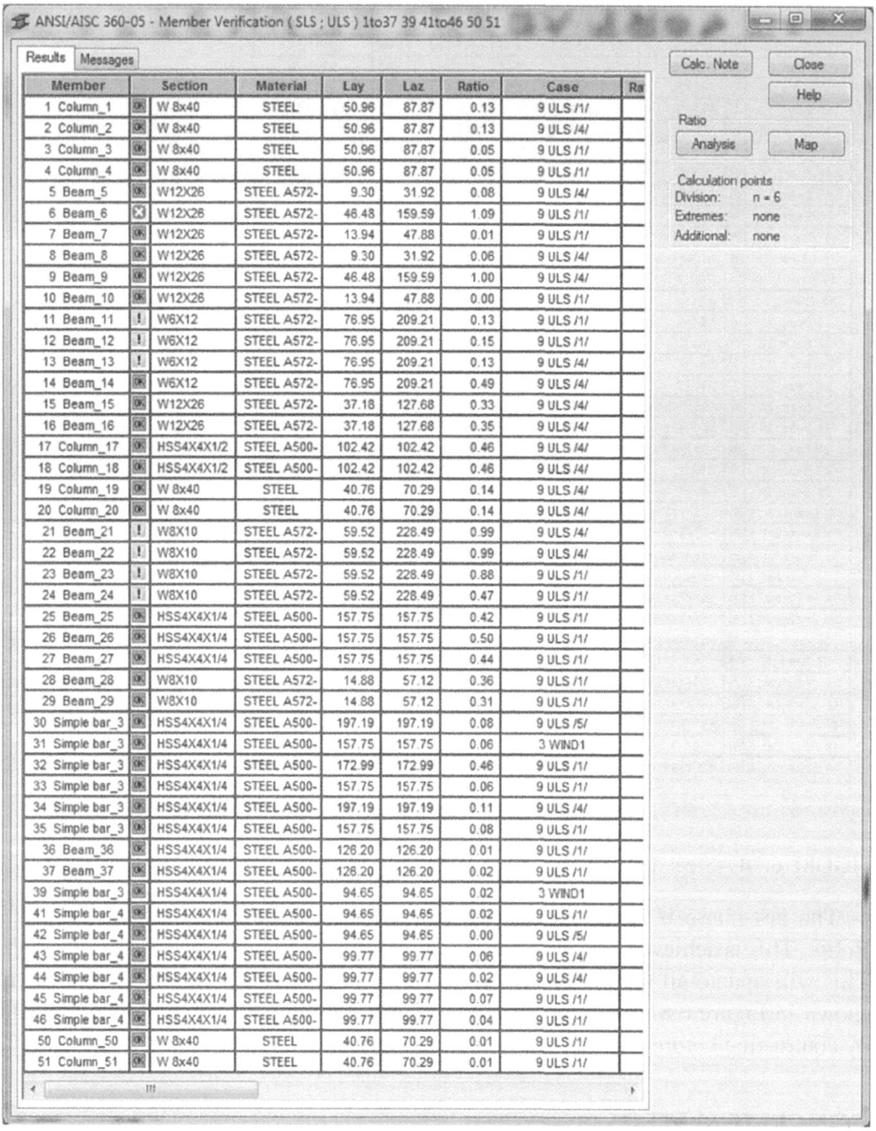

Member		Section	Material	Lay	Laz	Ratio	Case	Ra
1 Column_1	OK	W 8x40	STEEL	50.96	87.87	0.13	9 ULS /1/	
2 Column_2	OK	W 8x40	STEEL	50.96	87.87	0.13	9 ULS /4/	
3 Column_3	OK	W 8x40	STEEL	50.96	87.87	0.05	9 ULS /1/	
4 Column_4	OK	W 8x40	STEEL	50.96	87.87	0.05	9 ULS /1/	
5 Beam_5	OK	W12X26	STEEL A572-	9.30	31.92	0.08	9 ULS /4/	
6 Beam_6	X	W12X26	STEEL A572-	46.48	159.59	1.09	9 ULS /1/	
7 Beam_7	OK	W12X26	STEEL A572-	13.94	47.88	0.01	9 ULS /1/	
8 Beam_8	OK	W12X26	STEEL A572-	9.30	31.92	0.06	9 ULS /4/	
9 Beam_9	OK	W12X26	STEEL A572-	46.48	159.59	1.00	9 ULS /4/	
10 Beam_10	OK	W12X26	STEEL A572-	13.94	47.88	0.00	9 ULS /1/	
11 Beam_11	!	W6X12	STEEL A572-	76.95	209.21	0.13	9 ULS /1/	
12 Beam_12	!	W6X12	STEEL A572-	76.95	209.21	0.15	9 ULS /1/	
13 Beam_13	!	W6X12	STEEL A572-	76.95	209.21	0.13	9 ULS /4/	
14 Beam_14	OK	W6X12	STEEL A572-	76.95	209.21	0.49	9 ULS /4/	
15 Beam_15	OK	W12X26	STEEL A572-	37.18	127.68	0.33	9 ULS /4/	
16 Beam_16	OK	W12X26	STEEL A572-	37.18	127.68	0.35	9 ULS /4/	
17 Column_17	OK	HSS4X4X1/2	STEEL A500-	102.42	102.42	0.46	9 ULS /4/	
18 Column_18	OK	HSS4X4X1/2	STEEL A500-	102.42	102.42	0.46	9 ULS /4/	
19 Column_19	OK	W 8x40	STEEL	40.76	70.29	0.14	9 ULS /4/	
20 Column_20	OK	W 8x40	STEEL	40.76	70.29	0.14	9 ULS /4/	
21 Beam_21	!	W8X10	STEEL A572-	59.52	228.49	0.99	9 ULS /4/	
22 Beam_22	!	W8X10	STEEL A572-	59.52	228.49	0.99	9 ULS /4/	
23 Beam_23	!	W8X10	STEEL A572-	59.52	228.49	0.88	9 ULS /1/	
24 Beam_24	!	W8X10	STEEL A572-	59.52	228.49	0.47	9 ULS /1/	
25 Beam_25	OK	HSS4X4X1/4	STEEL A500-	157.75	157.75	0.42	9 ULS /1/	
26 Beam_26	OK	HSS4X4X1/4	STEEL A500-	157.75	157.75	0.50	9 ULS /1/	
27 Beam_27	OK	HSS4X4X1/4	STEEL A500-	157.75	157.75	0.44	9 ULS /1/	
28 Beam_28	OK	W8X10	STEEL A572-	14.88	57.12	0.36	9 ULS /1/	
29 Beam_29	OK	W8X10	STEEL A572-	14.88	57.12	0.31	9 ULS /1/	
30 Simple bar_3	OK	HSS4X4X1/4	STEEL A500-	197.19	197.19	0.08	9 ULS /5/	
31 Simple bar_3	OK	HSS4X4X1/4	STEEL A500-	157.75	157.75	0.06	3 WIND1	
32 Simple bar_3	OK	HSS4X4X1/4	STEEL A500-	172.99	172.99	0.46	9 ULS /1/	
33 Simple bar_3	OK	HSS4X4X1/4	STEEL A500-	157.75	157.75	0.06	9 ULS /1/	
34 Simple bar_3	OK	HSS4X4X1/4	STEEL A500-	197.19	197.19	0.11	9 ULS /4/	
35 Simple bar_3	OK	HSS4X4X1/4	STEEL A500-	157.75	157.75	0.08	9 ULS /1/	
36 Beam_36	OK	HSS4X4X1/4	STEEL A500-	126.20	126.20	0.01	9 ULS /1/	
37 Beam_37	OK	HSS4X4X1/4	STEEL A500-	126.20	126.20	0.02	9 ULS /1/	
39 Simple bar_3	OK	HSS4X4X1/4	STEEL A500-	94.65	94.65	0.02	3 WIND1	
41 Simple bar_4	OK	HSS4X4X1/4	STEEL A500-	94.65	94.65	0.02	9 ULS /1/	
42 Simple bar_4	OK	HSS4X4X1/4	STEEL A500-	94.65	94.65	0.00	9 ULS /5/	
43 Simple bar_4	OK	HSS4X4X1/4	STEEL A500-	99.77	99.77	0.06	9 ULS /4/	
44 Simple bar_4	OK	HSS4X4X1/4	STEEL A500-	99.77	99.77	0.02	9 ULS /4/	
45 Simple bar_4	OK	HSS4X4X1/4	STEEL A500-	99.77	99.77	0.07	9 ULS /1/	
46 Simple bar_4	OK	HSS4X4X1/4	STEEL A500-	99.77	99.77	0.04	9 ULS /1/	
50 Column_50	OK	W 8x40	STEEL	40.76	70.29	0.01	9 ULS /1/	
51 Column_51	OK	W 8x40	STEEL	40.76	70.29	0.01	9 ULS /1/	

(a)

FIGURE 6.79 (a) Steel design results from robot after changing column sizes. (b) Design calculations for the updated steel column.

(Continued)

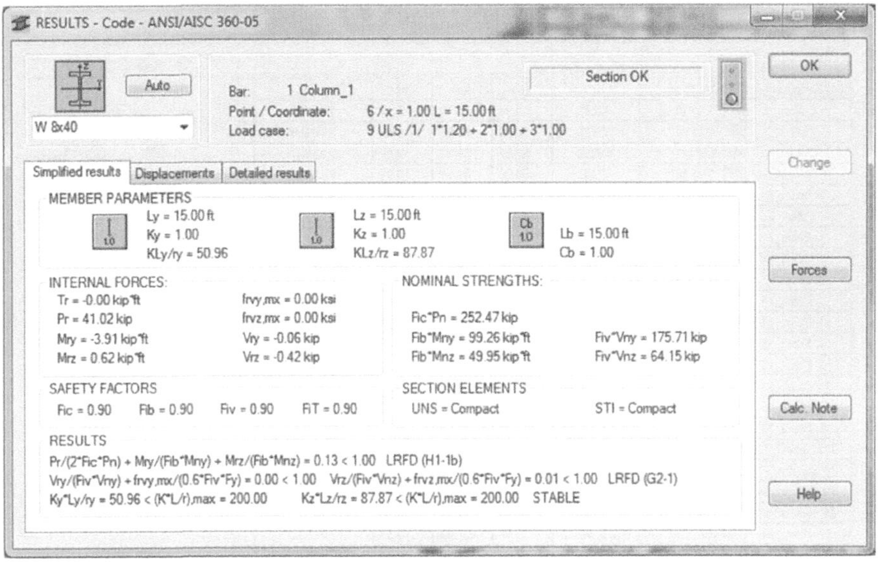

(b)

FIGURE 6.79 (*Continued*) (a) Steel design results from robot after changing column sizes. (b) Design calculations for the updated steel column.

Steps similar to the ones followed in the previous section for structural steel systems are adopted for concrete systems. The BIM model is exported to Robot through the direct link (Figure 6.84), and then load combinations are defined according to LRFD ASCE 7–10 in Robot. Computations for the structural analysis must be conducted before starting the reinforced concrete structural element design. Part of the analysis results is given in Figures 6.85 and 6.86.

A reinforced concrete design for the structure can be started by selecting the **Design** menu item from Robot interface, then choosing the **Required Reinforcement for Beams/Wall…** option. A dialog box for design computation options will pop up to allow for various code settings (Figure 6.87). Calculations are performed in this example according to the American Concrete Institute (ACI) 381–08 specifications. Results are presented in a tabular form that gives the required area of reinforcement at each location along the beam or column along with the rebar sizes (see Figure 6.88).

The next design details for each reinforced concrete element can be obtained by first selecting the element in the 3D model in Robot and launching **Provided Member Steel Reinforcement** from the **Design** tab. This will open an interface with various options. Specify the calculation options by selecting the icon for that purpose and then make the appropriate settings for the material grades and rebar sizes. For example, for Beam 6 (see Figure 6.88), the settings are given in Figure 6.89.

Design results are given through a number of tabs illustrating the internal forces diagrams, beam reinforcements, drawing details, reports covering the code computations, and design checks. Some of these results are depicted in Figure 6.71 for Beam 6.

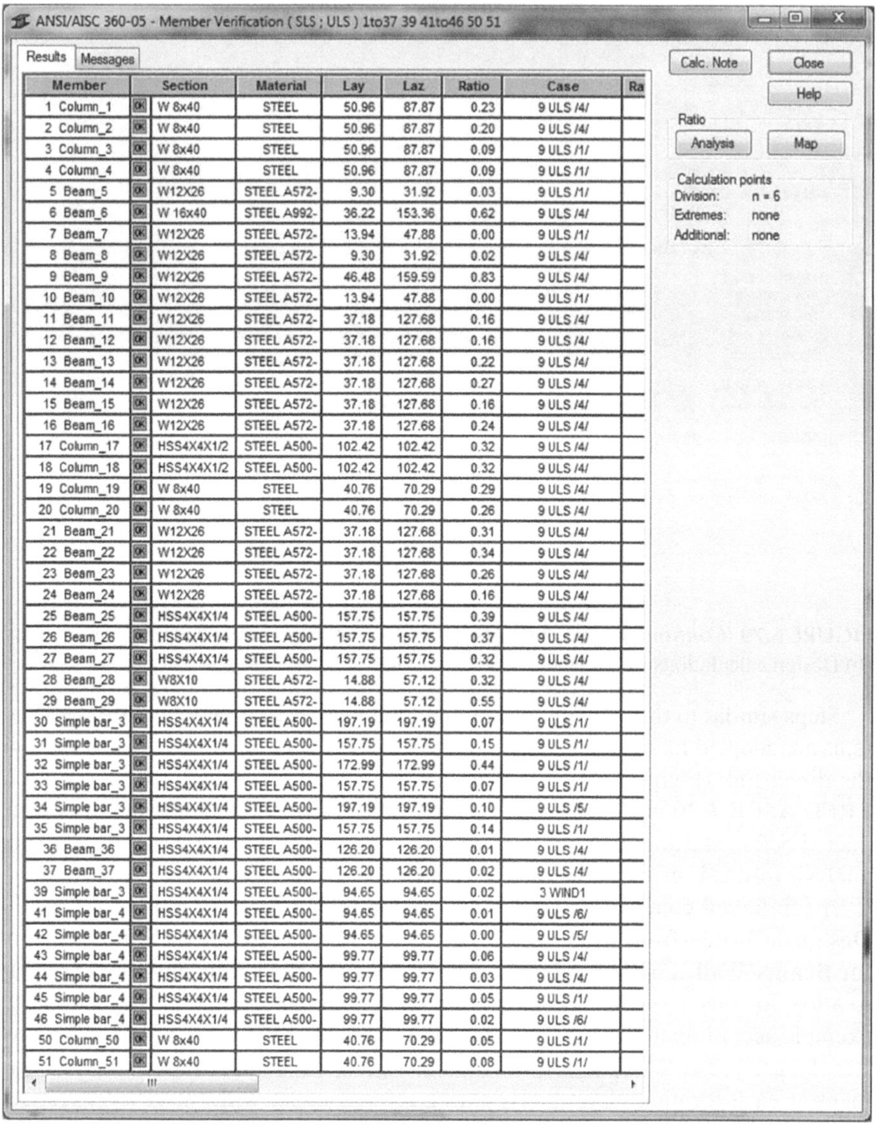

Member	Section	Material	Lay	Laz	Ratio	Case	Ra
1 Column_1	W 8x40	STEEL	50.96	87.87	0.23	9 ULS /4/	
2 Column_2	W 8x40	STEEL	50.96	87.87	0.20	9 ULS /4/	
3 Column_3	W 8x40	STEEL	50.96	87.87	0.09	9 ULS /1/	
4 Column_4	W 8x40	STEEL	50.96	87.87	0.09	9 ULS /1/	
5 Beam_5	W12X26	STEEL A572-	9.30	31.92	0.03	9 ULS /1/	
6 Beam_6	W 16x40	STEEL A992-	36.22	153.36	0.62	9 ULS /4/	
7 Beam_7	W12X26	STEEL A572-	13.94	47.88	0.00	9 ULS /1/	
8 Beam_8	W12X26	STEEL A572-	9.30	31.92	0.02	9 ULS /4/	
9 Beam_9	W12X26	STEEL A572-	46.48	159.59	0.83	9 ULS /4/	
10 Beam_10	W12X26	STEEL A572-	13.94	47.88	0.00	9 ULS /1/	
11 Beam_11	W12X26	STEEL A572-	37.18	127.68	0.16	9 ULS /4/	
12 Beam_12	W12X26	STEEL A572-	37.18	127.68	0.16	9 ULS /4/	
13 Beam_13	W12X26	STEEL A572-	37.18	127.68	0.22	9 ULS /4/	
14 Beam_14	W12X26	STEEL A572-	37.18	127.68	0.27	9 ULS /4/	
15 Beam_15	W12X26	STEEL A572-	37.18	127.68	0.16	9 ULS /4/	
16 Beam_16	W12X26	STEEL A572-	37.18	127.68	0.24	9 ULS /4/	
17 Column_17	HSS4X4X1/2	STEEL A500-	102.42	102.42	0.32	9 ULS /4/	
18 Column_18	HSS4X4X1/2	STEEL A500-	102.42	102.42	0.32	9 ULS /4/	
19 Column_19	W 8x40	STEEL	40.76	70.29	0.29	9 ULS /4/	
20 Column_20	W 8x40	STEEL	40.76	70.29	0.26	9 ULS /4/	
21 Beam_21	W12X26	STEEL A572-	37.18	127.68	0.31	9 ULS /4/	
22 Beam_22	W12X26	STEEL A572-	37.18	127.68	0.34	9 ULS /4/	
23 Beam_23	W12X26	STEEL A572-	37.18	127.68	0.26	9 ULS /4/	
24 Beam_24	W12X26	STEEL A572-	37.18	127.68	0.16	9 ULS /4/	
25 Beam_25	HSS4X4X1/4	STEEL A500-	157.75	157.75	0.39	9 ULS /4/	
26 Beam_26	HSS4X4X1/4	STEEL A500-	157.75	157.75	0.37	9 ULS /4/	
27 Beam_27	HSS4X4X1/4	STEEL A500-	157.75	157.75	0.32	9 ULS /4/	
28 Beam_28	W8X10	STEEL A572-	14.88	57.12	0.32	9 ULS /4/	
29 Beam_29	W8X10	STEEL A572-	14.88	57.12	0.55	9 ULS /4/	
30 Simple bar_3	HSS4X4X1/4	STEEL A500-	197.19	197.19	0.07	9 ULS /1/	
31 Simple bar_3	HSS4X4X1/4	STEEL A500-	157.75	157.75	0.15	9 ULS /1/	
32 Simple bar_3	HSS4X4X1/4	STEEL A500-	172.99	172.99	0.44	9 ULS /1/	
33 Simple bar_3	HSS4X4X1/4	STEEL A500-	157.75	157.75	0.07	9 ULS /1/	
34 Simple bar_3	HSS4X4X1/4	STEEL A500-	197.19	197.19	0.10	9 ULS /5/	
35 Simple bar_3	HSS4X4X1/4	STEEL A500-	157.75	157.75	0.14	9 ULS /1/	
36 Beam_36	HSS4X4X1/4	STEEL A500-	126.20	126.20	0.01	9 ULS /4/	
37 Beam_37	HSS4X4X1/4	STEEL A500-	126.20	126.20	0.02	9 ULS /4/	
39 Simple bar_3	HSS4X4X1/4	STEEL A500-	94.65	94.65	0.02	3 WIND1	
41 Simple bar_4	HSS4X4X1/4	STEEL A500-	94.65	94.65	0.01	9 ULS /6/	
42 Simple bar_4	HSS4X4X1/4	STEEL A500-	94.65	94.65	0.00	9 ULS /5/	
43 Simple bar_4	HSS4X4X1/4	STEEL A500-	99.77	99.77	0.06	9 ULS /4/	
44 Simple bar_4	HSS4X4X1/4	STEEL A500-	99.77	99.77	0.03	9 ULS /4/	
45 Simple bar_4	HSS4X4X1/4	STEEL A500-	99.77	99.77	0.05	9 ULS /1/	
46 Simple bar_4	HSS4X4X1/4	STEEL A500-	99.77	99.77	0.02	9 ULS /6/	
50 Column_50	W 8x40	STEEL	40.76	70.29	0.05	9 ULS /1/	
51 Column_51	W 8x40	STEEL	40.76	70.29	0.08	9 ULS /1/	

FIGURE 6.80 Final steel design results from robot.

The process for designing reinforced concrete columns using Robot is similar to concrete beam design. Initially, select the column element in the 3D model in Robot and then launch the **Provided Member Steel Reinforcement** tool from the **Design** tab. This will open an interface with various options. As previously discussed, you need to set up the calculation options by selecting the icon for that purpose, and then make the appropriate choices for the material grades and rebar sizes (see Figure 6.71).

Design results for reinforced concrete columns are presented through a number of tabs in Robot showing the axial load–bending moment interaction diagram for each

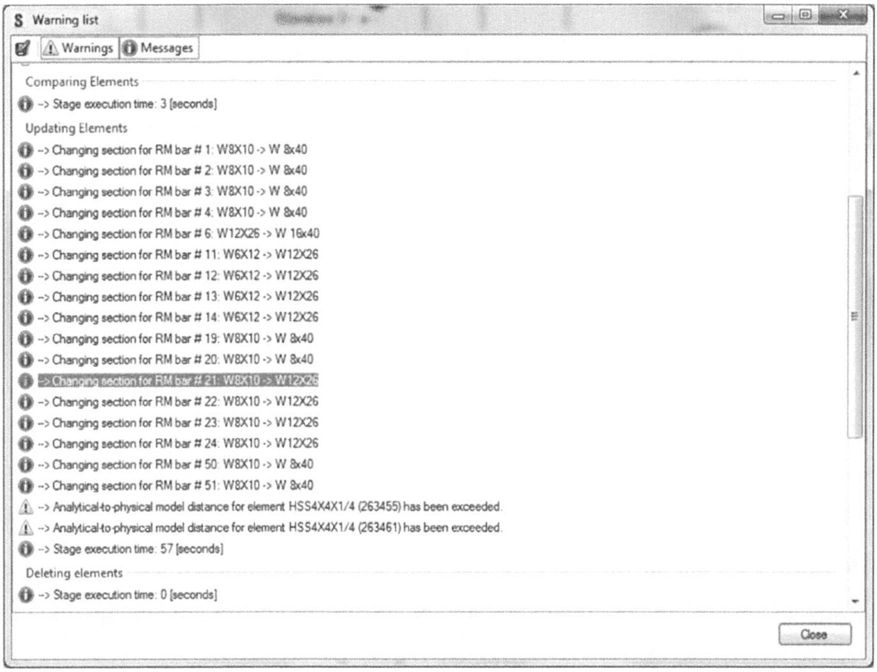

FIGURE 6.81 Revit report on the updated elements after importing from robot.

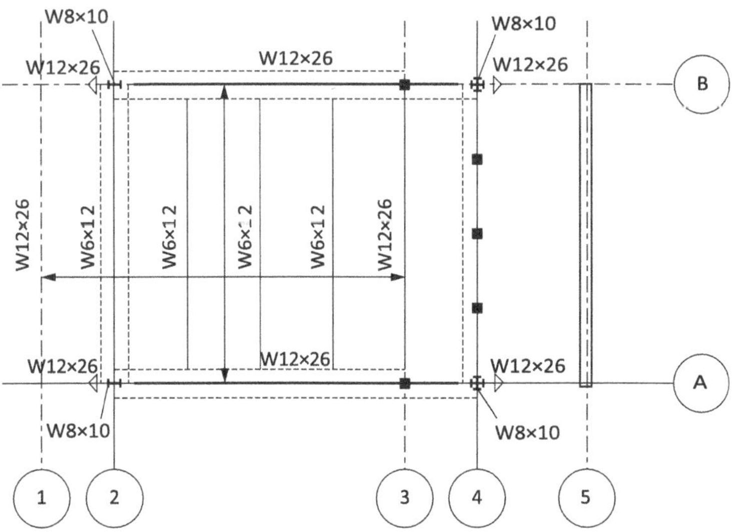

(a)

FIGURE 6.82 Floor-framing plan: (a) before the update from robot; (b) after the update from robot.

(Continued)

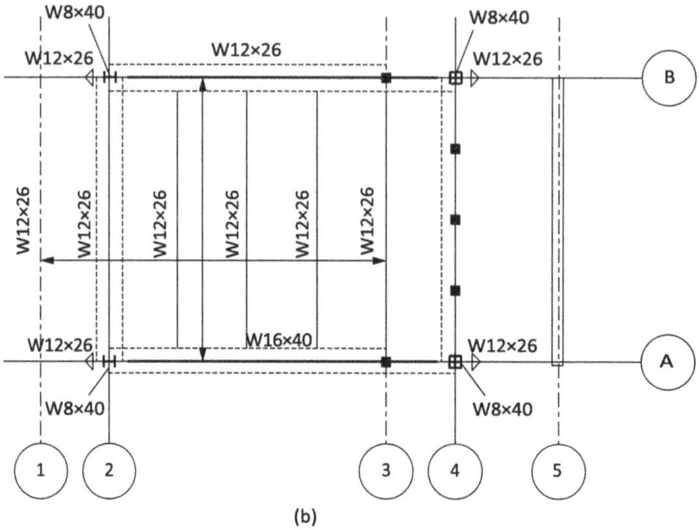

FIGURE 6.82 (*Continued*) Floor-framing plan: (a) before the update from robot; (b) after the update from robot.

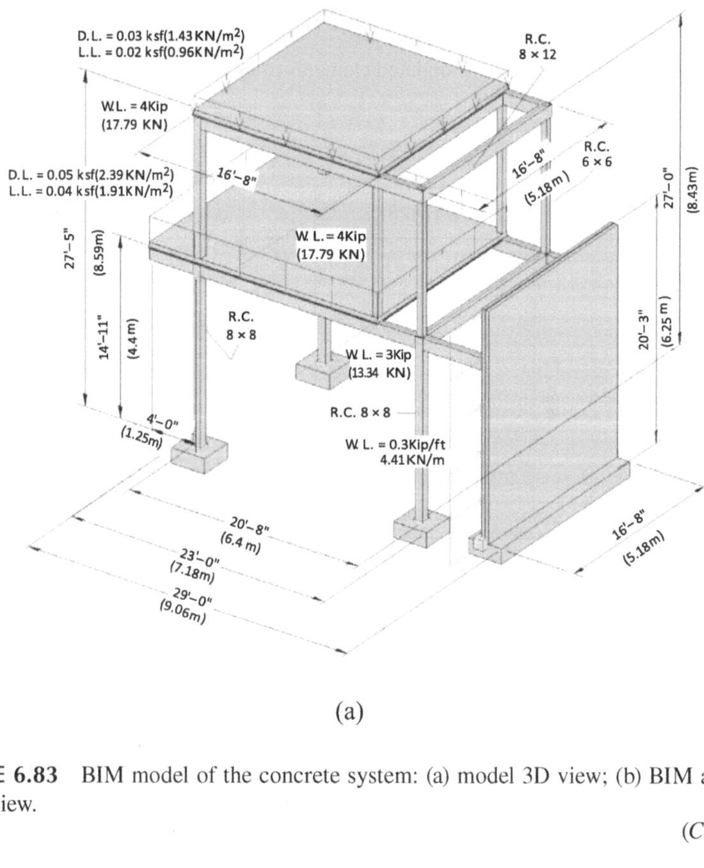

FIGURE 6.83 BIM model of the concrete system: (a) model 3D view; (b) BIM analytical model view.

(Continued)

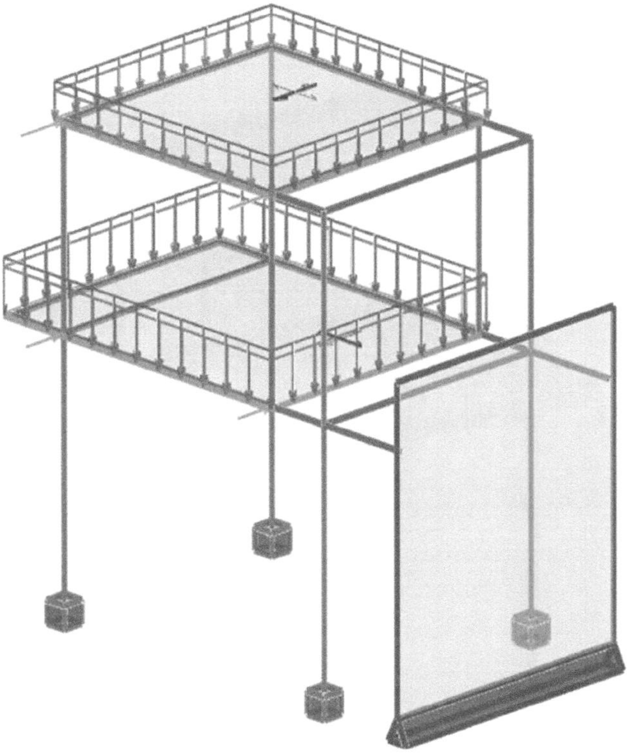

(b)

FIGURE 6.83 (*Continued*) BIM model of the concrete system: (a) model 3D view; (b) BIM analytical model view.

specific load combination (Figure 6.72), column reinforcements, drawing details, report covering the code computations, and design checks. Some of these results are depicted in Figures 6.72 and 6.73. Reports are available on the companion website (http://www.crcpress.com/product/isbn/9781482240436).

The design of planar elements is illustrated by selecting the floor slabs and the shear wall. To start the design computations, instantiate the **Required Steel Reinforcement of RC Slabs/Walls…** tool from the **Design** menu of Robot (Figure 6.74). After completing design calculations, select the type of result that needs to be shown on the screen (Figure 6.75). The required and minimum amounts of steel reinforcement according to ACI 318-08 can be displayed in each direction (Figures 6.76 and 6.77). Other options for displaying results include the number of bars and the spacing between the reinforcing steel bars.

The next step in the design process is to determine the amount of steel reinforcement in the slabs and create detailed drawings. This can be started by selected the icon for **Slab-Provided reinforcement**. This will provide further options about the steel reinforcement that the designer would like to use for the slab (see Figure 6.78).

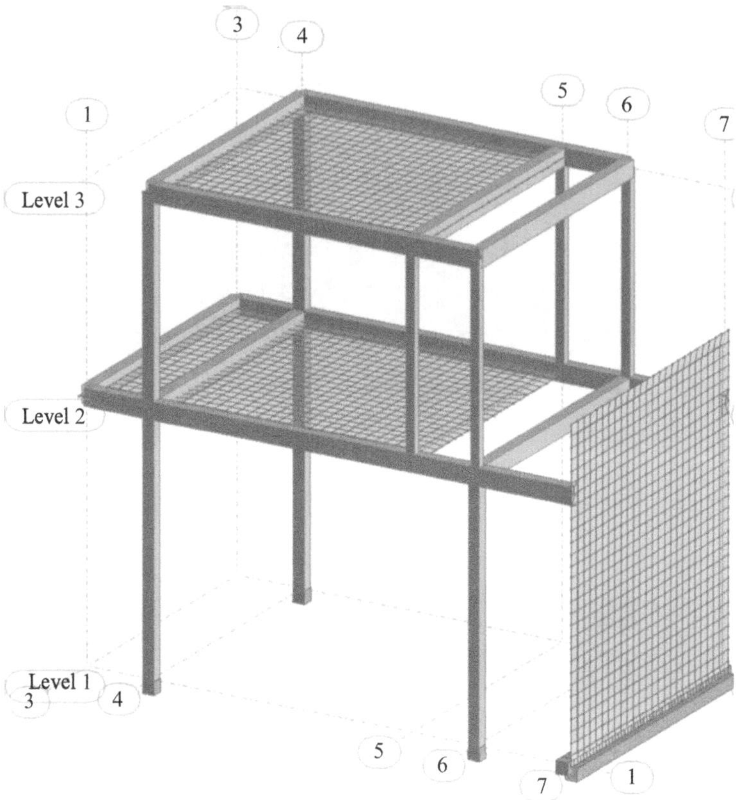

FIGURE 6.84 Exported BIM model in robot showing meshes for the planar element.

After performing the computation for the reinforcement provided based on the settings shown in Figure 6.78, the **Slab – Reinforcement** option can be selected. This will yield results for the selected slab. For example, for the first-floor slab, the reinforcement results are given in Figure 6.79. Also, you can create steel reinforcement drawing details by selecting the icon for **Drawing Details** in Robot (Figure 6.80).

A full report about the steel reinforcement can be created and checked for further verification. An example of this report is available on the companion website (http://www.crcpress.com/product/isbn/9781482240436). At this point, it is possible to update the BIM model in Revit with the steel reinforcement in each member. This can be accomplished by invoking the link to Robot from the **Analyze** tab and then checking the appropriate update options (see Figure 6.81). Steel reinforcement can be further adjusted in Revit by using the **Steel Reinforcement** tool from Revit Extensions (6.90–6.100).

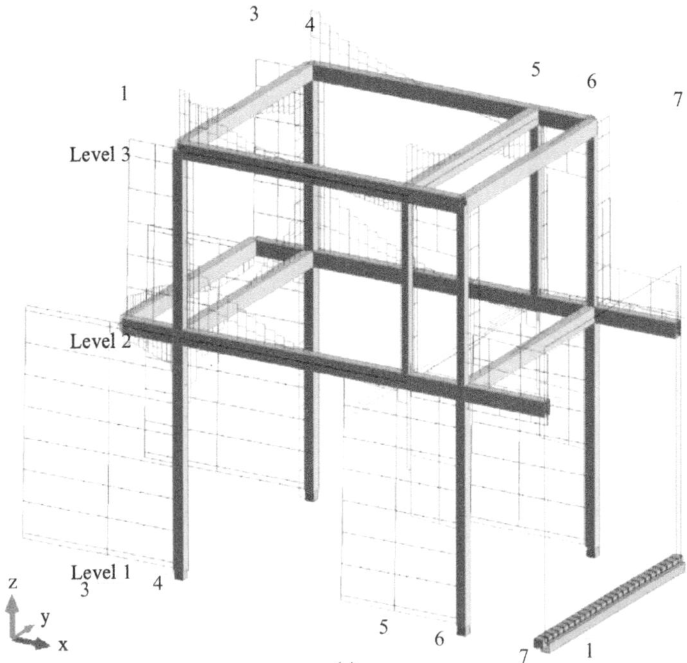

(a)

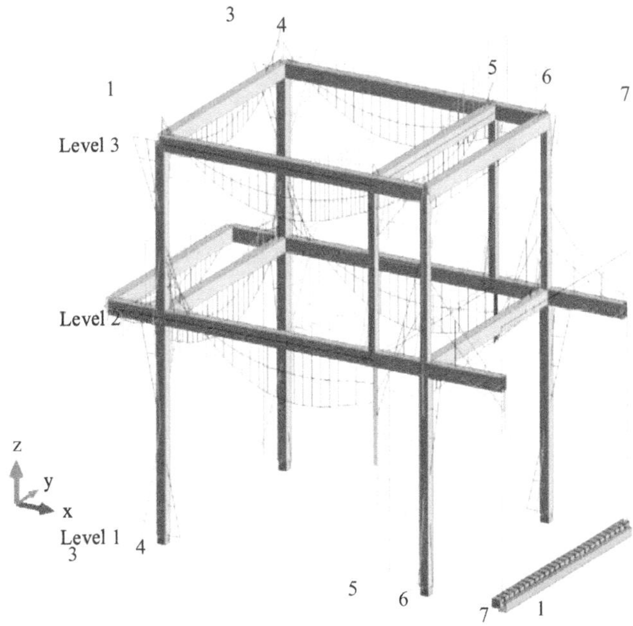

(b)

FIGURE 6.85 Examples of the analysis results for linear elements: (a) shear force diagrams; (b) bending diagrams.

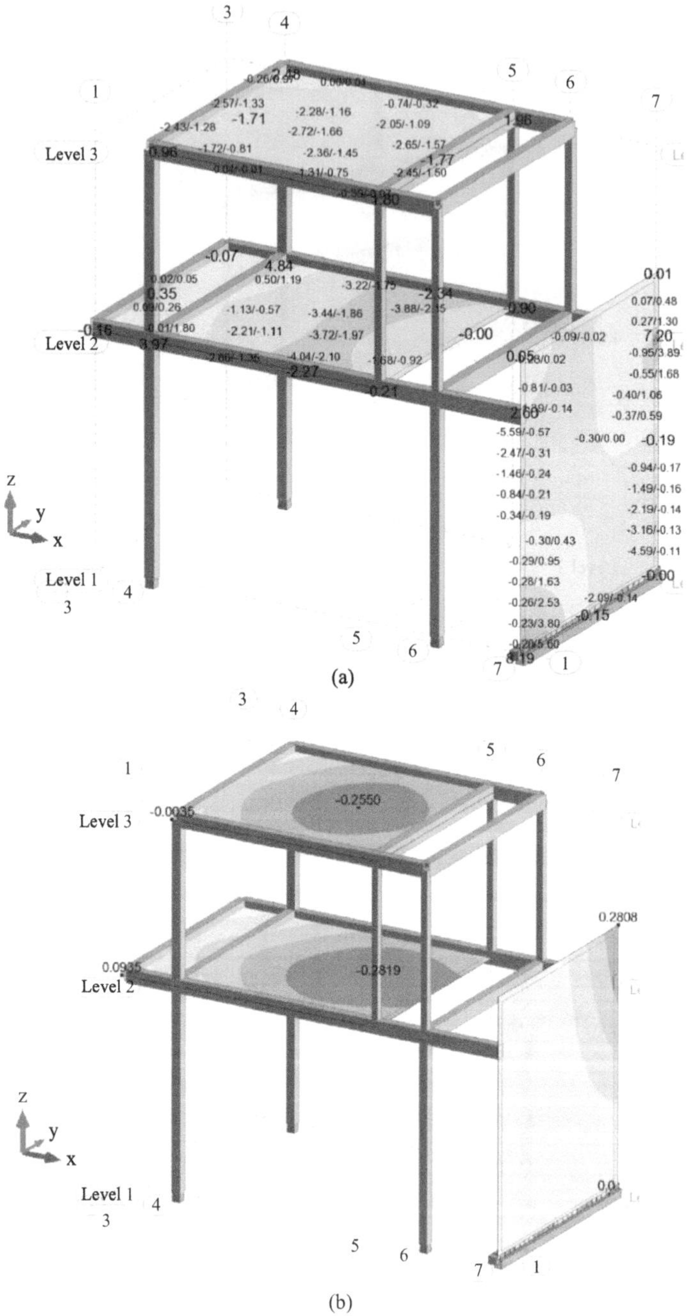

FIGURE 6.86 Examples of the analysis results for planar elements: (a) bending moment about the x axis M_x; (b) vertical displacement U_z.

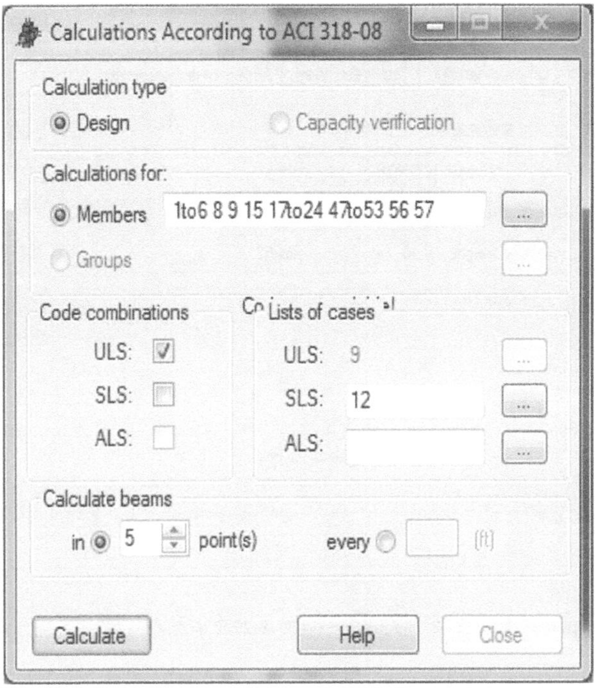

FIGURE 6.87 Design computation options in robot.

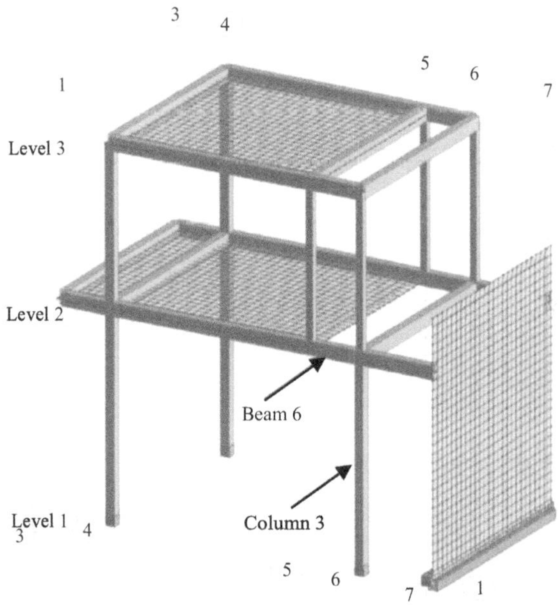

FIGURE 6.88 Beam 6 selected for design.

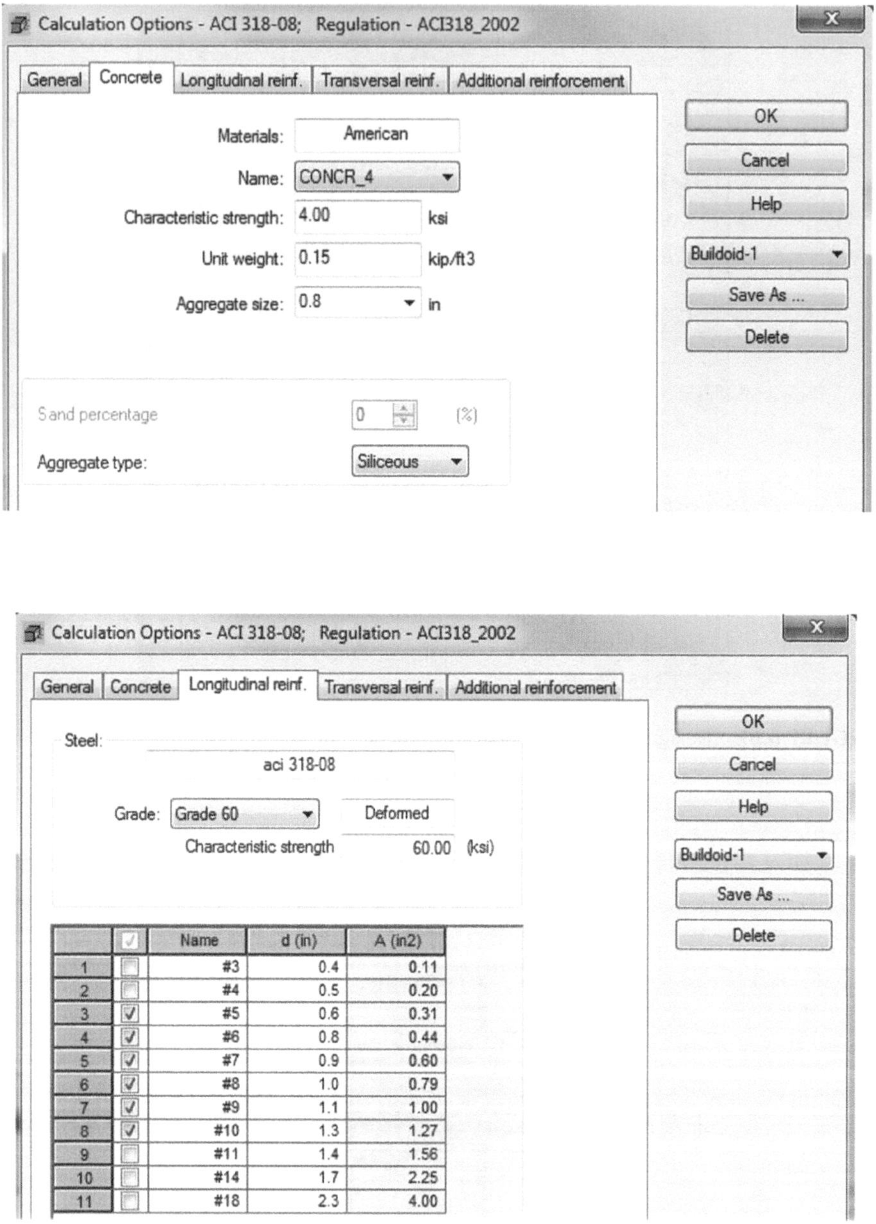

FIGURE 6.89 Settings for Beam 6.

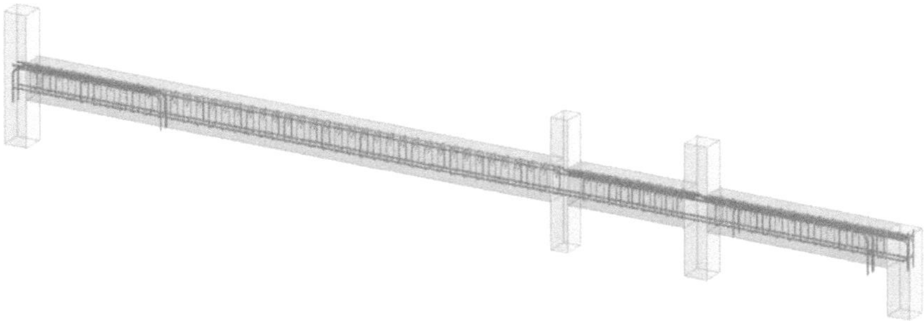

(a)

General | Detailed | Summary table | Spacings and areas

	No.	Reinforcement Type	Steel Grade	Diameter	Number	(ft)
1	1	main-top	Grade 60	#6	2	B = 4.41
2	2	transverse-main	Grade 60	#3	65	B = 0.40
3	3	main-bottom	Grade 60	#5	2	B = 17.93
4	4	assembly-top	Grade 60	#5	2	
5	5	main-bottom	Grade 60	#5	2	B = 6.39
6	6	assembly-top	Grade 60	#5	2	
7	7	main-bottom	Grade 60	#5	2	B = 6.88
8	8	assembly-top	Grade 60	#5	2	
9	9	main-top	Grade 60	#7	2	B = 9.18
10	10	main-top	Grade 60	#5	2	B = 5.23
*						

(b)

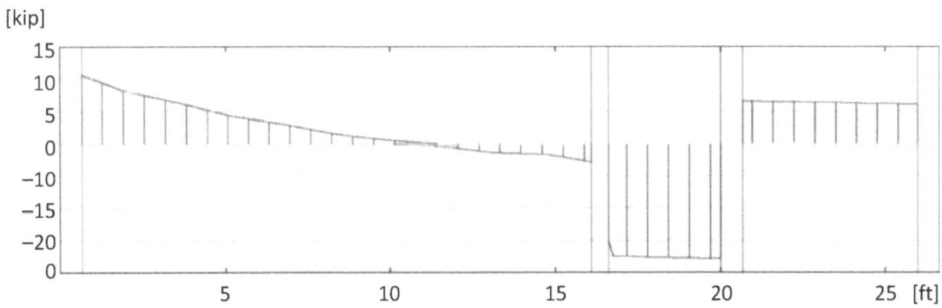

(c)

FIGURE 6.90 Design computation results for the selected beam: (a) Beam 6; (b) steel reinforcement for Beam 6; (c) critical shear force diagram for Beam 6; (d) critical bending moment diagram; (e) provided bending moment capacity as related to the required bending moment; (f) elevation view of the beam reinforcement; (g) beam sections; (h) steel reinforcement details.

(Continued)

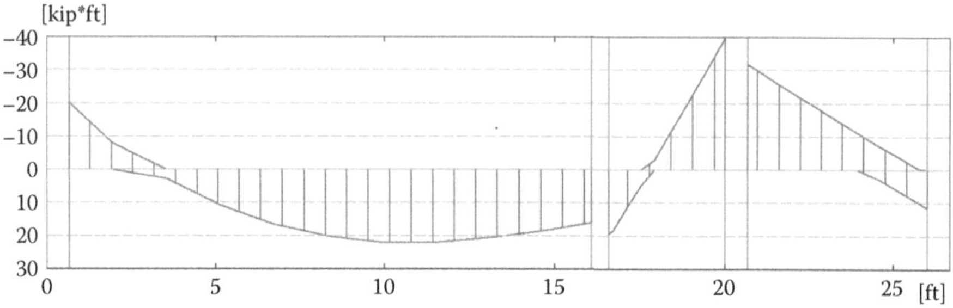

(d)

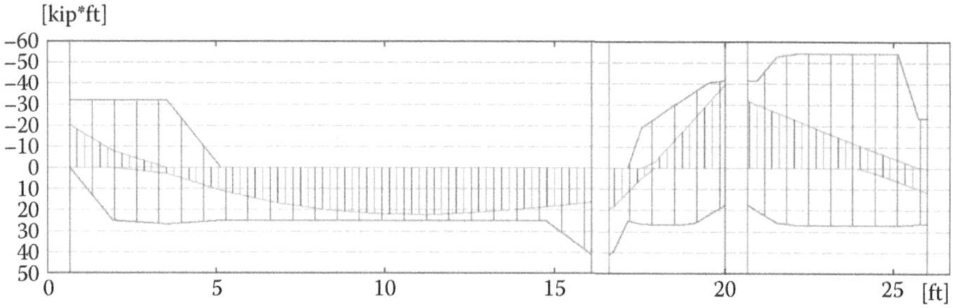

(e)

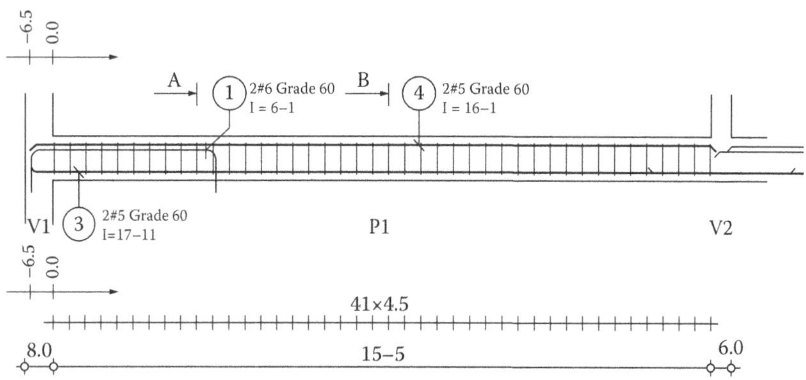

(f)

FIGURE 6.90 (*Continued*) Design computation results for the selected beam: (a) Beam 6; (b) steel reinforcement for Beam 6; (c) critical shear force diagram for Beam 6; (d) critical bending moment diagram; (e) provided bending moment capacity as related to the required bending moment; (f) elevation view of the beam reinforcement; (g) beam sections; (h) steel reinforcement details.

(Continued)

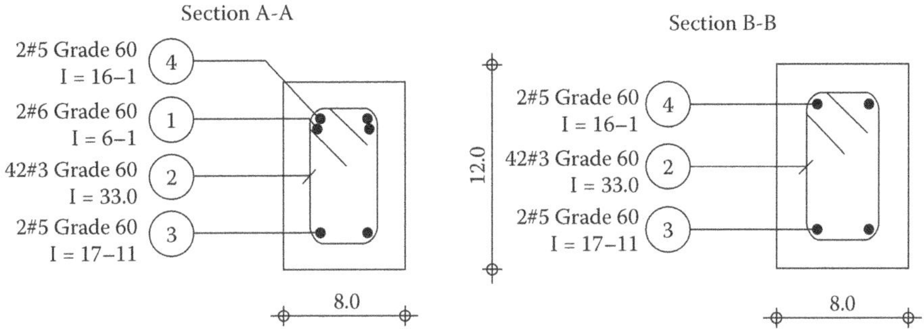

(g)

Pos.	Reinforcement		Shape
1	2#6 Grade 60	I=6–1	12.0 4.5 12.0
2	22#3 Grade 60	I=33.0	4.5 4.0 9.0
3	2#5 Grade 60	I=17–11	17–11.5
4	2#5 Grade 60	I=16–1	16–1

(h)

FIGURE 6.90 (*Continued*) Design computation results for the selected beam: (a) Beam 6; (b) steel reinforcement for Beam 6; (c) critical shear force diagram for Beam 6; (d) critical bending moment diagram; (e) provided bending moment capacity as related to the required bending moment; (f) elevation view of the beam reinforcement; (g) beam sections; (h) steel reinforcement details.

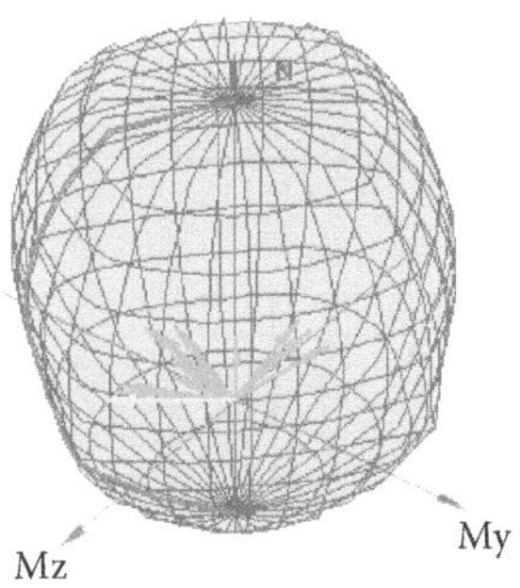

Description	N (kip)	My (kip*ft)	Mz (kip*ft)
1.20DL1+1.60WIND1 (B)	26.14	-4.75	8.40
1.40DL1 (A)	32.92	-5.78	2.24
1.40DL1 (C)	32.92	-4.27	0.90
1.40DL1 (B)	32.92	3.12	-1.12
1.20DL1+0.80WIND1 (A)	27.18	-1.18	-1.98
1.20DL1+0.80WIND1 (C)	27.18	-2.95	1.49
1.20DL1+0.80WIND1 (B)	27.18	-1.04	3.72
0.90DL1+1.60WIND1 (A)	19.08	3.84	-6.35
0.90DL1+1.60WIND1 (C)	19.08	-3.01	3.46
0.90DL1+1.60WIND1 (B)	19.08	-5.42	8.64

FIGURE 6.91 Reinforced concrete column interaction diagram for Column 3.

(*Continued*)

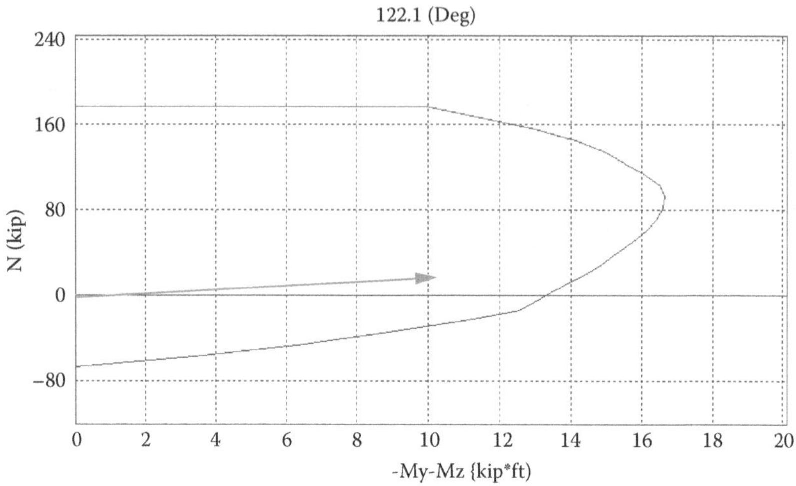

FIGURE 6.91 (*Continued*) Reinforced concrete column interaction diagram for Column 3.

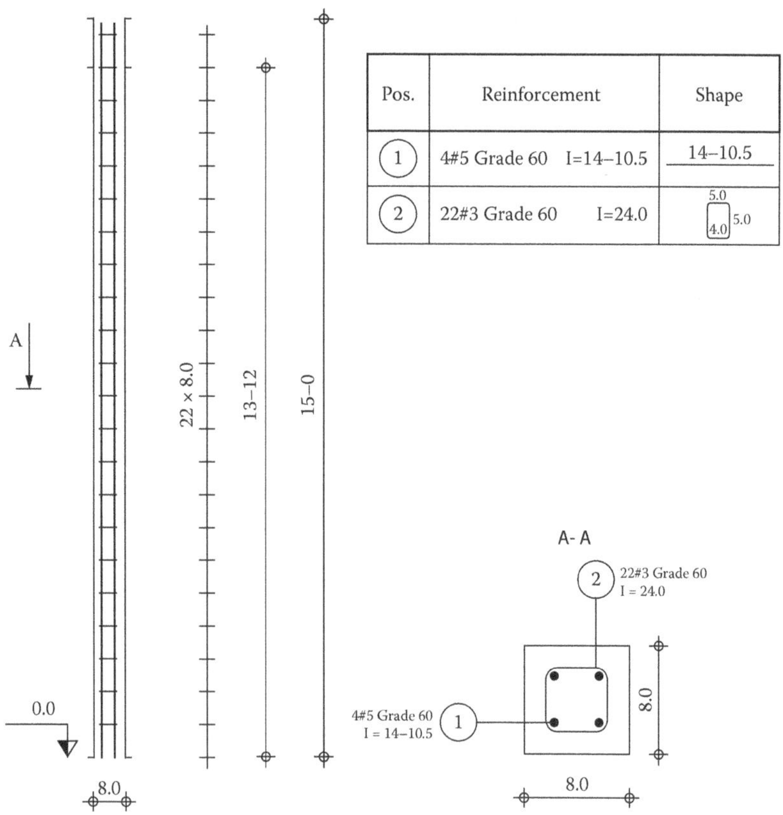

Pos.	Reinforcement	Shape
1	4#5 Grade 60 I=14–10.5	14–10.5
2	22#3 Grade 60 I=24.0	5.0 4.0 5.0

FIGURE 6.92 Column 3 elevation and section with steel reinforcement details.

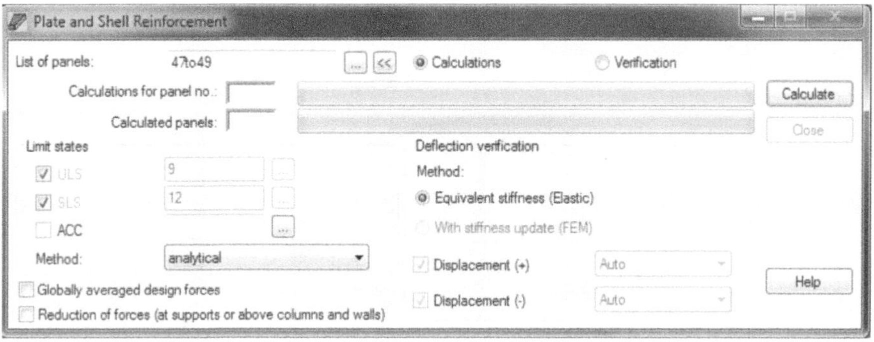

FIGURE 6.93 Design computation for structural concrete slabs and walls.

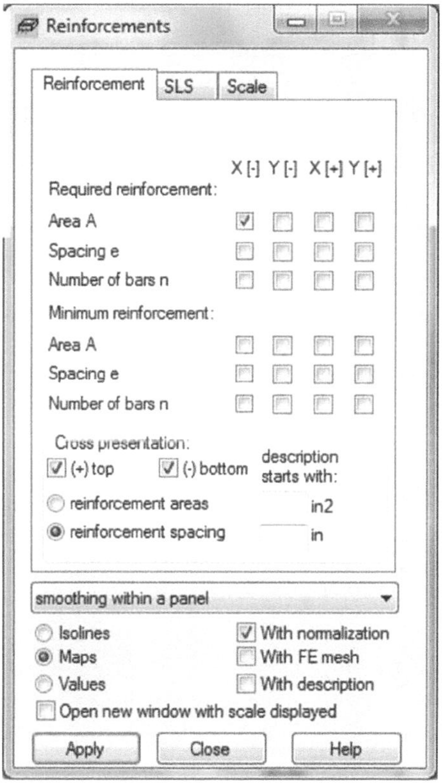

FIGURE 6.94 Types of design results for reinforced concrete slabs and walls.

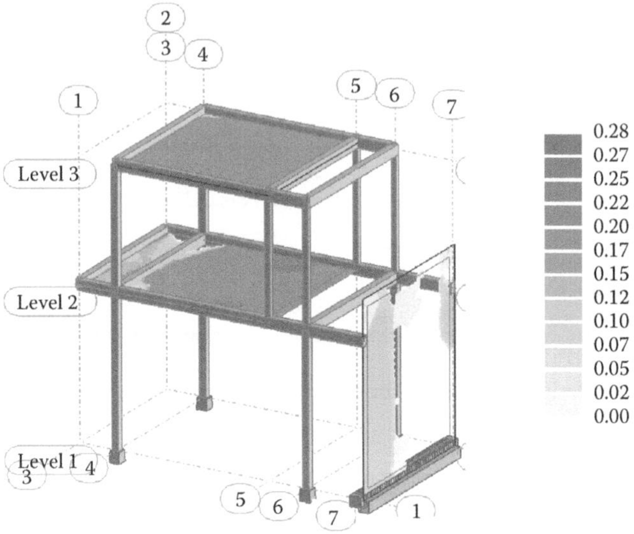

FIGURE 6.95 Required area of steel for bottom reinforcement due to M_x.

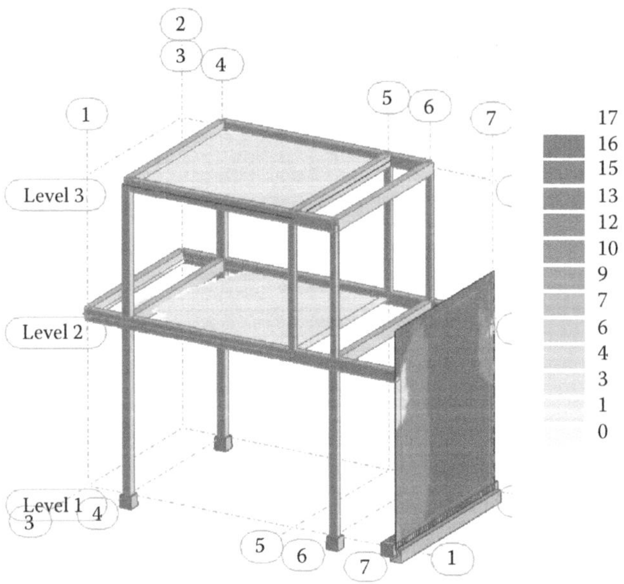

FIGURE 6.96 Minimum number of bars for bottom reinforcement due to M_x.

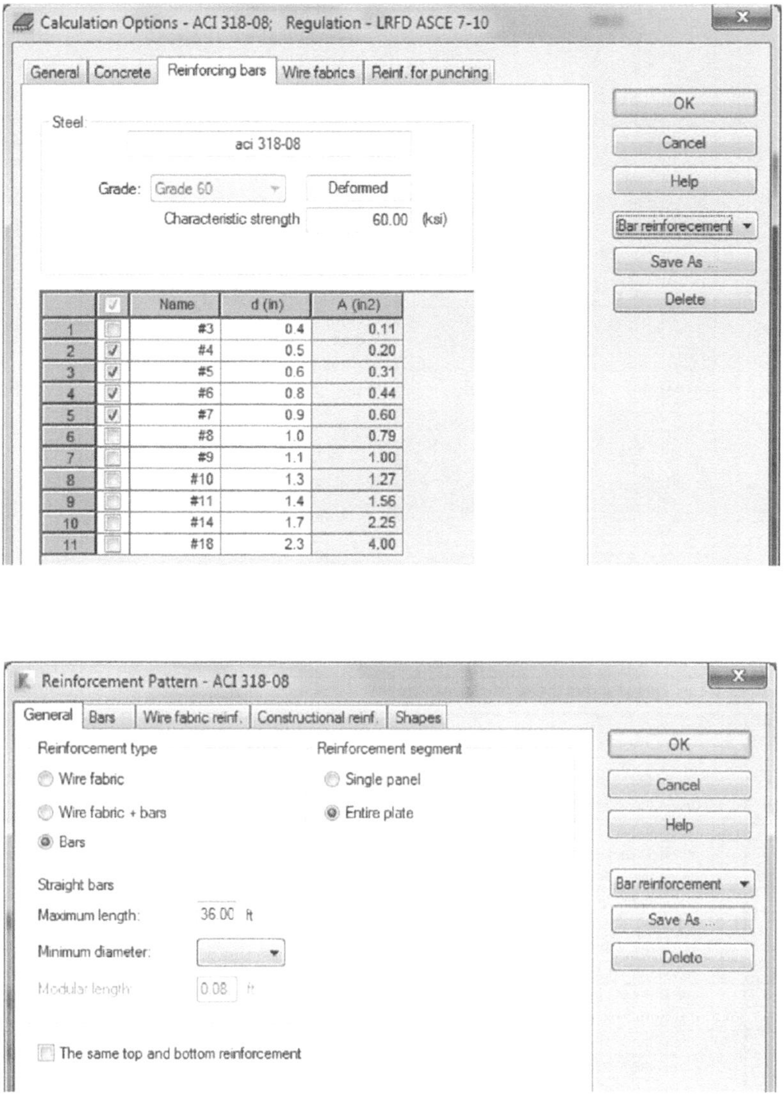

FIGURE 6.97 Options for provided steel reinforcement for slabs in robot.

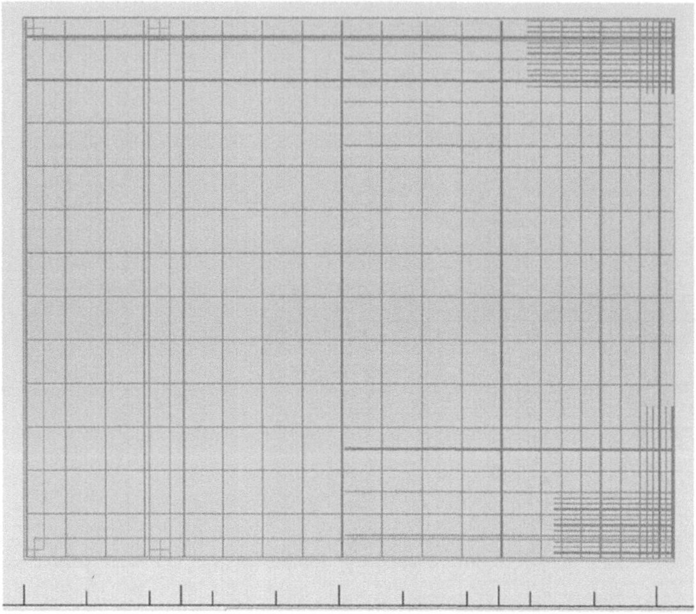

General | Detailed | Summary table | Wire fabrics - general | Wire fabrics - detailed

	No.	Reinforcement Type	Steel Grade	Diameter	Shape Code	Number	Sp
		Plate48					
1	1	top in the X direction	Grade 60	#4		20	
2	2	top in the X direction	Grade 60	#4		2	
3	3	top in the X direction	Grade 60	#4		4	
4	4	top in the X direction	Grade 60	#4		4	
5	5	top in the X direction	Grade 60	#4		16	
6	6	top in the X direction	Grade 60	#4		3	
7	7	top in the Y direction	Grade 60	#4		12	
8	8	top in the Y direction	Grade 60	#4		27	
9	9	top in the Y direction	Grade 60	#4		33	
10	10	top in the Y direction	Grade 60	#4		6	
11	11	top in the Y direction	Grade 60	#4		4	
12	12	top in the Y direction	Grade 60	#4		8	
[illegible]	13	bottom in the X direction	Grade 60	#5		[illegible]	

FIGURE 6.98 Steel reinforcement for floor slab.

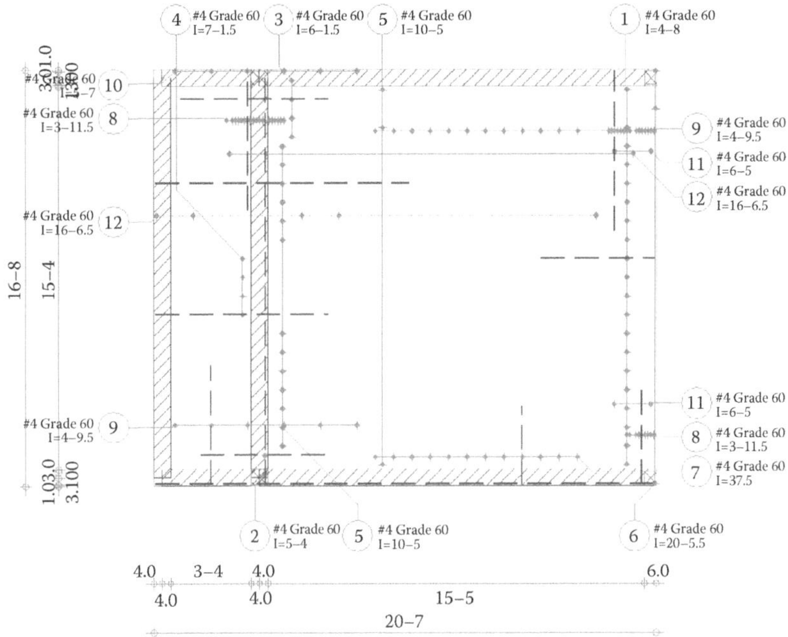

(a)

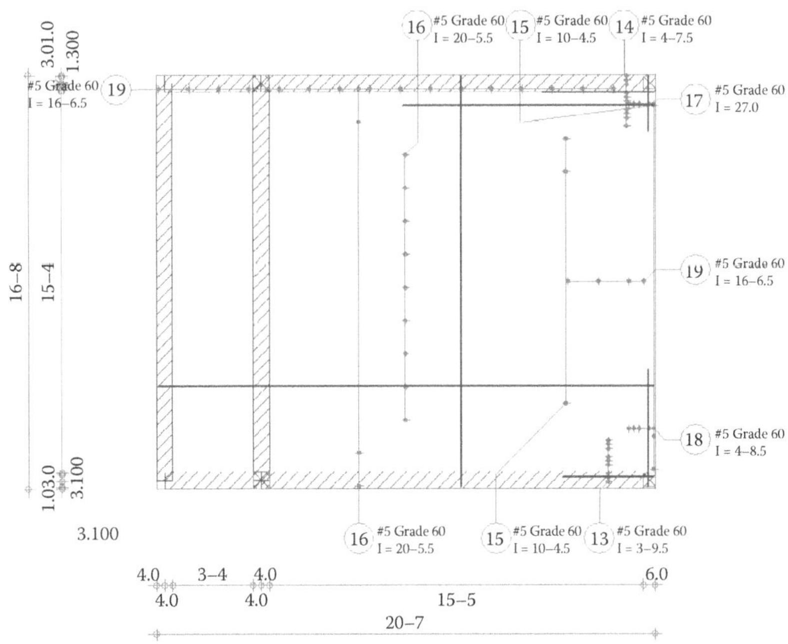

(b)

FIGURE 6.99 Top and bottom steel reinforcement details from robot: (a) top steel reinforcement for first-floor slab; (b) bottom steel reinforcement for first-floor slab.

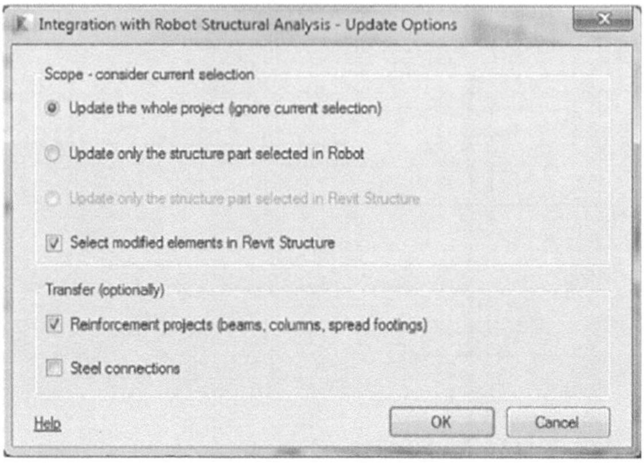

(a)

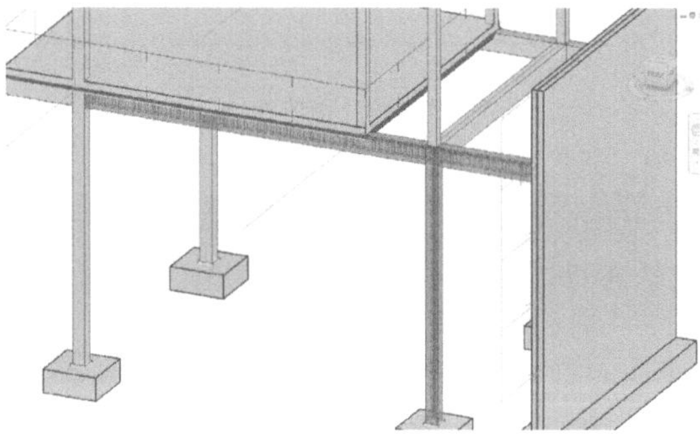

(b)

FIGURE 6.100 Imported data from robot into Revit model: (a) options for updating the BIM model from robot; (b) updated model showing steel reinforcement for the designed beam and column.

EXERCISES

6.1. Using the model developed for Exercise 5.3, apply the following loads and then study the load path using the **Load Takedown** tool in Revit Extensions: Floor dead load = 50 psf (2.39 kPa); floor live load = 60 psf (2.87 kPa); roof dead load = 30 psf (1.44 kPa); roof live load = 20 psf (0.96 kPa) (shown below).

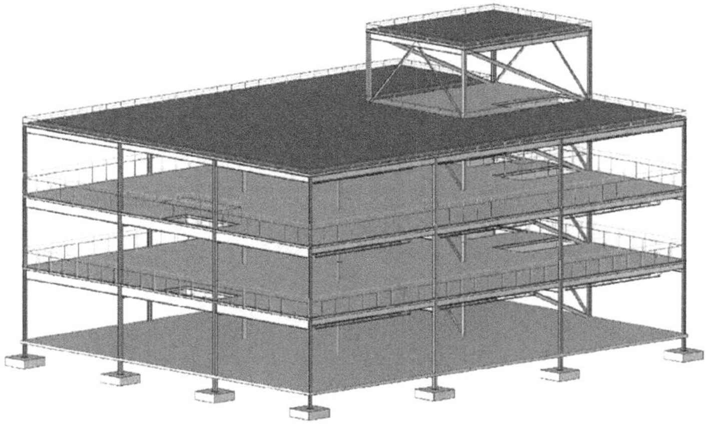

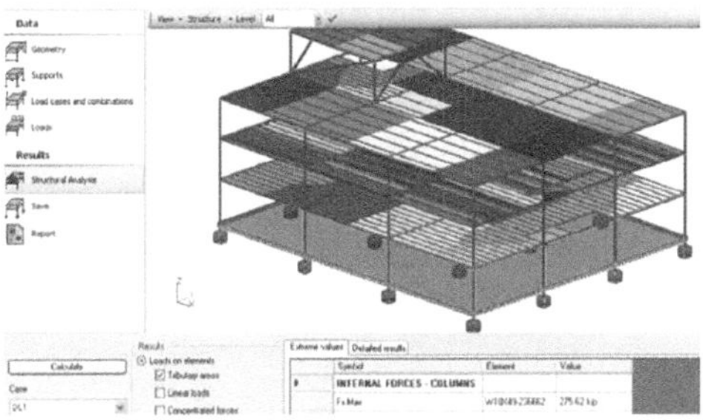

6.2. Create a BIM model for the structure shown below. Using the **Beam** tool in Revit Extensions, analyze the cantilever beam indicated as B1. Assume the weight of the concrete planter to be equal to 4.2 kips (17.68 kN).

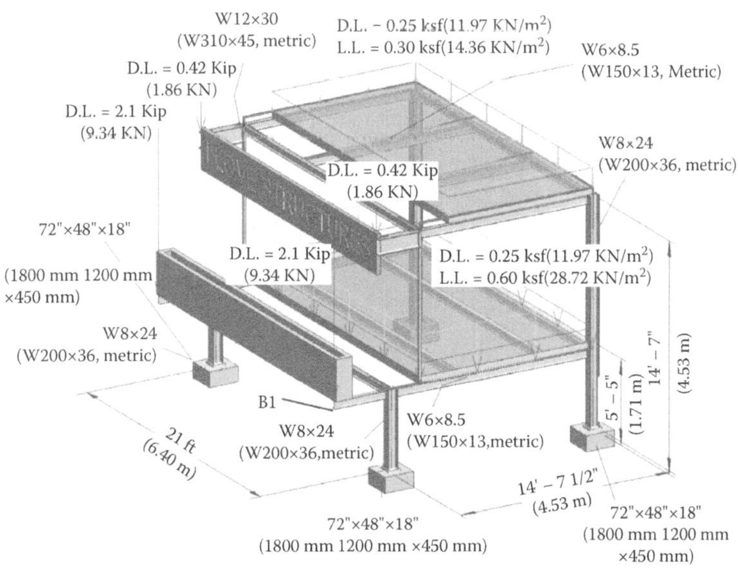

6.3. Using the structure that you modeled in Chapter 4 (see Figure below), analyze one of the trusses using the **Truss** tool available in Revit Extensions. Loads applied to the truss joints include point DL = 2 kips (8.90 kN) and live load = 3 kips (13.34 kN).

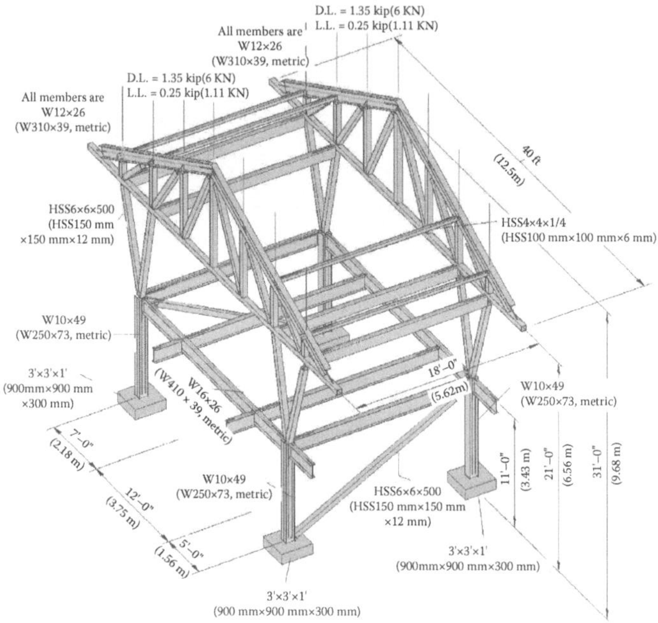

6.4. Download the BIM model shown in Problem 6.3 and then analyze the frame indicated as FR1 using the **Frame Analysis** tool from Revit Extensions (see Figure below).

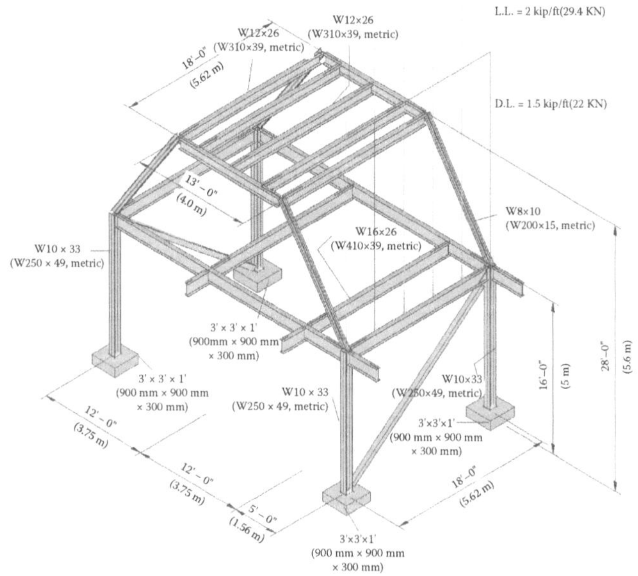

6.5. Model the steel truss shown below using Revit and then run the direct link to Robot to analyze and design the different members. List and mark the elements that did not meet the AISC 2005 specifications.

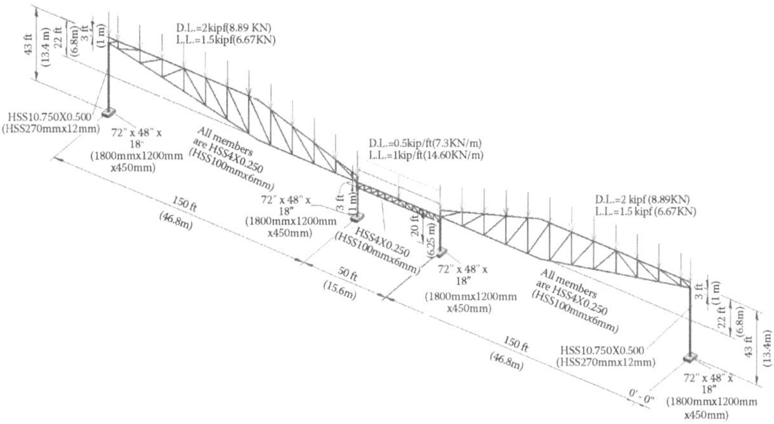

6.6. For the truss in Exercise 6.5, recommend structural elements that can successfully replace the failed ones. Verify your results by producing a full report of the truss analysis.

6.7. Model the structure shown below from the given foundation and first-floor framing plan. Then, analyze the slab using the **Slab** tool in Revit Extensions.

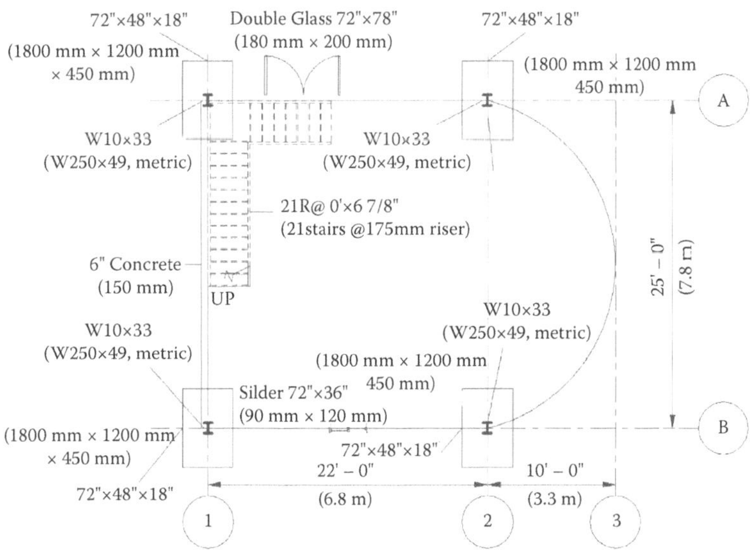

Foundation plan

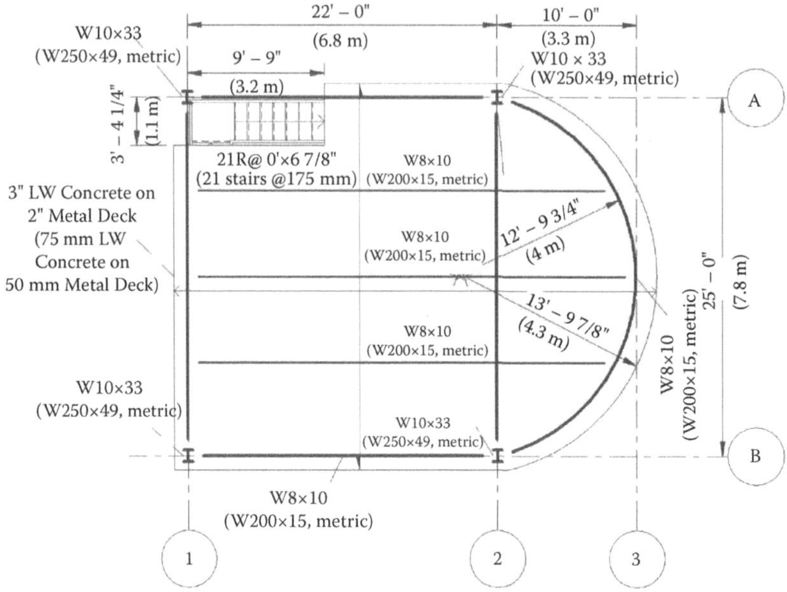

Floor framing plan

6.8. Create the mass shown below and then analyze it using the **Conceptual Form Simulation** tool. The mass is subjected to dead, live, and wind loads of 1.0 ksf (47.88 kPa), 0.5 ksf (23.94 kPa), and 0.05 ksf (2.394 kPa), respectively.

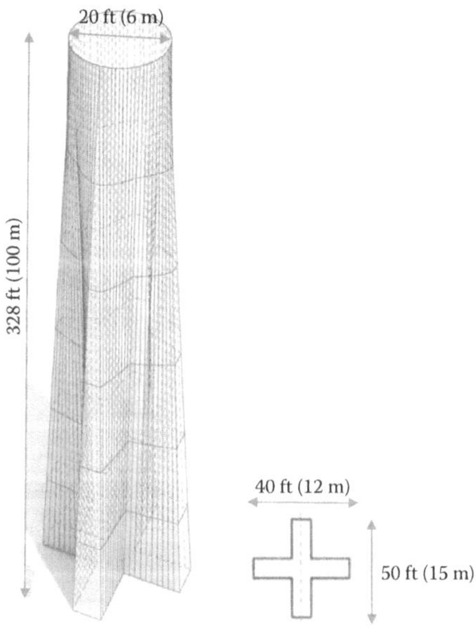

3D view Base cross-section

6.9. Analyze the entire BIM model developed in Exercise 6.2 using the direct link to RISAFloor and then RISA-3D. Update the model with the verified member sections from RISA.

6.10. Repeat Exercise 6.9 using the direct link to Robot.

6.11. Using a buildoid plan with an isosceles triangle, develop various arrangement patterns for lateral growth as suggested by the SAS framework. For each option, propose a reasonable program along with the designation for circulation.

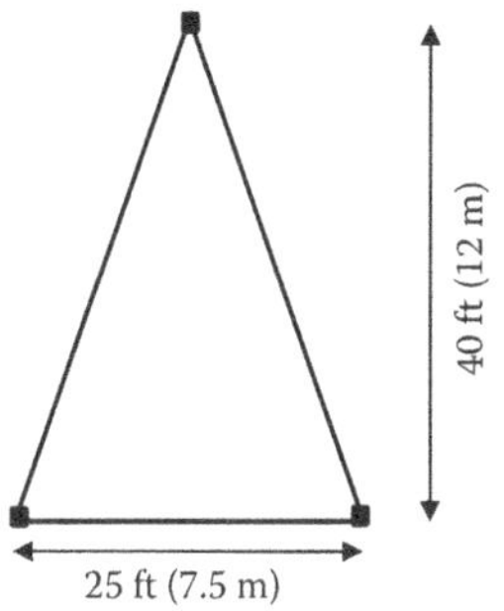

6.12. Design a buildoid for the plan given below using wood structural elements with interior and exterior spaces. Then, applying the SAS framework, develop lateral expansion (double the space) and vertical progression. The program can be a two-story residential unit. A complete architectural and structural model is required.

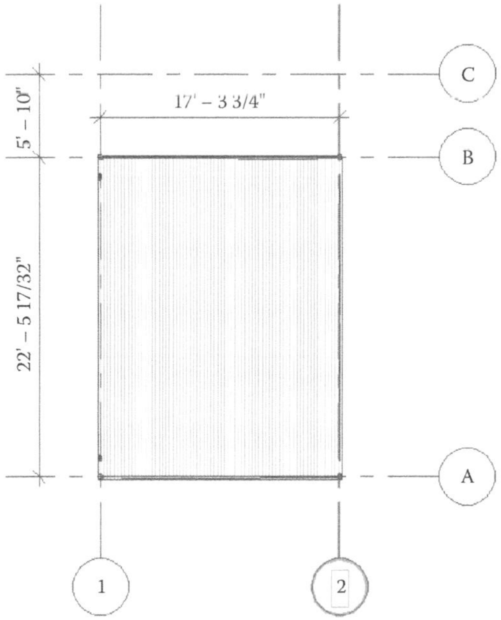

6.13. Analyze the developed structural solution for Exercise 6.12 using the direct link to Robot or RISA. Assume a floor dead load of 30 psf and floor live load of 40 psf. For the roof, assume a dead load of 15 psf and a live load of 20 psf. Apply a wind load of 30 psf. Provide a full report that verifies the adequacy of your design.

6.14. Repeat Exercise 6.12 for the moon crest plan shown below. The crest is formed by two circles having a diameter of 25 ft (10 m), and their centers are 12.5 ft (5 m) apart.

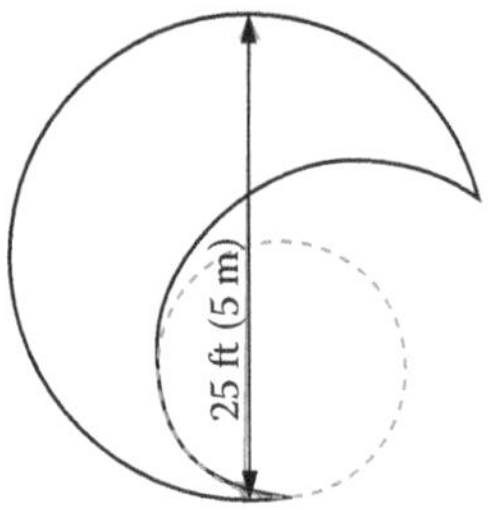

6.15. Using the principles of the SAS framework, design a buildoid for the plan given below. The plan gives the location for columns and footing. Then, develop expansion and growth patterns for doubling the space and tripling the functional space to fulfill a program of your choice. Use structural steel systems.

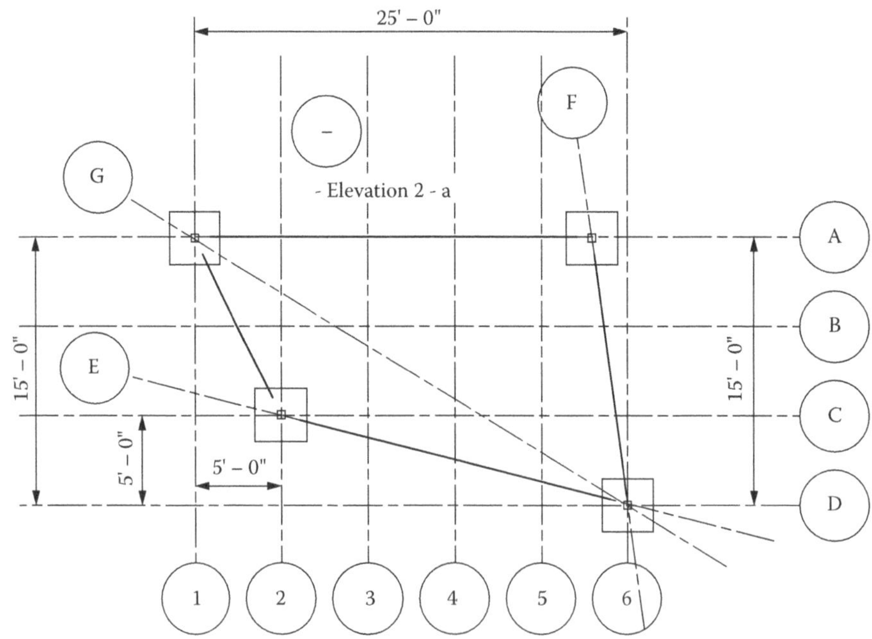

While developing your design, check the following criteria:

- The expression of the primary functional activities of the building through a rich, hierarchical composition of structural geometries.
- Express structural systems to reinforce formal and spatial objectives.
- Incorporation of coherent circulation patterns to provide clear routes and connections in and around the building.
- Development of spatial sequences to reinforce the circulation patterns and functional activities.
- Articulation of spaces in and around the building to enhance the public realm.

6.16. Analyze the developed structural solutions for (a), (b), and (c) in Exercise 6.13 using the direct link to Robot or RISA. Provide a full report that verifies the adequacy of your design.

6.17. Repeat Exercise 6.15 for the plan shown below:

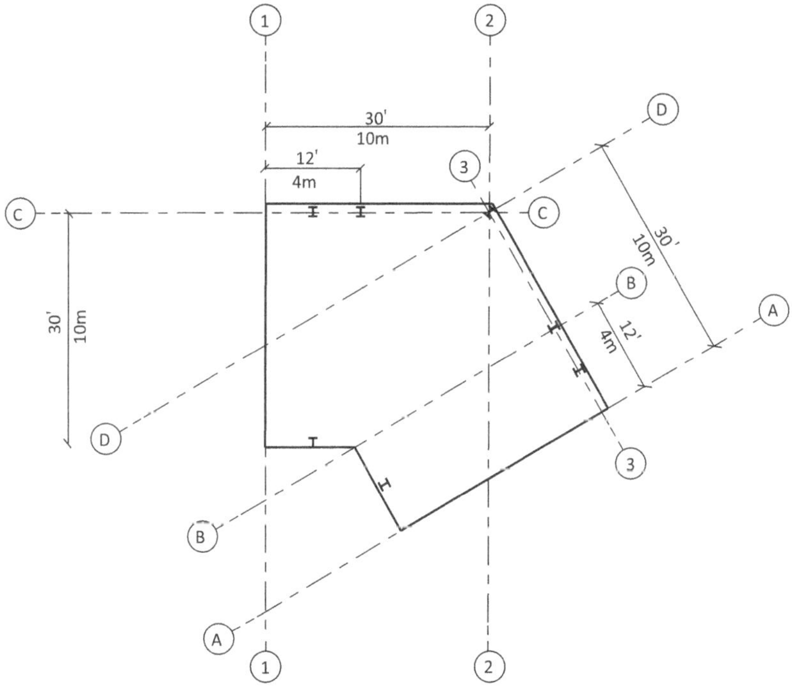

6.18. Beginning with the regular octagon plan shown with a plan area of 80 m^2 (861 ft^2), using the SAS framework design a community center to be used by an indigenous community. The design should address the following programmatic conditions:

Site: The site is at the banks of the Atrato River in the municipality of Vigía del Fuerte, Antioquia, Colombia.

Ecosystem: The location is a tropical rain forest on floodplains, one of the most humid and rainy places on Earth.

Human Environment: Embera indigenous communities and settlers of Spanish and African descent share this land under precarious conditions and substantial economic limitations.

Building: The project consists of a palafitte (stilt) open building that will allow the members of the Embera community to articulate to life in the municipality, receive education, meet with Embera of different villages, and offer their products to the community of settlers. Develop lateral expansion patterns to fulfill various requirements:

- Double the space for a grocery building.
- Three times the space for social living.
- Four times the space for a public restaurant.
- Five times the space for education.
- Six times the space for dormitories.

The site is in a rural area of Colombia (South America). Try to aim for simplicity of the structure while maintaining stability, functionality, and durability. Use wood structural systems for this assignment. No interior columns are allowed.

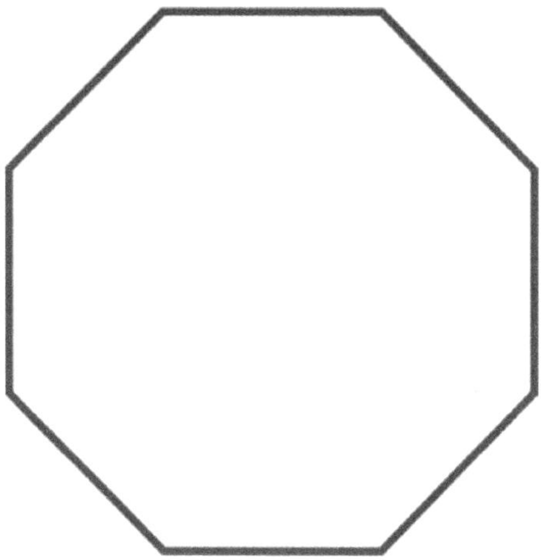

While developing your design solution, check the following criteria:

- The expression of the primary functional activities of the building through rich, hierarchical composition of structural geometries.
- Express structural systems to reinforce formal and spatial objectives.
- Incorporation of coherent circulation patterns to provide clear routes and connections in and around the building.
- Development of spatial sequences to reinforce the circulation patterns and functional activities.
- Articulation of spaces in and around the building to enhance the public realm.

Bibliography

Addis, B. (2001). *Creativity and Innovation: The Structural Engineer's Contribution to Design*. Oxford, UK: Architectural Press.

Addis, B. (2007). *Building: 3000 Years of Design, Engineering and Construction*. London: Phaidon.

Addis, W. (1990). *Structural Engineering—The Nature of Theory and Design*. New York: Ellis Harwood.

Addis, W. (1992). Engineering history and the formation of design engineers. *International Journal of Engineering Education*, 8, 6, 408–412.

Addis, W. (1998). Using the history of engineering design. In *Proceedings of the 20th SEED Annual Design Conference*, ed. American Society Autodesk. London: Imperial College, pp. 97–100.

AIA. (2008). *Document E202-2008: Building Information Modeling Exhibit*. Washington, DC: American Institute of Architects. https://www.aia.org/contractdocs/AIAB095713

AIA. (2013). *Document G202*. Washington, DC: American Institute of Architects. https://www.aia.org/aiaucmp/groups/aia/documents/pdf/ aiab099086.pdf

American Society of Civil Engineers, ASCE 7-10. (2010). *Minimum Design Loads for Buildings and Other Structures*. Reston, VA: American Society of Civil Engineers.

American Society of Civil Engineers, ASCE 7-22. (2022). *Minimum Design Loads for Buildings and Other Structures*. Reston, VA: American Society of Civil Engineers (ASCE).

Autodesk Inc. (2013). *Revit 2014 Wikihelp*. San Francisco, CA: Autodesk Inc. https://help.autodesk.com/view/RVT/2014/ ENU/?guid=GUID-ACC03901-54B0–45DF-9432-0C0C1A033BCB

Autodesk Inc. (2025). *Revit 2025*. San Francisco, CA: Autodesk Inc.
https://help.autodesk.com/view/RVT/2025/ENU/?guid=RevitReleaseNotes_2025release_html

Autodesk Inc. (2025). *Robot Structural Analysis professional 2025*. San Francisco, CA: Autodesk Inc.
https://help.autodesk.com/view/RSAPRO/2025/ENU/?guid=GUID-CDBFEAA4-0FFB-41AC-AB8A-79F5C413C221

Bard, J. (1990). *Morphogenesis: The Cellular and Molecular Processes of Developmental Anatomy*. Cambridge: Cambridge University Press.

Barison, M. B. and Santos, E. T. (2010). BIM teaching strategies: An overview of the current approaches. In *Proceedings of the International Conference on Computing in Civil and Building Engineering,* ed. W. Tizani. Nottingham, UK: Nottingham University Press, pp. 577–584.

Becerik-Gerber, B., Gerber, D. J., and Ku, K. (2011). The pace of technological innovation in architecture, engineering, and construction education: Integrating recent trends into the curricula. *ITcon. Journal of Information Technology in Construction*, 16, 411–432.

Bently Inc. (2014). Accessed on January 2014 from https://www.bentley.com/en-US/Products/Building+Analysis+and+Design/

Billington, D. (2003). *The Art of Structural Design—A Swiss Legacy*. New Haven, CT: Yale University Press.

BIMForum. (2023). *Level of Development 2023 Specification*. BIMForum: https://buildingsmartalliance.org/standards/bimforum/lod-specification/

Chappell, D. and Willis, A. (2000). *The Architect in Practice* (Running a project). Oxford, UK: Blackwell Science, 98–103.

Construction Specifications Institute. (2023). *OmniClass Construction Classification System*. Construction Specifications Institute: https://www.csinet.org/omniclass .

Dhonukshe, P. (2011). *Cell Biology*. Netherlands: Universiteit Utrecht. https://web.science.uu.nl/cellbiologylindex.php?option=com_content&view=article&id=42&Itemid=34

Eastman, C., Teicholz, P., Sacks, R., and Liston, K. (2011). *BIM Handbook: A Guide to Building Information Modeling for Owners, Managers, Designers, Engineers and Contractors*. 2nd ed. New York: Wiley.

Gallaher, P., O'Connor, A., Dettbarn, J., and Gilday, L. (2004). *Cost Analysis of Inadequate Interoperability in the U.S. Capital Facilities Industry*. NIST GCR 04-867. Gaithersburg, MD: U.S. Department of Commerce Technology Administration National Institute of Standards and Technology, Advanced Technology Program Information Technology and Electronics Office.

GraphiSoft Inc. (2014). Accessed on January 2014 from https://www.graphisoft.com/archicad/

Hasanain, F., and Nawari, N. O. (2022). BIM-based model for sustainable built environment in Saudi Arabia, *Frontiers in Built Environment*, Switzerland.

Hensel, M., Menges, A., and Weinstock, M. (2010). *Emergent Technologies and Design: Towards a Biological Paradigm for Architecture*. 1st ed. London: Routledge.

International Alliance for Interoperability (IAI). (2012). *BuildingSMART International*. Hershey, PA: Industry Foundation Classes (IFC). https://www.iai-international.org and https://www.iai-tech.org

Khan, Y. S. (2004). *Engineering Architecture: The Vision of Fazlur R. Khan*. New York: Norton.

Lawal, Oluwatoyin O., Nawari, N. O., and Omobolaji Lawal (2025). AI-enabled cognitive predictive maintenance of urban assets using city information modeling—systematic review, *Buildings,* 15, 5, MDPI.

Lawal, Oluwatoyin O., and Nawari, N. O. (2024). Blockchain-enabled city information modeling (CIM) framework for urban asset management, *Journal of Architectural Engineering* 30, 3, ASCE, USA.

Lawal, Oluwatoyin O., and Nawari, N. O. (2023). Blockchain and city information modeling (CIM): A new approach of transparency and efficiency, *Journal of Information Technology in Construction (ITCon)* – 8, 711–734, ISSN 1874-4753. https://doi.org/10.36680/j.itcon.2023.037

LEED v4.1 (2019). Leadership in Energy and Environmental Design and Green Building Certification System, U.S. Green Building Council's (USGBC). https://www.usgbc.org/leed/v41

McGraw-Hill's SmartMarket Report. (2012). *The Business Value of BIM in North America: Multi-Year Trend Analysis and User Ratings (2007–2012)*. Belford, MA: McGraw-Hill Construction.

National Design Specification. (2005). *(NDS®) for Wood Construction*. Leesburg, Virginia: American Wood Council (AWC).

National Design Specification. (2022). *(NDS®) for Wood Construction*. Leesburg, Virginia: American Wood Council (AWC).

National Institute of Building Sciences. (2007). *National Building Information Modeling Standard (NBIMS). Version 1, Part 1. Overview, Principles, and Methodologies*. Washington, DC: National Institute of Building Sciences. https://www.nationalcadstandard.org/

Nawari, N. O. (2020). Blockchain Technologies: Hyperledger Fabric in BIM Work Processes, *Proceedings of the 15th International Conference on Computing in Civil and Building Engineering (ICCCBE + CIB W78 2020),* San Paolo, Brazil, June 2020.

Nawari, N. O. (2019). BIM data exchange standard for hydro-supported structures, *Journal of Architectural Engineering (JAE),* ASCE, 25, 3, 04019015.

Nawari, N. O., and Ravindran, S.G. (2019). Blockchain technology and BIM process: Review and potential applications, *Journal of Information Technology in Construction (ITcon)* 24, 12, 2019, 209–238.

Nawari, N. O. (2018). *Building Information Modeling: Automated Code Checking and Compliance Processes*: CRC Press, Taylor and Francis Group. http://www.crcpress.com/; March 2018. ISBN-13: 978-1498785334

Nawari, N. O. (2011). Standardization of structural BIM. In *Proceedings of the 2011 ASCE International Workshop on Computing in Civil Engineering*. Miami, FL, pp. 405–412.

Nawari, N. O. and Kuenstle, M. (2011). *Building Structures: Fundamentals of Crossover Design*. San Diego, CA: University Readers, Inc. & Cognella Academic Publishing.

Nawari, N. O., Itani, L., and Gonzalez, E. (2011). Understanding building structures using BIM tools. In *Proceedings of the 2011 ASCE International Workshop on Computing in Civil Engineering*. Miami, FL, pp. 478–485.

NBS. (2015). *Uniclass 2015*. NBS (National Building Specification): https://www.thenbs.com/our-tools/uniclass-2015

Nervi, P. L. (1965). *Aesthetics and Technology in Building*. Cambridge, MA: Harvard University Press.

New Oxford American Dictionary. (2010). 3rd ed. S.V. Poetry.

Omar Bagasi, and Nawari, N.O. (2025). "BIM and AI in Early Design Stage: Advancing Architect-Client Communication," *Buildings, MDPI.*

Önür, S. (2009). IDS for ideas in higher education reform. In *Proceedings of the First International Conference on Improving Construction and Use through Integrated Design Solutions*. Espoo, Finland: VTT-Technical Research Centre of Finland, pp. 52–71.

Prowler, D. (2012). *Whole Building Design*. Washington, DC: The Whole Building Design Guide (WBDG), The National Building Institute. https://www.wbdg.org/wbdg_approach.php

Rafiq, Y. (2010). A radical rethink in educating engineering students. In *Proceeding of the 19th Analysis and Computation Specialty Conference, ASCE*. Orlando, FL, pp. 366–376.

Sacks, R. and Barak, R. (2010). Teaching building information modeling as an integral part of freshman year civil engineering education. *Journal of Professional Issues in Engineering Education Practice*, 136, 1, 30–38.

Sacks, R. and Pikas, E. (2013). Building information modeling education for construction engineering and management. I: Industry requirements, state of the art, and gap analysis. *Journal of Construction Engineering Management ASCE*, 139, 11, 04013016.

Salingaros, N. A. and Mehaffy, M. W. (2006). Rules of beauty and order in past times. In *A Theory of Architecture*, ed. J. Quillien. Solingen, Germany: Umbau-Verlag, pp. 29–30.

Sandaker, B. N. (2008). *On Span and Space: Exploring Structures in Architecture*. New York: Routledge, Taylor & Francis Group.

Schodek, D.L. (2004). *Structures*. 5th ed. Toronto, ON: Prentice Hall.

Schueller, W. (1995). *The Design of Building Structures*. Englewood Cliffs, NJ: Prentice-Hall.

Schueller, W. (2007). *Building Support Structures: Analysis and Design with SAP200 Software*. Englewood Cliffs, NJ: Computer and Structures.

Sharag-Eldin, A. and Nawari, N. O. (2010). BIM in AEC education. In *2010 Joint Structures Congress with the North American Steel Construction Conference*. Orlando, FL, pp. 1676–1688.

Sinnott, E. W. (1960). *Plant Morphogenesis*. New York: McGraw-Hill.

Viollet-le-Duc, E.-E. ([1854] 1990). *The Foundations of Architecture*, (Translated by Kenneth D. Whitehead from the original French.) New York: George Braziller.

Viollet-le-Duc, E.-E. (1990). *The Architectural Theory of Viollet-le-Duc: Readings and Commentary*, ed. M. F. Hearn. Cambridge, MA: MIT Press.

Waldrep, L. W. (2006). The education of an architect. In *Becoming an Architect: A Guide to Careers in Design*, ed. M. J. Malecha, Hoboken, NJ: Wiley.

Wong, K. A., Wong, K. F., and Nadeem, A. (2011). Building information modelling for tertiary construction education in Hong Kong. *ITcon. Journal of Information Technology in Construction*, 16, 467–476.

Zafar, R. (2017). Construction Costs and Health Outcomes Using LEED, Thesis, published by Texas A&M University. http://hdl.handle.net/1969.1/161349

Index

Note: **Bold** page numbers refer to tables and *italic* page numbers refer to figures.

C

D

E

N

O

P

W

X

Z